de Gruyter Studies in Organization 52

Technology Policy

de Gruyter Studies in Organization

Innovation, Technology, and Organization

This international and interdisciplinary book series from de Gruyter presents comprehensive research on the inter-relationship between organization and innovations, both technical and social.

It covers applied topics such as the organization of:

- R & D
- new product and process innovations
- social innovations, such as the development of new forms of work organization and corporate governance structure

and addresses topics of more general interest such as:

- the impact of technical change and other innovations on forms of organization of micro and macro levels
- the development of technologies and other innovations under different organizational conditions at the levels both of the firm and the economy.

The series is designed to stimulate and encourage the exchange of ideas between academic researchers, practitioners, and policy makers, though not all volumes address policy- or practitioner-oriented issues.

The volumes present conceptual schema as well as empirical studies and are of interest to students of business policy and organizational behaviour, to economists and sociologists, and to managers and administrators at firm and national level.

Technology Policy

Towards an Integration of Social and Ecological Concerns

Editors

Georg Aichholzer and Gerd Schienstock

Walter de Gruyter · Berlin · New York 1994

Georg Aichholzer
Senior Researcher, Institute of Technology Assessment, Austrian Academy of Sciences, Vienna, Lecturer at the University of Economics, Vienna, Austria

Gerd Schienstock
Senior Researcher, Academy of Technology Assessment, Stuttgart
Lecturer at the University of Stuttgart, Department of Social Sciences, Germany

With 34 tables and 23 figures

⊗ Printed on acid-free paper which falls within the guidelines of the ANSI to ensure permanence and durability.

Library of Congress Cataloging-in-Publication Data

Technology policy : towards an integration of social and ecological concerns / edited by Georg Aichholzer, Gerd Schienstock.
p. cm. – (De Gruyter studies in organization : innovation, technology, and organization ; 52)
Includes bibliographical references.
ISBN 3-11-013677-5 (alk. paper)
1. Technological innovations. 2. Technology and state.
I. Aichholzer, Georg. II. Schienstock, Gerd. III. Series : De Gruyter studies in organization ; 52.
HC79.T4T453 1993
338'.064–dc20 93-37994
CIP

Die Deutsche Bibliothek – Cataloging-in-Publication Data

Technology policy : towards an integration of social and ecological concerns / ed. Georg Aichholzer and Gerd Schienstock. – Berlin ; New York : de Gruyter, 1994
(De Gruyter studies in organization ; 52 : Innovation, technology, and organization)
ISBN 3-11-013677-5
NE: Aichholzer, Georg [Hrsg.]; GT

Typesetting: Converted by Knipp Satz und Bild digital, Dortmund – Printing: Karl Gerike GmbH, Berlin. – Binding: Dieter Mikolai, Berlin. – Cover Design: Johannes Rother, Berlin. – Printed in Germany.

Preface

In almost every Western industrialized country, technology policy has become a field central to public policy. This development is due to the fact that international competitiveness is heavily dependent on the extent to which a country succeeds in achieving a position of technological leadership. However, it has long since become generally accepted that a mere policy of promoting innovative activities can but insufficiently stimulate technological progress. Rather, successful political action in this area depends on the creation of structures conducive to innovation in such societal realms as education and the economy at large.

Technology policy has also experienced an extension in a second sense. The fact that a rather widespread 'technology euphoria' has been followed by a more sceptical assessment of technology and technological progress, has also brought its negative social and ecological effects more into the limelight. The prevention of such negative impacts creates a need for a whole new field of tasks. However, these cannot merely be added on to the traditional tasks of technology policy: they demand an integrative reorientation. Socially- and ecologically-oriented technology policy, as it is understood here, is aiming at a stronger orientation of other policy fields towards technological development as well as towards the prevention of the more negative effects of technological progress.

As a whole, the contributions in this volume have a dual focus. The papers in the first part are mainly devoted to investigating from a theoretical point of view the possibilities and opportunities as well as the problems and weaknesses of a socially- and ecologically-oriented technology policy. The purpose of the second part is to inform on the mode and extent to which a social and ecological orientation has already found its way into national technology policies. As will be seen, the existence of an integrative policy conception in a really advanced version is still far from realization anywhere in the world. However, a number of different and promising beginnings can be discerned which do deserve attention and further efforts.

Our thanks go to everyone who has participated in the creation of this book. We would especially like to thank all our authors for their cooperation. Their patience has been sorely tested by the considerable difficulties and delays which such an international project is bound to bring along with it. We would also like to acknowledge the support received from the Institute for Advanced Studies, Vienna, where we started this work as well as from our present employing institutions, the Institute of Technology Assessment (ITA) at the Austrian Academy of Sciences and the Academy of Technology Assessment of Baden-Württemberg, respectively. We are especially grateful to Gertraud Stadler, who has cared for the book in a most committed way from its beginning to the final editing. Our gratitude goes also to Suzanna Stephens for her linguistic checking, as well as to all those supporters and colleagues at various public institutions in Austria who have in various ways contributed to the realization of this project.

Georg Aichholzer and Gerd Schienstock
Vienna/Stuttgart, December 1993

Contents

Technology Policy in the Process of Change: Changing Paradigms in Research and Technology Policy?

Gerd Schienstock

1 Introduction

There is no modern industrialized country not pursuing research and technology (R&T) policy in one way or another. Although state interventionism in general is viewed more and more sceptically and despite an increasing shift towards favouring the regulating power of the market, hardly any call has been voiced thus far for the state's withdrawal from the fields of research, innovation and technology. To the contrary: not only is public R&T policy considered to be as indispensable as ever for stimulating economic growth, but it has acquired an even growing importance within the scale of state activities. It may be expected that, as knowledge about the complexity of technological innovation increases, state R&T policy will gradually develop into an independent policy field (see Edquist, this volume).

Rather than asking for an expansion of technology policy along existing lines, critics refer to the lack of attention devoted to the social shaping of technological development, as well as to the social compatibility of technological change. Moreover, it is pointed out that there is a lack of ecological precautions and environmentally-benign technology – a deficit whose solution is also expected from the state. What one has in mind, though, is not an expansion of R&T policy by complementary measures which take into account social and ecological aspects. Rather, there is a generalized call for either an integrative or systemic technology policy. According to OECD, there is a need for complete re-orientation within this policy field. Its substance must be newly defined, new spheres of intervention must be identified and new structural arrangements must be developed (OECD 1988). In the following, some of the tendencies shall be pointed out of such a re-orientation in state R&T policy.

2 Traditional Motives for Public R&T Policy

Discussions on the justification of state R&T policy are generally based on the assumption that economic growth is the central indicator of social well-being and, as such, constitutes a high-ranking societal goal. In turn, decisive growth impulses are said to emanate from technological innovation. Technological development is typically depicted as a linear phase scheme; while here, for reasons of simplification, the three stages of research, innovation and diffusion shall be distinguished. As disruptions in the course of technological development constitute a danger to economic growth, there is therefore a major societal interest in overcoming them (see Braun, this volume).

There is a reason for attributing such an important role in the process of technological innovation to the state: the structural problems arising within the various stages are often not solved with equal efficiency by the market. The market process, oriented along the self-interests of private actors as far as the goal of economic growth is concerned, leads to an insufficient level and/or to an inappropriate structure of national economic innovation. The central task of state R&T policy is considered to be to correct or even to avoid any such deficiencies by influencing innovation. Moreover, the latter should basically continue to follow market economic principles (Littmann 1975:45). Thus, in justifying state R&T policy, crucial importance is attached to the aspect of market failure in achieving economic growth – regardless of how great this failure is considered to be.

The following enumeration includes the major structural problems of the process of technological innovation: According to the theory of economics, these are solved unsatisfactorily, if at all, by market-oriented economies: the availability of sufficient knowledge in the natural sciences, the technical or economic indivisibility of large-scale projects, and the lack of transparency of the market (Ewers/Wein 1989). As a consequence, the state is assigned a central role in coping with these problems.

Originally, the justification of state intervention in the process of technological innovation was primarily based on two central arguments. It is generally acknowledged that, at the enterprise level, research is connected with considerable external effects. Especially in the case of fundamental research in the natural sciences, third parties may use new findings without any market transaction having taken place. As in the case of fundamental research, the market economic principle of elimination fails: there is no incentive for enterprises, according to wide theoretical agreement, to invest more heavily in terms of resources in this field (Arrow 1962). This argument is less valid for applied research: in principle, patent protection – as well as the chance to lead in know-how in the long run – provide sufficient incentives for investment by private industry.

Subsequently, as far as research is concerned, market economic regulation provides an unsatisfactory allocation of resources for attaining the goal of economic growth. This shortcoming is used to justify the fact that, above all, fundamental research is not controlled by the market but rather by the state – i.e. that it is either carried out by state institutions or at least supported by these, with results accessible to all market participants (Ewers 1990:150).

This is all the more true if technological progress is claimed to be based on the "trickling down model" (Hetzler 1970). This serves as the second argument for state R&T policy. This model implies that the extent of fundamental research substantially influences the opportunities for technological innovation – which, in turn, determines the growth rate of the social product. In the case of an inadequate level of fundamental scientific research, the redistribution of resources towards this stage of technological progress makes it possible to initiate a process of national economic growth.

Although there is no doubt that an increase in scientific knowledge raises a society's innovative potential, its utilization is by no means automatically guaranteed. This is why, in connection with the support of fundamental research, the concept of "conditions providing policy" is used to distinguish it from R&T policy as such (see Boxsel, this volume). Transforming new scientific findings into technological innovation, is basically a matter of those decisions taken by business on how to utilize them. Actions taken at the enterprise level, however, quite often prove detrimental to national economic growth. Therefore, it seems obvious to put the emphasis of state R&T policy on guaranteeing the transfer of technological knowledge, rather than on encouraging and supporting fundamental research. Such a strategy proves rational, especially for small national economies unable to provide the enormous financial means necessary for carrying out fundamental research. Even in Japan, the transfer of modern technology from the leading industrialized nations was originally promoted by the state primarily to initiate a process of economic recovery. Only recently, after Japan's establishment as a leading industrialized nation, has the promotion of fundamental research gained in importance within state R&T policy (see Kawasaki, this volume).

In the case of large-scale technologies, the transformation of scientific findings is also becoming a problem hardly manageable by the market alone. The massive financial requirements resulting from their technological and/or economic indivisibility, quite often exceed the capacities of private companies. In such fields as transport technology, space technology, nuclear energy and computer technology, the governments of most of the large industrialized nations subsequently feel compelled to extend their engagements in technological innovation to fields of applied research. These fields are closer to the market and thus avoid the risk of innovative activities being curtailed due to the financing problems of private industry. However, the following must be admitted as well: in view of the enormous research and development budgets of multinational corporations, the failure of the market in terms of technological and/or economic indivisibility can be used only to a limited extent to legitimatize state research and technology policy (Ewers 1990:151).

The additional danger of fundamental scientific knowledge being transformed into technological innovation, only exists if there is a lack of information on the probability of the success of these innovations. This problem is gaining more and more in importance: the profitability of investment in high-tech research and development is hard to estimate due to the long 'running-in' periods involved. Although public subsidies may not reduce the risk of failure, they may well reduce related costs. It is argued that, if companies' research expenditure is reduced by state subsidies, then companies will engage more readily in financial commitments difficult to calculate in advance (see Braun, this volume). Public support in these cases, however, is by no means an uncontentious issue: it implies that public institutions are indeed more capable than industrial enterprises to estimate future market risks and, accordingly, that the companies' risk precautions are not justified in case of an appropriate transparency of the market. Reality has proved otherwise: public institutions' estimations of financial risks in the development and marketing of new technologies are quite frequently inadequate. The result of this is the permanent necessity of their adopting a role as permanent donors of subventions.

As discussions have shown, a public R&T policy based on economic targets is rather limited in its scope of legitimatization. State intervention in the process of technological innovation, however, may not be justified by the failure of the market argument alone. One should also consider the necessity of gaining and maintaining international competitiveness (Rothwell/Zegveld 1980). Competitive enterprises in so-called key technology areas are considered to be inevitable if technological dependencies detrimental to the national economy are to be avoided. Here, the perspective of state support does not only refer to fundamental research but also increasingly to the stage of development at the enterprise level.

Public technology policy in Japan is especially characterized by such an orientation towards key technologies. According to McMillan, "The Japanese have learnt the lessons of the post-industrial and information revolution and turned them to competitive advantage" (1984:64). What is characteristic of Japanese R&T policy, is that the economy is regarded as a bundle of sectors, branches and industries. All of these are analyzed by the state with regard to such criteria as prospects for the future, net product or world market potential. Also, an optimal composition of the economy is aimed at by special state R&T policy measures encouraging such promising industries as information and data technology, new material technology, biotechnology, automatic manufacturing systems, etc. (see Kawasaki, this volume). In the meantime, however, a number of other countries have also adopted an R&T policy concentrating on the support of key technologies.

This means that national R&T policies are becoming more and more similar. All states promote the development of the same key technologies, without taking into account national specificities or the characteristics of national innovation systems. At present, a vividly-discussed topic is the fact that the semiconductor development is characterized by considerable structural deficiencies in European countries – which bears the risk of a European technological dependence on the US and Japan. This is why, again and again, the massive subsidization of European companies has been demanded in the development of silicon semiconductors.

Here again, however, the legitimacy of public research and technology policy seems to be questionable at best. For one thing, the argument of the expediency of international division of labour may be used against it. For another, in view of the growing globalization tendencies of large corporations and the increasing number of strategic alliances being formed, it has become difficult to ascertain which are those enterprises worthy of support in terms of maintaining national competitiveness. In the case of semiconductor development, cooperation between a German and a US corporation at the expense of a European one, raises doubts about the existence of national or Europe-wide support of key technologies. As shown by Edquist, this dilemma of public technology policy may frequently be encountered in small countries.

Traditional research and technology policy, as well as its bases of legitimatization, have been subject to criticism in various respects. First, there is considerable doubt about the basic assumptions of the "trickling down model". It may by no means be clearly proved whether the support of fundamental research over various stages of the innovation process, finally leads to the development of new marketable products and processes – and thus, to economic growth. The substantive growth processes which took place in the highly-developed industrialized countries after World War II, may not even predominantly be attributed to state innovation policies. It certainly seems equally justified to name a variety of other factors as having caused them.

By the same token, there is no definite empirical evidence that international competitiveness may be guaranteed by means of state policy. It is true that public research and development subventions are provided massively in those areas where individual industrialized nations hold a particularly strong position in international competition (Krupp 1987). A clear-cut assignment of cause and effect, though, is hardly possible. Also, one may consider the thesis of plausible state research and development policy as being attracted by economic success rather than as causing it. In any case, it is difficult to prove the success of traditional innovation policy with regard to the objective of initiating economic growth.

However, traditional R&T policy is criticized not only because of the uncertain prospects of success. It is also questioned because of its traditional understanding of technology development and policy. Economic competitiveness is exclusively related to technological progress; thus, a material conception of technology prevails (see Braun, this volume). According to this narrow definition, technology is understood in terms of machines and production facilities. Accordingly, public R&T policy concentrates on the support of scientific findings and the development of material technology. However, there is much room for doubt as to whether the competitiveness of enterprises on the world market depends exclusively on the modernity of material technology. Conversely, it might be the other way around:

as shown by the example of Japan, organization and management aspects play an equally important role. This implies that the traditional, technologically-oriented R&T policy must accept the argument regarding the limits of its purely technological approach towards an optimal path to economic growth. At the same time, the question is raised again, though from another perspective, in how far state interventionism is expected to interfere in market-oriented stages of the technological innovation process.

3 From Material Technology to Technological Practice

In recent debates on Fordism, it has become clear that the progress in productivity connected with its diffusion was not due to the mere technical innovation of the conveyor belt: rather, this was possible only in combination with an appropriately adapted work organization and enterprise culture (Boyer 1991). Mechanization of production, Taylorist principles of work organization and a bureaucratic management concept form an entity, so to say; it is their interaction that can be considered the secret behind the marked stage of growth after World War II. Hence, Fordism as a new technological paradigm was not only characterized by technological progress, but also by changes in work organization and enterprise culture.

This is an expression of a changed basic understanding of technology. The material technology concept is replaced by a relational one (Rammert 1989). Technology is defined as an organized system of men and machines in which scientific findings and experience are applied in the solution of practical problems (cf. also Dufourt 1991, quoted in Abdelmalki/Kirat, this volume). In this connection, Pacey talks about technological practices (1983:6) which, apart from a technical dimension, also include both an organizational and a cultural dimension. Technological progress is thus a process of combined techno-organizational and socio-cultural innovation.

Immediately evident when studying the specific structure of modern information technology, are both the significance of a technology concept extended by the element of practice and, simultaneously, the problematic nature of traditional policy. Due to the software element which it incorporates in addition to the hardware, this is characterized by a significantly higher degree of 'influenceability' by social factors other than traditional technology. This means, in other words, that technological progress may not be reduced to the development of material technology; rather, the way of structuring the working process via the development of software, is an endogenous and functionally necessary constituent in the development of technological systems (Naschold 1986:232).

Serving as the basic assumption is, on the one hand, the autonomous development of the *technological* paradigm in the form of hardware; and on the other, the *social* paradigm in the form of software (Dosi 1982). The utilization of scientific findings may already have resulted in the development of concrete technologies; however, the systemic combination of these hardware components with relevant software programmes, can only be considered to be in an experimental stage. The same is true, for that matter, for the supervisory structure, the regulation of working conditions and certain aspects of organization culture. In any case, it cannot yet be foreseen which combination of hardware and software components, work-organizational norms and enterprise culture patterns may finally be considered as the most promising technology line with respect to productivity and growth development (see Badham/Naschold, this volume).

The susceptibility towards a variation of technological system configurations becomes obvious in the case of the conception of CIM systems. There, at least two technological practices have developed. According to Brödner (1985), there is a technocentric and anthropocentric development line; and according to Badham (this volume), a neo- and a post-Fordist one. The two totally contrary systems – both of which differ in technological structure, machine programming and regulations of work organization – are above all based on different appreciations of human labour. A similar argument is used by Kern and Schumann, who discover in their analysis of new production concepts a management philosophy totally different from Taylorism. They are influenced by quite specific organizational culture patterns (Kern/Schumann 1984).

The concept of technological practice suggests that the productivity potential of new technologies may be fully exploited only in connection with organizational and socio-cultural innovations. Due to their systemic or paradigmatic character (Brandenburg et al. 1975; Mensch 1977), the implications of new technologies – especially if based on advanced electronics – go beyond the sphere of work at the enterprise level. To allow for their efficient application, changes in a variety of societal institutions have also become necessary. According to this basic idea, new technologies represent a scientific development so powerful that it puts into motion entire whole chains of technological and social innovations (see Braun, this volume). In this connection, Nelson and Winter talk about generalized natural trajectories of technology (1977, 1982).

In her concept of "techno-economic paradigm", Perez also emphasizes the importance of the connection between technological progress and societal change (1987). Successful techno-economic paradigms are the result of, on the one hand, the interaction between new lines of development; and on the other, both selective mechanisms of the economy and the natural and social environment. Before a new technological paradigm can lead to any substantial productivity rise, it is argued, a crisis of structural adaptation must be overcome. One should add that old institutions and cultural patterns as well become obsolete; they correspond to the requirements of the outdated technological system. A mis-match occurs between new technologies and the old social model of production. This mis-match must be overcome by means of institutional change if the new technologies' productive potential is to be fully exploited. By institutional change, Perez refers not only to work organization and management aspects but above all to the educational system, industrial labour relations, societal value conceptions, etc. Crucial importance is attached to the reform of the system of vocational education in connection with the change in paradigm due to the introduction of new information technologies. Their efficient use substantially depends on the degree to which employees are in a position to enter into active dialogue with the technological systems (Loikkanen/Seppälä, this volume).

The example of the automobile lends itself once again to demonstrating the consequences of paradigmatic technological change. Apart from representing a technological, work-organizational and – as can easily be checked with Ford – enterprise-cultural change in paradigm, the development and acceptance of the new transport technology has been similarly and decisively influenced by massive public investment programmes, as well as the rise in mass consumption via wage and social policies. To carry this example further, the oil crisis has not only encouraged new engine development but has also led to the social innovation of the "car-free weekend". Finally, massive technological development initiatives may be expected in connection with legal regulations for the solution of the recycling problem.

According to what has been said so far, social, cultural and institutional aspects of work organization at the level of both companies and society may be similarly decisive as technological innovation for productivity and economic growth. This seems to be taken into account by shifting emphases within public R&T policy. While the support of fundamental research continues to be uncontested, the limits have been extended within which public activities are considered to be appropriate (OECD 1988). In view of the central importance of technological practices and systemic technology structures for productivity development, it is argued that effective public innovation policy must comprise considerably more functions than the mere financial support of research and development or the pilot application of new technologies. It must be related much more closely to all those factors either hampering or favouring its development or use (Bräunling 1986:264). Essentially, two groups of factors may be discerned here: socio-cultural factors of work organization at the enterprise level, on the one hand, and institutional factors at the societal level on the other (see Badham/Naschold, this volume).

The direct or indirect subsidization of technology development may be of less importance for national economic growth than the elimination of barriers to innovation at the enterprise level or the support of factors stimulating innovation. This fact has become uncontested as a basis of legitimatizing public innovation policy. Accordingly, there is consensus in principle regarding the state's active role in establishing both infrastructure and social institutions supporting technological change. This is the case, for instance, with such economic measures as guaranteeing a low interest level; eliminating protective tariffs; providing efficient traffic, transport or modern communication systems. In the same way, this extends to an economical energy supply, far-reaching technological standardization or even the extension of research and educational systems. However, the question of how far the educational system – including continuing education – is to be adapted to the requirements of technological progress, is being discussed quite controversially. Even stronger reservations exist when it comes to state intervention for the purpose of re-organizing such social institutions as the system of industrial relations, property rights or social security.

The central importance of social institutions for technological change leads to the demand not only for extending but also for centralizing public research and development policy. As a consequence, there has been a demand for the systematic integration into public policy of all those policy spheres more or less unintentionally influencing technological progress, i.e. transforming implicit into explicit policy (see Edquist, this volume). Independently of the fact that such a policy perspective is accompanied by enormous coordination problems (see Badham; Aichholzer et al., this volume), the question also arises of how far any such dominance of technological progress – as compared with other societal objectives – is desirable at all. The concept of national innovation systems (see Dalum, this volume), stressing the very orientation of public R&T policy towards national institutions and cultural patterns, is becoming increasingly widespread in the literature. This suggests at the very least that the relation between technological progress and the establishment of social institutions should not be understood as a process of unilateral adaptation – and accordingly, that there is due justification for maintaining the independence of individual public policy fields from R&T policy. This in no way means, however, the discontinuing of considerations for coordinating public policies.

Even more controversial than the state's role in institutional adaptation processes and in providing infrastructure, is its intervention in designing production and work processes at the enterprise level. There is basic agreement, however, that the social shaping of new technological systems is of eminent importance under the very aspect of productivity; and this consensus also extends to the market as an appropriate medium for the implementation of social models of work organization within a limited range (Naschold 1986:236). From this perspective, there is quite a significant scope of intervention for state technology policy.

Objections to the state's active role in conceiving the social model of new technologies, primarily relate to the low impact of supporting projects of an exemplary nature as a means for public R&T policy. This refers not only to the structural conservatism of business enterprises (Child et al. 1987) but also to the problematic nature of the transferability of forms of social organization at the enterprise level – and thus, the model nature of publicly-subsidized projects. Secondly, the specific sensitivity to state intervention in this field, is due to the fact that intervention into enterprises' production and working processes may not be seen under the mere aspect of productivity growth. Inevitably raised are questions of social compatibility and ecological consequences. Taking into account such problems and thus incorporating non-economic goals as determinants of public

R&T policy means nothing more than questioning the traditional growth-oriented policy pattern and suggesting a complete strategic and institutional re-orientation.

4 The Crisis of Growth-oriented R&T Policy

Thus far, the government policy objective of increasing social welfare has quite indiscriminately been equalled with growth and productivity progress. In the discussions in this chapter so far, however, the decisive argument has been that a lasting increase in productivity cannot be achieved by technological innovation alone: changes in enterprises' technological practices and in societal institutions are also necessary. The technological innovations encouraged by state innovation policy, however, do not lead to productivity increases only; their development, as well as their application, involve considerable risks of both a social and an ecological nature. This has given rise to an increasing scepticism towards the traditional objectives of public R&T policy (see Braun, this volume). Social damage may by far outweigh productivity growth – leading, as a result, to a decline in social welfare.

It is important to note that social and ecological risks are the result not only of the very nature of material technology as such, but also of its use under specific circumstances. As has been shown quite impressively by Perrow (1984), the ecological risks of modern large-scale technology are primarily the result of a lack of mutual adjustment between social organization and technological structures. In addition, the hazards to human health as a central social risk are in many cases not due to specific mechanical properties, but rather to work-organization structures and excessive performance standards.

Due to a specific selectivity, the capacity of the market to eliminate social and ecological risks is considered to be extremely low. Consequently, there are a number of further examples of market failure which might possibly contribute to considerable shortcomings of production steering (Littmann 1975:52): external effects of production and consumption, neglect of social and human standards; as well as differences between private and general economic time spans.

The external effects of production and consumption increase dramatically in accordance with economic and technological development. Environmental damage – such as water and air pollution, the destruction of the ozone layer, as well as the gigantic growth in waste quantities – make it clear that many people's quality of life is considerably diminished by technological progress. Impairing the quality of the free goods of water and air as a consequence of industrial production, raises serious doubts about the positive effects of productivity increase and economic growth on the general welfare.

Scarce resources may be misdirected if environmental protection and improvement are left to the market. This becomes easy to explain when considering as the entrepreneurial goal a ratio as favourable as possible between private profits and costs. In line with this goal, the substitution of private costs for societal ones appears to be quite rational – at least from the entrepreneurial point of view. This applies to the emission of toxic gases in the same way as for discharging contaminated waste water into the rivers. As long as the wasteful use of scarce resources does not reduce profits, there is nothing – from the entrepreneurial perspective – that speaks against the exploitation of nature. Such an externalization of costs may mean technological progress from the single entrepreneur's point of view; while at the level of society at large, such behaviour results in reduced social welfare.

Specific market selectivity becomes also obvious when considering human-oriented aspects in the conception of technological practices. A decrease in operational costs as a consequence of work intensification due to the use of new technologies, is still considered as technological progress in the traditional sense – even if this goes hand in hand with physical and psychological strain. This assessment, though, cannot be shared on the societal level: there are the considerable social costs, for instance, of medical care and human capital lying idle. Again, costs accrued on the enterprise level during the production process are externalized and must be borne by society as a whole.

Insufficient knowledge of the future scarcity of natural resources is said to be the reason for yet another form of market failure. Market prices, it is argued, are steering mechanisms related to the present, indicating only current scarcities of goods rather than signalling future developments. Hence, such a short-term perspective of the market inevitably leads to inappropriate resource allocation and a misdirection of the innovation process from the point of view of social welfare. One argument along this line refers to the insufficient consideration of the needs of future generations (Ewers 1990:151).

Such failure of the market is demonstrated by the way of dealing with non-renewable natural resources. The fact that their availability is limited, is hardly reflected in market prices as long as the supplied quantities correspond to demand. In this way, the necessary adjustment processes are deferred – which may lead to considerable friction and slow-down in growth at a later stage. Only if processes of adaptation to future conditions are initiated at a very early stage, can the expected social conflicts be even somewhat alleviated (Littmann 1975:57).

The market mechanism contributes to environmental damage, the violation of human standards and the neglect of future situations of scarcity. This fact suggests the need for further perspectives on public innovation control. This does not necessarily suggest renouncing the objective of growth, but rather – and at the very least – a re-orientation in the policy of growth.

There is the widespread view that the development and application of the new EDP-based technologies may solve not only economic but also ecological and social problems. These technologies are not only labelled as clean: they are also supposed to lead to the simultaneous optimization of economic and social objectives. Stimulating technological progress by supporting research in new technologies – and possibly in their pioneering application as well – is thus considered as a comprehensive form of state intervention no longer requiring any additional orientation towards ecological and social objectives. Rather, there should be emphasis on maintaining the basic pattern of technology support.

Daneke criticizes the idea that such a conception of new technologies makes them a kind of panacea. He argues, however, that these are not unproblematic from the ecological point of view. In addition, they place high demands on enterprise organization and public infrastructure (see Daneke, this volume) while by no means leading automatically to the creation of socially-compatible working structures (see Badham, this volume). If the welfare-enhancing function of technological progress is not to be impaired by any negative effects, ecological and social objectives must explicitly be made the hallmarks of public R&T policy.

5 The Role of Technology Policy: Support or Regulation?

At first sight, a simple idea seems to provide a sensible orientation for a type of R&T policy aiming at economic, ecological and social objectives. Littmann describes this basic idea as follows:

> Economic-technological innovations which would not be carried out under the conditions of a (distorted) empirical price system, being considered unprofitable from the private enterprise point of view, shall be stimulated provided that they contribute to working towards the socio-economic optimum. On the other hand, those types of innovation must be discriminated against which seem promising of success merely from the financial point of view of private industry but, taking into consideration social costs and profits, would result in no gain or even in a loss of people's economic welfare (Littmann 1975:58).

This basic idea seems to suggest a dual R&T policy incorporating both support and regulation. "The two main thrusts of technology policy are the support of technology and the regulation of technology. The former stems mainly from the desire to strengthen the national economy, while the latter is mainly necessary in order to reduce health and environmental hazards caused by the use of technology" (Braun, this volume). Badham (in this volume) refers to the partial inadequacy of this role assignment of the two instruments. Thus, the development of new technologies – as mentioned above – contains certain aspects of regulation; while on the other hand, there are considerable difficulties in eliminating health hazards via legal provisions regulating working procedures. It is rather the support of new technologies which might lead to more effective results in this area.

Putting the above-mentioned idea into practice poses considerable difficulties as well. Above all, the point in time at which the state should intervene by either supporting or hampering the process of innovation, seems to be an open question. Mayntz/Scharpf give a very good illustration of the problematic nature of such a simple recipe as far as fundamental research is concerned. In politics, according to the authors, it is not possible "to nurse the vegetables and pull out the weeds: the same roots feed both the potential for increasing social welfare and for jeopardising it" (1990:61). The knowledge about nuclear fission, genetic engineering, or information storage as such, may not a priori be categorized as supportive or detrimental to social welfare. Thus, the aim of welfare optimization by public R&T policy may not be achieved by supporting fundamental research only in those selected fields in which new findings are expected to be conducive to welfare, while withholding support in all those areas where new findings conjure up ecological and social risks.

It obviously causes insurmountable problems to attempt to exclude any negative effects of technological progress, be they of an ecological or a social nature, through state control over fundamental research. It does, however, seem realistic to orient this research towards finding solutions to existing problems. It is possible to find out which type of knowledge is necessary for solving a given social or ecological problem. Of course, this by no means guarantees that an adequate remedy to the problem will actually be found. On the one hand, knowing which kind of scientific knowledge is required, does not automatically guarantee success – even if massive financial means are available for the necessary fundamental research. On the other hand, new scientific findings are not immediately transformed into the necessary forms of technology utilization. In spite of these imponderabilities, it seems appropriate to assign to state policy three main tasks (Mayntz/Scharpf 1990:63):

- generating knowledge to find technological solutions to social problems which have not been caused by technology itself;
- generating knowledge about possible negative side effects of currently-practiced or prospective technological solutions (see also Edquist, this volume); and
- generating knowledge for minimizing such side effects and solving problems arising from former technological practices.

What the authors have in mind here, is primarily fundamental research in the field of ecological systemic relationships. However, their policy conception may easily be transferred to social problems. So, for instance, fundamental research may be aimed at finding out where technological practices do not comply with the criterion of social compatibility. Also, research into the adverse long-term effects of work practices can be made the object of fundamental research, in quite the same way as investigating possibilities for confining or totally eliminating such adverse effects. The authors are well aware, however, that the incentive structure currently existing in fundamental research, is hardly adequate for directing scientific interest towards the investigation of ecological cause-and-effect relationships. To the contrary, it is systematically designed to neglect interdisciplinary research. In view of this, it can be seen as the essential task of the state to initiate the kind of institutional change in fundamental research that is necessary for embarking on interdisciplinary problems (Ewers 1990:159-160). In this respect, some of the developments in the Netherlands seem to be interesting. In the course of the 1970s, special programmes and courses were established at universities and other research institutes. All of these were devoted to investigating the relationships between science, technology and society (see van Boxsel, this volume).

There exists a dilemma of not being able to exclude the detrimental effects of technological advances through controlling fundamental research. Its solution is often considered to lie in a specific model of dualistic R&T policy. There have been attempts to reach the goals of stimulating a level of productivity growth conducive to raising social welfare, as well as that of maintaining national competitiveness in the international context by supporting fundamental research. At the same time, knowledge application is controlled by a variety of forms of state regulation, with a view to avoiding undesired social and ecological consequences of technological progress – or at least of limiting possible damage (Braun 1984:123). According to Littmann, it might certainly be much easier to correct any undesired aspects of technological progress on the user level rather than trying to establish an optimally-structured development path for the economy as a whole via the distribution of research funds (Littmann 1975:59). In this connection, Collingridge (1980) has drawn attention to the central dilemma of control. This is due to the fact that, during the early stages of development, technological innovation may not be controlled because of insufficient predictability; while later, as social and ecological consequences may appear, it is possible to do so. Control at this stage, however, becomes increasingly difficult, since any changes are extremely costly due to the amount of technological, financial, institutional and cultural investments already made.

State regulation may take the form of either laying down concrete legal provisions or of influencing price-setting: for instance, via the levy of taxes or other charges. This is also described by the concepts of normative or economic regulation. Public regulation is discussed above all as a means of achieving ecological objectives.

A great deal of controversy surrounds the issue of which form of regulation is more efficient for implementing non-economic goals in economic life. By means of economic instruments such as taxes and other charges, producers and consumers are supposed to bear those external costs which they themselves have caused and which so far have been paid for by the community (see Simonis, this volume). "Shadow prices" for causing damage to the environment, it is assumed, will make the rationally-acting subjects of economic life use clean technologies and environmentally-benign products so as to avoid the costs arising from legal requirements and price hikes. Logically speaking, state intervention in the price setting mechanism will lead to the elimination from the market of unclean technologies to be replaced by environmentally-benign product and process technologies.

As criticized in the contribution by Abdelmalki and Kirat, this argumentation has some weaknesses. One major difficulty in fixing 'shadow prices' lies with determining the origins of a specific environmental damage – and accordingly, discriminating against the technologies causing the damage. In most cases, it is argued, damage is of complex origins and appears only after some time. Moreover, levies and charges by no means invariably lead to the substitution of production technologies and of products depriving the state and the environment of their clean counterparts. This is primarily a question of how high such shadow prices are. The problem is that fixing exact price levels is not possible due to insufficient information on damage caused by a specific technology. They tend to be the result of biased negotiations in which typically rather moderate prices are fixed, mostly providing little incentive for a change in technological practices or buying habits. In addition, the case of future generations is not taken into consideration in this procedure. Of special importance, though, seems to be the argument that trying to fix a price for environmental damage in principle suggests that nature-destructive processes are reversible. However, in most instances, this is not the case. Therefore, ecologically-oriented government regulation by way of price is in principle not suitable for solving the problem of preventing environmental damage (Söderbaum 1990:3).

Legal regulation also has a number of disadvantages. It has proved to inhibit innovation insofar as there is no special incentive for producers to improve the 'state-of-the-art' beyond maximum permissible emission levels. The latter are generally fixed by the state aiming at the development of cleaner technologies. Those emissions remaining below the standard fixed by the state, thus do not incur any costs (Ewers 1990:156). Moreover, the problem with setting emission standards is that this must be done at a point in time when the extent of the individual technologies' harmful effects is still largely unknown (see Simonis, this volume). It comes as no surprise that emission standards tend to be too low: as a rule, the expected damage is underrated (Nowotny/Eisikovic 1991). In addition, proof needs to be supplied that the foreseen legal regulations actually lead to the desired ecological effects.

Various authors therefore display a generalized doubt regarding the possibility of efficient state control of knowledge application at the enterprise level, be it via price or legal regulations (Mayntz/Scharpf 1990:65-66). This scepticism may be substantiated by a number of arguments. For instance, it cannot be excluded that – even by far-reaching state control of knowledge application – it is still impossible to prevent the unintended negative side effects of a basically desirable innovation. After all, technological contexts have become increasingly complex and unpredictable. This is also true, for example, with the catalyzer, whose use is certainly less harmful to the environment but which poses substantial, formerly neglected problems of ultimate disposal.

In addition, the thesis of controllability of knowledge application is not very plausible because of the fact that new scientific findings by their very nature constitute a public good applicable on a world-wide basis. State regulations, though, do not extend beyond national borders – thus being incapable of preventing certain technological developments as such.

Not even within national borders is it possible to invariably prevent any damage attributable to technological development by state control of knowledge application. As can be seen from the example of nuclear energy, such a policy is powerless as soon as the damage is due to causes lying beyond national borders. This is true in the same way for all those types of emission which cross national borders via air or water pollution. In this case, a technology policy based on regulation can be successful only if supranational bodies check the compliance with technological standards and are also empowered to impose sanctions.

Finally, there are those technologies which are unwanted from the supranational point of view but which are of enormous advantage to individual states. This is especially true for the military sector. In these fields, national technology policy oriented towards controlling knowledge application cannot achieve any satisfactory results. Even if there is a ban on the military use of nuclear technology in individual countries, this cannot hamper the further development of nuclear weapons.

A series of quite similar problems would pose themselves to a state technology policy attempting to achieve social objectives by means of controlling knowledge application at the enterprise level. 'Shadow prices' for the use of labour in ways detrimental to health, for instance – independently of the difficulty of setting them – would be problematic primarily because of the assumption of health as a renewable good. Legal regulation, on the other hand, is confronted with the problem of proving the causality between damage and specific technologies – an obstacle hardly to be overcome. Chronic diseases are primarily the consequence of complex cause-and-effect relationships in which the coping potential of the person concerned plays a substantial role: in addition, they appear in many cases only towards the end of a person's working life. Even greater difficulties arise if indicators other than health hazards are taken as a yardstick for assessing the criterion of social compatibility: for example, that of self-realization.

Apart from the various forms of regulation, public contract-placing is considered an especially suitable tool for implementing social and ecological aspects by means of public R&T policy (see Edquist, this volume). By virtue of the mere choice from among several technologies available on the market, the state might conceiveably influence technological development, provided that the level of state demand is of any significance. For instance, vehicles for state institutions might be purchased from the point of view of specifically ecological considerations. Also, private enterprises whose production organization is considered as exemplary with regard to social criteria might be given preference when placing public contracts. Admittedly, however, this is only an indirect way of steering technological progress.

Of much greater significance is the placing of public contracts with the object of providing specific research and development results. There are special advantages to this way of exerting public influence: the expected research and development results may be specified in detail; the instrument of contract-placing may be used in a highly flexible way; and the state, through compiling the contracts, has the possibility of securing itself rather far-reaching rights to controlling the innovation process (Littmann 1975:210). However, if this instrument is to be used successfully, there must be a guarantee of interest on the part of private enterprises to take on contracts from the state. What is decisive here, is whether the state monopolizes patent rights to research and development results or whether those are also at the disposal of the contracting enterprise.

The chances for putting greater emphasis on social and ecological aspects by means of placing research and development contracts are relatively good within public R&T policy. There are, however, some problems with using this instrument which must not be overlooked. Let us mention here only the difficulty of a technically-adequate specification of research and development contracts by public institutions. As a rule, it cannot be taken for granted that there is sufficient qualified staff with the specific knowledge required for carrying out this task satisfactorily.

Therefore, in specifying contracts and defining the criteria of ecological and social objectives, the state usually has to rely on external know-how. This, however, involves the risk of expertize being provided from the point of view of a later participation in a call for tenders. In order to minimize such a conflict of interests, it seems reasonable to split the process of placing public contracts into several steps. Still, the problem remains that this instrument might be of quite limited efficiency for orienting technological innovation towards ecological and social objectives, due to the insufficient technical qualification of state actors.

6 R&T Policy as a Social Experiment

In view of the problems pointed out so far, it seems necessary for public R&T policy to adopt new approaches. Undoubtedly of but limited success would be a technocratic policy trying to achieve ecological and social aims by means of interfering in the market mechanism or the improvement of regulatory practices. Such a re-orientation can only be initiated by a change in political decision-making culture and by the development of new political decision-making structures and institutions.

Regarding the change in political decision-making culture primarily with a view to taking into account ecological objectives, a number of demands have been expressed (Ewers 1990:156-157; Simonis, this volume):

- the obligation of the emitter to prove as non-harmful the effect of emissions;
- the acceptance of heuristic methods of decision-making, since cause-and-effect relationships are insufficiently known;
- the individual obligation to automatically reduce permissible standards; and
- giving priority to less optimistic prognoses.

Quite similar demands can be made on state policy to take into consideration social objectives of technology development. To give only one example, enterprises should be obliged to prove that specific technological practices are not detrimental to health. In addition, abandoning the proof of causality of work-related illness, may be conceived as a basis for political decisions. There is some doubt, however, as to the feasibility of actually implementing such changes in political decision-making culture. This explains an approach which, unlike the call for a change in public decision-making culture, does not aim at an automatic expansion of state control or regulation over the dynamics of technological development. Rather, it chooses to aim at a redefinition of the conditions for legitimatizing regulatory policies (Nowotny/Eisikovic 1991).

It would certainly be asking too much to assign to the state the role of central societal control – including the definition of basic technological development lines, the establishment and administration of comprehensive research and development programmes, and the control of knowledge application at the level of society. One factor would be the immense coordination problems involved in such a process. In addition, as previously mentioned, it is these very technical qualification barriers that presently speak against such state dominance in the technological innovation process. Thus, a re-orientation of public R&T policy must in the first place be accompanied by a redefinition of the state's role in the technological innovation process. This should primarily be based on coordination, integration and information. There is more or less unanimous agreement in professional literature that public R&T policy can by no means remain restricted to either supporting technological knowledge or to controlling its application. It seems to be the state's central task to create a consensus on socially-desired and undesired lines of technological development – not least because of the far-reaching social consequences primarily due to the systemic character of new technologies (Bräunling 1986). At the same time, this means that new decision-making structures should not only aim at integrating external expertise but also at more incorporating democratic elements.

The establishment of technology assessment centres is generally considered an essential institutional innovation aiming at simultaneously pursuing economic, ecological and social ends within the scope of public R&T policy. However, the days seem to be over when one witnessed the heyday of a type of technology assessment oriented towards risk assessment in mere terms of quantity. There are several reasons for this. Firstly, risk assessment by various experts – especially in the early stages of the process of technological innovation – have proved to be widely diverging. This makes it clear that risk assessment is highly dependent on the subjective interpretations and interests of experts. In addition, risks of a social or ecological type are often connected with specific technological practices rather than with a specific material technology. Here, however, a purely technocratic risk assessment is impossible. Finally, the traditional form of technology assessment – as a rule characterized by centralization, bureaucratization and expert-orientation – lacks the democratic element. This is due to the serious social impact of new technology systems, one decisive for an R&T policy based on consensus.

The impossibility of objectivizing technological risks involves the danger of specific interests being imposed, even if unintentionally, by way of technology policy. There are various decision-making approaches designed to avoid this problem. Especially in the USA, public R&T policy is based on the instrument of concurrent expertise. One has to admit, though, that those views perceived as too critical are often excluded from the opinion-forming process: either they lack the convincing lobby or the necessary resources for scientifically sustaining their ideas. To compensate for this, public R&T policy would have to be committed to increased support for 'alternative research institutes'.

The problem of orienting the technological innovation process towards social and environmental concerns, may not be solved by means of expert advice: this type of technology assessment continues to be based on the idea of direct regulatory-interventionist control by the state. However, this means that there is no direct link between technology assessment and the development of new – or the improvement of existing – technologies. There is no direct input of knowledge on social and ecological risks into the technological innovation process; only an indirect one exists, via government measures. Moreover, technological practices largely defy the logics of state interventionism. These include, apart form technical, also organizational and cultural aspects. Thus, public R&T policy must watch out for alternatives to the classical centralist technology assessment model based on expert knowledge.

Such new forms of technology assessment have already started to appear on the horizon. At the enterprise level, those directly concerned are increasingly assigned an expert role in the techno-organizational restructuring of production and work processes. Their experience with regard to social and ecological impacts of technology is thus immediately integrated into the technological innovation process (Naschold 1986; Badham/Naschold, this volume). At the level of society, the institutionalization of a democratic dialogue between various social groups and institutions constitutes a form of technology assessment (Nowotny/Eisikovic 1991). The idea of "Constructive Technology Assessment", as spread by the Netherlands Organization for Technology Assessment (NOTA) in the Netherlands, is another indication of change in the forms of technology assessment (see van Boxsel, this volume).

Characteristic of this change is the transition from public regulation to the self-regulation of technological innovations (Latniak/Simonis, this volume). The state – or rather, the institutions established by the state – confine themselves primarily to the role of providing the necessary conditions for experiments of a socially- and environmentally-oriented technology development and application. Thus, for instance, it may be regarded as the central element of constructive technology assessment, to provide for the networking of researchers, innovators and other social groups according to the idea of a socio-technical – as well as to strengthen the mechanisms of communication and cooperation. Organizing such a dialogue pursues the aim of integrating a maximum of social perspectives and interests into the innovation process at a stage as early as possible. Accompanying investigations on socially- and ecologically-oriented technological innovations, shall certainly provide additional input into the social dialogue. In this way, this is at least the intention, a continuous learning process will be set into motion. Moreover, by means of the widest possible diffusion of the knowledge thus gained, there will be a heightening in social awareness of the social and ecological problems of technological progress.

Quite similar aims to the concept of constructive technological assessment are pursued by the North Rhine-Westphalian programme 'Man and Technology – Socially Acceptable Shaping of Technology (SoTech)' (Latniak/Simonis, this volume) as well as by the 'Integrated Environment Programme' developed by the Environmental Protection Agency (EPA) in the USA (Daneke, this volume). As far as the scope of expansion of such R&T policy is concerned, there is due cause for scepticism. Many state-supported social and ecological experiments are far from having model character, and there is hardly ever a diffusion throughout society at large. Nevertheless, such experiments seem to contribute to a cultural change, in increasing people's awareness of the social and ecological aspects of technology development and utilization.

7 Conclusion

Since its onset, public R&T policy has undergone considerable change. The new understanding of technology may serve as initial evidence of this. While state intervention was originally aimed at material aspects only, now organizational and cultural aspects are increasingly taken into consideration. The object is no longer pure material technology but rather technological practices. Moreover, to an ever-increasing extent, social and ecological aspects are integrated into the objectives of public R&T policy, despite its original exclusive orientation towards economic growth. Rather than being predominantly oriented towards fundamental research, the scope of state interventionism increasingly includes those stages of the technological innovation process that are more closely linked to the market.

While public R&T policy was originally part of economic policy, it is gradually developing into an independent policy field closely interlinked with such fields as educational, financial, industrial and labour market policy – to name just a few. Formerly implicit technology policy is thus more and more developing into explicit technology policy, which means that technological aspects are also considered in the form of state influence on these policy fields. Finally, the role of the state in the technological innovation process is also undergoing change. While the state originally understood itself to be the central actor trying to impose specific technological development lines by means of support and regulation, it has recently rather turned into a facilitator and coordinator providing institutional arrangements for the self-regulation of technological innovation. This can also be called a transition from direct to context control. In the following overview, some significant characteristics of traditional and modern technology policy shall once again be compared.

Table 1: Characteristics of traditional and modern technology policy

	Traditional R&T policy	Modern R&T policy
Object	Material aspects (material technology)	In addition, organizational, institutional and cultural aspects (technological practices)
Objective	Economic growth	In addition, social and ecological compatibility
Stage of technological innovation process	Stages of little bearing on the market (primarily fundamental research)	Also stages closer to the market (technology transfer)
Policy integration	Part of economic policy (largely implicit R&T policy)	Independent policy-field closely interlinked with other policy areas (increasingly explicit technology policy)
Role of the state	Central actor of technological innovation process	Facilitator and coordinator of the self-regulation of the innovation process
Instrument	Support, regulation	Provision of infrastructure
Policy type	Direct control	Context control

It goes without saying that this is a presentation of ideal types. In reality, R&T policy in the individual countries is a combination of traditional and modern elements. However, one may clearly observe that such aspects as context control, explicit policy, the combination with other policy fields, integration of social and ecological objectives, greater closeness to the market, and technological practices are all indeed increasingly becoming constituent elements of the R&T policies of individual countries. Nevertheless, the development of a new type of public intervention in the technological innovation process cannot yet be concluded.

References

Arrow, K. (1962) The Economic Implications of Learning by Doing, in: *Review of Economic Studies* 29 (80): 155-173.

Boyer, R. (1989) New Directions in Management Practices and Work Organization, paper at the OECD Conference "Technological Change as a Social Process", Helsinki.

Brandenburg, A.G., P. Brödner, H.W. Hetzler, G. Schienstock (1975) *Die Innovationsentscheidung. Bestimmungsgründe für die Bereitschaft zur Investition in neue Technologien*, Göttingen: Otto Schwartz.

Braun, E. (1984) *Wayward Technology*, London: Pinter Publishers.

Bräunling, G. (1986) Ansätze, Konzepte und Instrumente staatlicher Technologiepolitik, in: H.-H. Hartwich (ed.) *Politik und die Macht der Technik*, Opladen: Westdeutscher Verlag, 264-277.

Brödner, P. (1985) *Fabrik 2000 – Alternative Entwicklungspfade in die Zukunft der Fabrik*, Berlin: Edition Sigma.

Child, J., H.-D. Ganter, A. Kieser (1987) Technological Innovation and Organizational Conservatism, in: J. Pennings, A. Buitendam (eds.) *The Development and Diffusion of Microelectronics*, Cambridge, MA: Ballinger, 87-115.

Collingridge, D. (1980) *The Social Control of Technology*, New York: Martin's Press.

Dosi, G. (1982) Technological Paradigms and Technological Trajectories, in: *Research Policy* 11: 147-162.

Dufourt, D. (1991) Les politiques technologiques: une nouvelle rationalité de l'intervention de l'État dans le système productif?, in: J. De Brandt, D. Foray (eds.) *L'évaluation économique de la recherche et des innovations technologiques*, Paris: Edition du CNRS, 351-380.

Ewers, H.J. (1990) Marktversagen und Politikversagen als Legitimation staatlicher Forschungs- und Technologiepolitik, in: H. Krupp (ed.) *Technologiepolitik angesichts der Umweltkatastrophe*, Heidelberg: Physica, 147-160.

Ewers, H.J., T. Wein (1989) Gründe und Richtlinien für eine Deregulierung, Discussion Paper No. 139, Berlin: Technische Universität.

Hetzler, H.W. (1970) *Soziale Strukturen der organisierten Forschung*, Düsseldorf: Westdeutscher Verlag.

Kern, H., M. Schumann (1984) *Das Ende der Arbeitsteilung? Rationalisierung in der industriellen Produktion*, München: Beck.

Krupp, H. (1987) Forschungspolitik und internationale Wettbewerbsfähigkeit, in: *Fraunhofer-Gesellschaft. Jahresbericht 1986*, München: Fraunhofer-Gesellschaft, 64-66.

Littmann, K. (1975) *Die Chancen staatlicher Innovationslenkung*, Göttingen: Schwartz.

Mayntz, R., F.W. Scharpf (1990) Chances and Problems in the Political Guidance of Reseearch Systems, in: H. Krupp (ed.) *Technologiepolitik angesichts der Umweltkatastrophe*, Heidelberg: Physica, 61-83.
McMillan, C. (1984) *The Japanese Industrial System*, Berlin, New York: de Gruyter.
Mensch, G. (1977) *Das technologische Patt*, Frankfurt/Main: Fischer.
Naschold, F. (1986) Politik und Produktion. Thesen zu Politik und Technologieentwicklung, in: H.-H. Hartwich (ed.) *Politik und die Macht der Technik*, Opladen: Westdeutscher Verlag, 231-241.
Nelson, R., S. Winter (1977) In Search of a Useful Theory of Innovation, in: *Research Policy* 6 (1): 36-75.
Nelson, R., S. Winter (1982) *An Evolutionary Theory of Economic Change*, Boston: Harvard University Press.
Nowotny, H., R. Eisikovic (1991) *Entstehung, Wahrnehmung und Umgang mit Risiken*, Bern: Schweizerischer Wissenschaftsrat.
OECD (1988) *New Technologies in the 1990s: A Socio-economic Strategy*, Paris: OECD.
Pacey, A. (1986) *The Culture of Technology*, Cambridge, MA: MIT.
Perez, S. (1987) Micro-Electronics, Long Waves and World Structural Change, in: *World Development* 13 (3): 441-463.
Perrow, C. (1984) *Normal Accidents: Living with High Risk Technologies*, New York: Basic Books.
Rammert, W. (1989) Technisierung und Medien in Sozialsystemen. Annäherung an eine soziologische Theorie der Technik, in P. Weingart (ed.) *Technik als sozialer Prozeß*, Frankfurt/Main: Suhrkamp, 128-173.
Rammert, W. (1990) Telefon und Kommunikationskultur. Akzeptanz und Diffusion einer Technik im Vier-Länder-Vergleich, in: *Kölner Zeitschrift für Soziologie und Sozialpsychologie* 1: 20-40.
Rothwell, R., W. Zegveld (1980) *Innovation and Technology Policy*, London: Pinter.
Söderbaum, P. (1990) National and Private Environmental Policy, paper at "Annual Conference of the E.A.E.P.E.", Florence.

From Socio-economic to Socially Oriented Innovation Policy

Richard Badham

1 Introduction

In the mid-1970s, Ernst Schumacher (1974:1) proclaimed that "The problem of production has not been solved". By this he meant that despite the appearance of affluence, the modern path of development was undermining its own foundations by destroying two of its key requirements: its natural and human resources. An economics "as if people mattered" would, he argued, broaden the criteria used by this "dismal science" to include these two key elements. In recent years, assertions such as this have moved out of the arena of what could be called the culture of protest and into the culture of power (Mathews 1988). Rather than being interpreted as a social protest against industrialism per se, the renewed concern with ecology and "human" work is commonly understood as a key requirement for the successful transition to a new stage of industrialism.

Unlike the 'post-industrial' rhetoric, this new stage is correctly seen as involving a radical reassessment of the nature of manufacturing – not a gradual decline in its significance. Although many of the 1970s social concerns of technology policy have been subordinated to the commitment to "reindustrialization", these concerns have emerged anew as a major consideration in economically oriented policies. Thus, in a recent overview of the technology policy context, Schmandt (1989) emphasizes that the assembly line and the organized R&D laboratory were the two key technical innovations of the Second Industrial Revolution, in which industries based on steel, electricity, chemicals and the internal combustion engine were of central priority. While economic policy was concerned with the overall regulation of demand for mass-produced goods, science and technology policy was oriented towards the 'engine of technological change' by stimulating R&D to up to 3% of GNP. In contrast, the present third stage of industrial development is led by the new science-based industries (microelectronics, nuclear physics, information science, genetics). Key focuses of technical concern are the improvement of materials and the conservation of scarce natural resources; and the combination of computer with telecommunications technologies in revolutionizing the generation, transmission and use of information. In addition, the capacity and demand for non-standardized high-tech production results in organized R&D being displaced from its past role as primarily a 'front end' input into the innovation process, to a new role, continually intervening in the innovation process in closer collaboration with production and marketing. The result of these developments is a role for economically oriented technology policy that incorporates environmental and energy concerns as well as the creation of innovative work and organizational forms for the new 'informated' workplace and integrated scientific-industrial establishment.

In the face of these developments, technology policies are slowly changing to meet the demands and help shape the new era. However, as we are currently seeing in Eastern Europe, institutional changes can take on a dynamic that leaves ideas and policies far behind. Even more significantly for the present dilemmas facing technology policy, I would argue, is that quite often ideas can change and become partially embodied in the conventional wisdom, long before they become embodied in concrete governmental institutional and policy changes. In a sense, I believe, this is the situation today. In the face of the changes outlined, we have witnessed the collapse in the intellectual realm of two key underpinnings of traditional policy analysis: the *R&D Push Model* of technological development and the *Dualistic Model* of technological policy as divided into two separate areas of techno-economic *promotion* and socially oriented *regulation*. This is indicated by the numerous calls for an increase in funding for diffusion and technology transfer, the adoption of a more "systemic" approach to science and technology policy, the establishment of radically new links between the education and industrial systems, the promotion of new "organizational technologies", the integration of energy-saving concerns with economic-oriented technology promotion, and the establishment of more "proactive" forms of technology assessment, etc.

Yet the recognition of the importance of such issues has only begun to be accompanied by the adoption of coherent policies and institutional mechanisms for their promotion. As described in the other chapters in this book, however, there are a number of important and interesting initiatives currently underway. These face a number of obstacles, even active sources of opposition, that have to be overcome for such developments to proceed and expand. Some of these are created by the traditional divisions drawn up between ministries, others by the politically controversial nature of the changes required, and others by the intrinsic policy difficulties of adopting a broader approach to technology policy.

The aim of this paper is to examine these developments in the context of the human resources rather than the natural resources area. The focus is on technology policy concerns directly related to the changes wrought by the combination of computer technology with communications. In particular the focus of my discussion will be on the promotion of changes in manufacturing production organization and skills in order to assist technological innovation and diffusion. In its broadest sense this could be referred to as "socio-economic innovation policy" in the manufacturing sphere. In the view of many – from, amongst others, work humanization programmes, departments concerned with education and industrial relations, and trade unions – this represents an important opportunity to explicitly orientate technological innovation towards social ends. If successful, such initiatives could be described as a move towards "socially oriented innovation policy". In the past references have been made to similar initiatives in England and Germany as "socially directed innovation" (Elliott 1986) and "socially oriented research and development policy" approaches (Naschold 1980). This chapter examines these approaches in the context of current developments in technology policy in Europe, with a particular focus on experiences in the promotion of "human centred" or "anthropocentric" production systems.

2 Socio-economic Innovation Policy

2.1 Conventional Wisdom

If we take statements from management journals, consultants recommendations, governmental reports, and trade union or industrial association publications as an indication of industry's 'conventional wisdom', it would appear that the importance of the human factor in successful innovation is now fully recognized (Hayes/Jaikumar 1988; Walton/Susman 1987; Weiner/Brown 1989; OECD 1988b; ACTU 1987; BMFT 1990). Similarly to Margaret Thatcher's claim that "the Cold War has now been won", there seems to have been victory in the human factors battle, at least at the level of public displays of conventional wisdom. As detailed by a leading US information technology expert, US firms have an annual 12% increase in their data processing budget, but only 2% productivity increases – they are therefore being directed to spend more attention to the 'human element'. This has been apparent at four different levels.

Firstly, it is widely asserted that the *effective use of new technologies* requires, in the words of the Harvard Business Review, "People Policies for the New Machines". As a result of the *cost*, *interdependence*, and *flexibility* of new microelectronics based equipment, its effective use requires an increase in the skills, responsibility and motivation of system users. Sophisticated and flexible technology, it is commonly said, requires sophisticated and flexible users. This issue is confirmed in a variety of opinion surveys of managers, who list the lack of skills as the primary obstruction to effective technology introduction (Daniel 1987; McLoughlin/Clark 1988).

Secondly, it is often stated that the trend towards computer-integrated manufacturing is primarily an *organizational based* rather than an equipment-based change. In a modification of the GIGO rule ("garbage in, garbage out"), there is an awareness that if the business organization is inefficient, the automation of inefficiency will provide little benefit. It is frequently commented by firms that have introduced the new structures that it was the organizational changes that provided the most benefit. Not only is there an acceptance of the slogan "No CIM without HIM" (or HER) but the organizational component of integrated manufacturing is seen as the most important but the most difficult to implement.

Thirdly, in technology policy circles the processes of technology *diffusion* or transfer are now widely recognized as a crucial component of technological development. In particular, many countries, especially the smaller ones, emphasize the significance of technology adoption rather than technology leadership strategies, i.e. the importance within technology policy of an emphasis on promoting national capabilities in the area of the evaluation, adaptation and development of foreign technologies. Moreover, within this area importance is attached to the organizational and skill components of effective diffusion. The OECD 1988 *Science and Technology Outlook* stated, for example, that policy in these areas should

> include (all) steps taken to increase the economic efficiency with which the new technology is utilized. These might include reorganization of factory work and materials flows (such as just-in-time programmes), improved management practices and increases in education and skills in the firm's labour force. In contrast to tangible investments in physical capital, these represent investments in organizational and human capital. They may be critical to capturing the economic potential of new technology, even though they have not traditionally been viewed as directly connected to science and technology policy (OECD 1988b:51).

This concentration on the diffusion process does not exclude R&D funding, however, but rather directs this funding to the adaptation and development of received technologies. For this reason the term 'socio-economic innovation' is used in this paper to refer to the institutional processes involved in the selection, adaptation and development of commercially available technologies.

Fourthly, as represented in the calls for a *systemic* approach to technology policy (Sweeney 1985), a broader understanding has developed of the range of factors influencing the diffusion process. It is recognized that the direct or indirect funding of technology promotion or diffusion activities may be of less significance than the modification of environmental variables that either obstruct or facilitate diffusion. These include such familiar factors as the high real cost of capital, lack of technical standards which increase adoption costs, long depreciation periods for capital equipment, government regulations that promote the use of inferior domestically-produced technology, the shortage of technical personnel acquainted with technology, the unfamiliarity of management with the potential of technology, and entrenched organizational and work practices. As the OECD *Outlook* commented, however, possibly "the most commonly cited barriers relate to inadequate human capital and concern the scarcity of appropriately skilled technical and managerial personnel" (OECD 1988b:76).

2.2 Institutional Implementation

Despite these changes in the conventional wisdom, however, the practical implementation of these ideas is frequently less dramatic. Firstly, as discerned in a number of industry studies, personnel and industrial relations managers in firms are frequently not involved in early discussions of technological innovation, and are consulted only late on in the process. Secondly, ergonomists or human factors engineers are not only regarded as of little significance by designers but their weak organizational location prevents them from imposing their ideas on the design process. Thirdly, trade union involvement in technological change, especially as measured by their involvement in technology agreements, is frequently 'too little, too late'. Even where they are not actively excluded from intervention at an early stage, they are frequently constrained by inadequate time, limited resources, and the restricted focus of many of their members on job security and working conditions. Fourthly, management – particularly line management – still has a strong interest in and focus on the control of the workforce, and this is frequently embodied in the development of system elements that enhance technical control even at the cost of restricting organizational learning. Such obstacles have proved to be extremely formidable when combined with other sources of resistance created by inadequate management strategic planning, the imbalanced conflict between financial and design/R&D management and production management, the opposition of middle management and planning department to the devolution of functions, etc.

At the governmental policy level the obstructing factors are no less significant. An examination of the science and technology policy funding in OECD countries, reveals a clear predominance of invention/innovation funding rather than innovation/diffusion. The OECD has attributed this to a variety of institutional obstructions, including the fact that

> technology development policies for the supply of technology are administratively more attractive than those for diffusion and demand. Technical objectives can generally be specified more easily than market objectives and are perhaps more consonant with those who have technological rather than economic backgrounds. It is also likely to be easier, both administratively and politically, to manage public funding for the supply of R&D than the demand for innovations. The latter can involve sensitive areas such as tariff dismantlement, restructuring of tax regimes and investment incentives, changing management attitudes, reducing rigidities in the labour force and so on (OECD 1988b:54).

Institutional difficulties are even more pronounced when an emphasis is placed upon the organizational and skill components of diffusion. Science and technology policy departments often have little experience with dealing with education, training and industrial relations issues, while the other departments concerned with these issues have a similar lack of expertise in technological innovation and diffusion processes. As a result, programmes for the promotion of skills and organizational change for new technologies often take the form of technology policy department initiatives oriented to the direct practical problems experienced by industry with little integration into the overall structure of education and training, while education and training department initiatives represent lengthy processes of transforming educational structures in a manner that is often condemned for its lack of knowledge of or concern with immediate industry requirements.

This general concern with diffusion and its organizational and skill components compounds the traditional problem of coordinating a central technology policy with the activities of other ministries that control important areas of technology. In order to systematically address these issues, integration is required between all those ministries with responsibilities for factors that bear upon this process – and that includes, in particular, education and training ministries and industrial policy ministries, as well as other sometimes key ministries with responsibility for areas of science and technology such as transport, communications, energy, public industries, etc. Yet each of these ministries have their own histories, traditions and interests, and while both industry and education/training ministries have programmes that have implications for diffusion, and sometimes directly address these issues, they are often inadequately coordinated with technology ministries and innovation/diffusion policies. As a result the challenges of a systemic technology policy that directly addresses the requirements of socio-economic innovation or diffusion processes are immense. This is especially significant given the existence within industry of the powerful factors, outlined above, that obstruct this process.

2.3 Socially Oriented Technology Policy

The attempt by technology policy-makers to grapple with these issues of socio-economic innovation occurs within a quite specific context. In the face of the changes outlined in the introduction, and the direct threat of "deindustrialization", the general emphasis of technology policy has shifted – from the more "socially" oriented initiatives in the late 1960s and 1970s, to a greater concern with adapting the science and technology system to the radically new "demands" of "reindustrialization". For some critics this had led to a "new politics of science" that, by promoting indirect funding and using tax relief mechanisms, substantially reduces public control over science and technology (Dickson 1984). Other critics have observed that it involves a move towards a "techno-corporatism" that creates an expert elite stratum of policy-makers that reduces the traditional role of interest group bargaining in the formulation of policy (Fischer 1988).

Drawing on a number of experiences in countries such as the United States, Great Britain and Australia, these perspectives regard government claims to be taking social factors into consideration in the innovation process is at best a techno-economically driven enterprise, and at worst a cynical ideological exercise designed to legitimate changes to organizational and work practices that worsen employees working conditions (Couchman 1990). In contrast, an alternative perspective has emerged referring frequently to initiatives in Scandinavia, West Germany and sometimes Japan. This more optimistic version argues that the decline of explicitly "social" policies has been accompanied by the arrival of a second industrial divide that provides the techno-economic preconditions for a repoliticization and democratization of technology policy, a new era of human capital, and new forms of socially directed innovation (Elliott 1986; Mathews 1989a; Piore/Sabel 1984; Reich 1983). In either case by pointing to the importance of the promotion of technology for non-economic social ends and the limitations of "reactive" forms of regulation, both perspectives demand a reconsideration of the Dualistic Model of technology policy as separate areas of promotion and regulation.

3 The Dualistic Model of Technology Policy

In the classic terms characteristic of the Dualistic Model view, Professor Ernest Braun stated in 1984 that

> The current interest in technology policy arises out of a paradoxical twin concern: on the one hand the wish by each nation not to be left behind in the race for technology-based prosperity and on the other hand multi-faceted fears about undesirable social and environmental consequences of technology. The belief that better technology means higher economic efficiency and international competitive advantage is a strong driving force behind attempts by most governments to support, foster and accelerate technological innovation. The fears about technology, on the other hand, force governments into a range of regulatory activities (Braun 1984:123).

This distinction is philosophically based on distinctions between facts and values, and techno-economic and social issues. In its institutional form, different commentators from across the political spectrum have made major distinctions between promotion and control institutions, stimulation and regulation institutions, and accumulation and legitimation activities of the state (O'Connor 1973; Johnston/Gummett 1979).

This distinction has, however, been undermined by theoretical developments and practices within both the 'promotion' and 'regulation' areas. From within the promotion area, innovation studies have simultaneously uncovered the socio-economic processes involved in establishing alternative technological trajectories, regimes or paradigms, and revealed the "paradox of automation", i.e that as automation succeeds in taking over a multiplicity of human tasks, it increases the significance of those remaining and establishes the basis for new learning opportunities and developments. In combination, these insights reveal both the social construction of paths of technological development, and the importance of social and organizational innovations as constituents of the innovation process.

Moreover, as Piore and Sabel point out in their classic work *The Second Industrial Divide* (1984), social changes of this kind do not emerge solely from the market or represent a simple social adaptation to the demands of technology.

Models of production, they point out, originate from within specific political and institutional constellations. They outline in particular the conditions affecting the European craft model and the United States mass production model. As is now widely discussed, such models are diffused from one country to another, with each country adopting 'successful' models but adapting them to their own specific circumstances (Boyer 1989). This process of adaptation involves changing customs and practices in a manner that imposes different costs and benefits on various groups. When new technologies as significant as microelectronics diffuse throughout the economy, and organizational models arise such as new Japanese management methods, the adoption and modification of these techniques in different production systems brings about a change in conditions of work and organization that have become accepted as a 'natural' part of economic life and even a key component of cultural identity. The resulting paths of socio-economic innovation are, consequently, inherently political and socially structured processes. As a result the "technical" promotion of technology for "economic" ends is inevitably subject to political shaping, however much 'technocrats' or what Professor Self has called "econocrats" attempt to support "the belief that there exist fundamental economic tests or yardsticks according to which policy decisions can and should be made" (Self 1975:15). The problem that this inevitably creates is how such political issues should be addressed, particularly within institutional organizations that have a self-understanding of their work as primarily technical or economic.

From within the so-called regulation area, the study of the domination of technological development by large scale systems has been a major factor leading to the explicit recognition of the "social shaping" of all technological artefacts and systems. Most of the traditional plethora of regulatory institutions (monitoring committees, public inquiries, technology assessment bodies, etc.) have found it extremely difficult to intervene effectively in the paths of development promoted by these systems, and have been restricted to a primarily "reactive" role. There has, consequently, been an appeal for a more "proactive" or "constructive" approach to technology assessment that attempts to intervene earlier on in the innovation process – thus attempting to 'promote' alternative paths of development. All such initiatives are grappling with what David Collingridge has called "the dilemma of control", i.e."attempting to control a technology is difficult, and not rarely impossible, because during its early stages, when it can be controlled, not enough can be known about its harmful social consequences to warrant controlling its development; but by the time these consequences are apparent, control has become costly and slow" (Collingridge 1980:19). If, more appropriately, the word 'shaping' is substituted for 'control', one of the central dilemmas of all socially oriented technology policy is revealed.

A third area of technology policy in which the distinction between promotion and regulation mechanisms has been broken down is in the sphere of development policy. In this sphere, a greater concern for the development of productive technologies 'appropriate' for radically different cultures has led to a greater awareness of the artificiality of the Dual Model. This has not, however, been a dominant emphasis within economic policy in many developing countries. In the recent publication *One World or Several* from the Development Centre of the OECD, for example, Federico Mayor commented that

> The recognition that expenditures on human resources development are not only necessary in themselves and go to the very heart of development, but that they also yield essential returns of an economic nature, has not yet been fully incorporated into economic development planning, nor into analyses of the allocation of increasingly scarce resources available to the social sectors (Mayor 1989:185).

One attempt to develop appropriate concepts has been stimulated by the work of Stevan Dedijer as witnessed in the collection *From Research Policy to Social Intelligence* (Annerstedt and Jamison 1988). This attempts to provide a concept of "social intelligence" as the organizational ability of a country to avoid mimicking overseas models of development and adapt to a changing world by combining the effective acquisition, evaluation and use of information with planned policies and activities for development.

4 Socially Oriented Innovation Policy

4.1 The Limitations of Socio-economic Innovation Policies

Despite this general breakdown of the Dualistic Model, in the area of innovation policy the claim that human and organizational considerations should be incorporated as a key factor in socio-economic innovation is frequently treated with suspicion. Although there is an increasing recognition of the need to take such factors into account, attempts to promote their incorporation are often seen by business and government policy makers as the inappropriate insertion of "soft" factors into the "hard" processes of technical and economic development. In response, the rhetoric of socio-economic innovation promoters strongly emphasizes the economic necessity of the organizational changes they advocate, often downplaying or actively denigrating the role of human or social considerations as goals of the process of change (den Hertog/Schöder 1989). Where greater emphasis is placed on human skills, teamwork and organizational decentralization, this is represented in traditional 'socio-technical' terms as a fortunate overlap between economic requirements and social desirability. Unlike the 'socio-technical' tradition, however, this view often involves a form of 'soft technological determinism' as these changes are portrayed as an inevitable requirement for the effective use of new microelectronic based technologies. It thus becomes difficult to orient innovation policy towards a consideration of alternative socio-economic innovation trajectories, and the stimulation of more 'humane' paths of development.

A second restriction on socio-economic innovation policy, as promoted by science and technology ministries, is a tendency to focus on traditional areas of specialization. This is exemplified by the response to the need for greater human skills, knowledge and intra-firm learning with policies for stimulating the number of R&D personnel. Thus an authoritative report on innovation policy by Rothwell and Dodgson, after arguing that "There exists strong evidence to show that an SMF's ability to access, transfer and assimilate external knowledge depends to a large extent on its level of in-house technical expertise" emphasizes the importance of "a scheme to subsidize the costs of hiring R&D personnel in SMFs" (Rothwell/Dodgson 1989:16). Despite the importance of such schemes, it is crucial that the level of "in-house expertise" is also taken to refer to developmental and reflexive processes of "learning by doing" on the part of system users. Thus, at a Japan Institute of Labour conference, for example, a broader approach to *The Role of Labor and Management in Technology Transfer* is adopted, defining the "technology transfer triangle" as a systemic interlinking of government, management and workers (Pock Too 1987). This emphasis is supported by evidence of the increasing concern of management not only with the computer skills of engineers but, centrally, with their human relations and industrial relations abilities (Williams 1988). A systematic consideration of such issues, however, takes innovation policy beyond the traditional province of many science and technology ministries.

4.2 The Nature of Socially Oriented Innovation Policies

As the term is used here, socially oriented innovation policies are intended to transcend both the restrictions outlined above. On the one hand the stimulation of socio-economic innovation is not understood purely in terms of facilitating the establishment of the 'social' conditions necessary to speed up technological innovation and maximize economic profitability. It is, rather, concerned with *investigating the range of socio-technical innovation alternatives available for stimulating economic advance, and promoting paths of development in accordance with the production culture and social priorities of different countries.* On the other hand this activity cannot, and should not, be seen in terms of the direct or indirect promotion of R&D funding and personnel but, rather, as *a systemic approach to intra-firm and inter-firm innovation processes and governmental innovation policy.*

The form in which these intra-firm and inter-firm innovation processes occur will differ substantially between firms, industries and countries. As a result, there can be no simple prescription of 'the' model of humane paths of development or appropriate policy mechanisms. As revealed in a number of studies of different production cultures, the development and use of new technologies is determined by the "social constitution" (Hildebrandt 1989) or "inherited characteristics" (Child 1987) of user firms. These may be related to their individual history, the nature of the industrial sector in which they are located, or their position within a national system of skills formation and industrial conditions. In order to develop effective socio-economic innovation policies and orient these towards social goals, these contexts need to be given central priority. Moreover, countries clearly vary considerably in the nature of their science and technology system and relevant governmental structures. The form taken by systemic approaches to socio-economic innovation will therefore differ in accordance with the nature and degree of established government intervention measures, the method of interest group representation, and negotiated forms of inter-departmental co-ordination.

4.3 Neo-Taylorist and Skill-Based Manufacturing Trajectories

Despite these qualifications about the context specific character of socio-technical innovation, there has been sufficient research conducted on international variations in the introduction of new microelectronics-based technologies in manufacturing to develop some general guidelines concerning the range and desirability of organizational options in this area. Due to the flexible and integrated character of the new technologies, the traditional conditions of both craft and mass producers are being changed. From the point of view of stimulating workers skills and autonomy, however, the form that this takes can vary between two extremes, variously described on the one hand as "neo-Taylorist", "technocentric" or "neo-Fordist" regimes and "skill-based manufacturing", "anthropocentric" or "post-Fordist" regimes on the other (Badham/Mathews 1989; Badham/Schallock 1991a).

The *neo-Taylorist* tendency is characterized by the creation of work systems that involve:

- a broader range of employee tasks but the restriction of job enrichment opportunities or 'vertical' skill attainment;
- an increase in skills and training but a simultaneous growth in system monitoring and pacing of the workforce;
- the tendency to develop and reward the skill of key employees and operators but a restriction of this group to a minority of the workforce;
- the attempt to motivate employees responsible for costly and interdependent equipment but the restriction of autonomous learning processes using the new information possibilities of new process technologies; and

- the training of workers in the new 'technical' skills required to operate the equipment but a lack of consideration to the use and integration of traditional skills in the design of the new systems.

In contrast, the *skill-based manufacturing* tendency involves:

- increasing job enrichment as design and planning tasks are devolved to system operators;
- a growth in the autonomy and responsibility of the workforce in the exercise of local initiative and the developmental "learning by doing" processes;
- a reduction in the division of labour between "skilled" and "unskilled", and an increase in job rotation and team-based organizations;
- the design of new system software, architecture and organization in order to facilitate the use and development of traditional skills and utilize system development capabilities.

The degree to which different firms, industries and countries will be able to, or will attempt to, approximate towards one or other end of the spectrum will vary. Moreover, it is only to be expected that in the series of negotiations and trade-offs involved in establishing the new production systems, no single model will emerge of 'the' desirable path of development. The challenge of socially oriented innovation policies is to adapt policies to these contexts while formulating and implementing criteria in a process of economic *and socially* oriented intervention. Although this may be regarded as a difficult task, it is important to recognize that the social structuring of paths of development will inevitably happen, even if this is carried out as a "practice without a name" while different interest groups promote their view as "the" "efficient" and "inevitable" form of socio-economic innovation. The advantage of socially oriented innovation policies is that they make this social structuring explicit, and directly relate negotiations to competing visions of humane and desirable working conditions.

5 Socially Oriented Innovation Policy Institutions

At the level of industry policy a wide variety of schemes have been suggested for integrating social, economic and technology policies in order to stimulate the process of socio-economic innovation. Influenced by the success of innovative regional economies, Piore/Sabel (1984) have recommended integrating financial investment, training and political initiatives in the stimulation of collaborative regionally based innovative economies. This has stimulated further widespread international interest in regional firm networks and methods for their promotion. Drawing upon experiences in Japan and Scandinavia, Robert Reich (1983) places central emphasis on the large corporation as the centre of innovation, and advocates government funding of corporate socio-technical innovation through integrating investment subsidies, job-retraining schemes, and social welfare funding. For others emphasizing the central role of a democratic industry policy, Scandinavian countries are often set up as a model with initiatives extending from work environment and codetermination legislation to nationally coordinated business retraining schemes and the use of pension funds to finance tripartite investment decisions (Mathews 1989b).

At the level of specific initiatives to stimulate socio-economic innovation in a direction that explicitly incorporates social as well as economic goals, a number of common general measures have been developed or suggested by proponents in different countries. In practice there are often some difficulties in distinguishing between more restricted socio-economic initiatives from socially directed ones, as the public rhetoric of many of the former include references to "social goals" (see section 3.1). However, some of the main policy mechanisms and institutional forms that could be used to stimulate socially oriented innovation are listed below.

5.1 Technology Transfer and Diffusion Mechanisms

Many of the technology transfer and diffusion mechanisms developed over the last decade may be utilized to promote social objectives in the innovation process. The foundation of such moves can be the incorporation of social objectives in the charters or mission statements of these programmes or institutions, composing a Board structure that incorporates at least tripartite representation and ensuring that suitable personnel and funds are made available. Some of the key areas in which such innovations can be made are:

Government-Assisted Technical Consultancy and Training Institutions: These institutions can assist in the selection, implementation and development of skill-based manufacturing systems. Activities may range from providing assistance in the development of appropriate technology and human resources strategies to the training of system users. A key component of such institutions could be the operational definition and application of appropriate system design and implementation methods for the new systems.

Government-Assisted Design and Innovation Centres: These centres may be responsible for software and system design to support skill-based manufacturing systems. A key function of these centres should also be the marketing of these systems, and diffusing knowledge of their advantages into industry.

University Research and Consultancy Units: The substantial resources of the higher education and university system can also be utilized to address the same issues as the above two types of centres. This can be assisted by the establishment of education-based centres for research and consultancy with similar guidelines to the above.

Industrial-Sector Restructuring Plans: Many countries now have sectoral industrial plans to assist ailing traditional industries and promote the development of rising high-tech industries. This assistance can be administered by tripartite bodies, and involve the funding of investment dependent upon firms involving unions in the establishment of business plans and the design of new work practices.

Government Procurement Policies and Offsets Arrangements: Government purchasing practices can be influential in two ways: the software and systems purchased for use in public manufacturing enterprises can be required to conform to skill-based manufacturing design guidelines; and government institutions may require that firms from which they buy goods and materials move towards skill-based manufacturing working practices. In addition, where goods are purchased from overseas companies a requirement can be made for these firms to transfer skill-based manufacturing and knowledge to collaborating firms in the home country.

Investment Funds, Programmes and Corporations: Publicly assisted investment funds may range from venture capital enterprises through development corporations to state regulated pension funds. These may be directed towards the funding of products or firms that enhance or are moving towards skill-based manufacturing processes, or be dependent upon initiatives being made in this direction.

5.2 Management and the Development of Human Capital Incentives

As revealed in a number of international studies, managements in different countries are influenced in their personnel policies by the structure of national systems of skills formation. The effect of these systems can, however, be strengthened or weakened by direct government actions. Initiatives to enhance the importance attached by management to upgrading the skills and responsibility of their workforce, and correspondingly their working conditions, include a variety of measures of which some are:

- Apprenticeship training grants
- Short term job training or retraining schemes
- Nationally coordinated training funds financed by employer contribution
- Employment insurance or other legislative measures to discourage external labour market strategies towards the hiring and firing of employees.

5.3 Workers Representatives and Concrete Codetermination Support

In the formulation of detailed skill-based manufacturing designs for specific firms, workers representatives play a crucial role in both determining the social criteria of central importance to the workforce and in ensuring the effective participation of employees in system selection, implementation and development. Consequently, the two major arguments for providing assistance to workers organizations and representatives are: firstly, the imbalance of resources between capital and labour; and, secondly, the difficulties that workers and their representatives have in adequately considering the implications of new systems and examining the choices available. Some possible government initiatives are:

Government Funding of Union Technology and Industrial Democracy Units: These can act as a source of information, training and advice for firm-based worker representatives, as well as initiating projects in the definition and implementation of skill based manufacturing systems.

Legislative Support for Information and Consultation Requirements in the Introduction of New Technology: A range of measures may be involved here, but an important consideration is the provision of arrangements for workers representatives to hire outside consultants, allocate sufficient paid time to considering or making innovation proposals, and participate in union/public institution training courses to improve their knowledge and skills in the area of innovation.

Legislative Guidelines on Work Environment Conditions That Must Be Met in the Introduction of New Technology.

A Requirement for Worker Involvement or Acceptance when Firms Review Government Assistance for R&D and Investment.

5.4 Nationally Based Skill Formation Restructuring

The degree of interest that individuals, firms and unions have in the upgrading of skills and employee responsibility reflects the national skill formation structures within which they are embedded. Many of the initiatives outlined above have an influence on these structures; indeed, in a sense they partially constitute it. However, government actions in at least three other areas have an important influence and are hence of crucial significance in shaping alternative paths of socio-economic innovation. These three are public education systems, government collective bargaining institutions, and public sector enterprises.

Public Education Systems: In order to facilitate both horizontal and vertical upskilling of employees, new broader-based training courses are required in addition to modular course structures that allow for retraining and upgrading. These course structures are part cause and part effect of career paths and job structures in industry, and consequently require modification to support skill-based manufacturing development.

Government Collective Bargaining Institutions: The nature and role of governments in the collective bargaining process clearly vary greatly between countries. However, these mechanisms can, where possible, be utilized to ensure that the successful resolution of conflicts over wages and working conditions are related to the introduction of a number of labour market characteristics that support skill-based manufacturing. Some of the most crucial are: the broadbanding of job classifications, pay for knowledge wage systems, and the establishment of skill-based career paths.

Public Sector Enterprises: The influence of the government on public sector enterprises can be used to assist the development of skill-based manufacturing processes by introducing the changes outlined above.

The schematic nature of this list necessarily provides only a general outline of possible socially oriented innovation policy measures. Any coherent and effective socially oriented innovation policy would require the considered development of an integrated package of such measures. In practice, however, technology policy programmes oriented towards social ends in the area of process innovation have predominantly taken the form of direct funding of research and development or company projects in this area. As illustrated in Table 1, there has been a proliferation of such government programmes to stimulate socio-economic innovation in Europe over the last two decades. Within these programmes one strand of research and promotion has developed that is explicitly directed towards socially oriented innovation – those concerned with the promotion of "human centred", "anthropocentric" or "skill-based" manufacturing systems. Despite the relatively small size of such projects and programmes when compared to overall government technology policy expenditures, their novelty and originality deserves some further consideration as an important set of initiatives to explicitly conduct a socially oriented innovation policy. The purpose of the next section of this chapter is therefore to briefly outline the range of programmes involved, describe the general character of the socio-technical trajectory promoted by these programmes, and discuss some of the problems and challenges that have faced these initiatives. As commented above, an effective socially oriented innovation policy requires a far broader and more extensive set of initiatives. However, as exercises in the use of direct promotion mechanisms to stimulate socially oriented innovation, the experience of human centred projects and programmes provide an important insight into the difficulties and dilemmas facing technology policy in this area.

6 European Promotion of Human Centred or Anthropocentric Systems

6.1 European Projects and Programmes

Government promotion of human centred or anthropocentric systems was prominently represented by a number of exemplary projects carried out in the early 1980s in Sweden, Germany and the UK. In the UK, at the University of Manchester, an interdisciplinary project funded by the British research council SERC was undertaken by the world-renowned control engineer, Professor Howard Rosenbrock, to create a human-centred lathe which left the creative programming decisions in the

Table 1: European government programmes promoting technology, work and organization

Government programmes	R&D	Implementation & diffusion	Greater technology orientation	Greater education and organization orientation	Contact
UK: Science policy research unit, ESRC funded programmes (PICT, New Technologies in the firm)	X	X	X		R. Coombs Manchester University
France: Ministry of S&T interdisciplinary programme of research on technology, work, employment & ways of life (PIRTTEM),	X				National Centre for Scientific Research
people, work and technology programme		X			S&T Ministry
Netherlands: Dept. of education & science: Technology work & organization (TAO) programme	X	X		X	Dr. Hertog MERIT Institute Maasstricht, also TNO Institute & Erasmus University
Sweden: Work fund programme (MDA, LOM, development)	X	X		X	B. Shepherd Nyman, W.E. Fund
Norway: HABUT programme		X		X	HABUT
Germany: Ministry for research & technology: work & technology (AuT)	X	X	X		BMFT AuT (P. Oehlke) Projektträger
manufacturing technologies (PfT)	X		X		BMFT PfT (T. Martin) Projektträger
programme, Bremen work & tech (AuT) programme	x	X			G. Simonis, AuT Institut, Gelsenkirchen, Prof. F. Rauner, Bremen Univ.
Belgium: Flemish Foundation for Tech Assessment (STV)	X	X			STV
EC: FAST 11:	x	X		X	W. Wobbe, FAST 11
ESPRIT, BRITE programmes	X		X		ESPRIT CIM Human Aspects Advisory Group (P. Kidd)

X: Represents relative emphasis compared to other programmes; they do not represent an exclusive concentration. Where absent, there is no comparatively dominant specialization.

x: Represents less significant emphasis.

X and x are intended as rough guides only.

Source: Adapted from den Hertog/Schöder (1989)

hands of the operator, with the computer performing repetitive and routine functions. In Sweden, from within the Centre for Working Life, a computer scientist with sociological training, Pelle Ehn, coordinated the UTOPIA computer-based page set up project, partially funded by the Work Environment Fund. The aim of the project was to develop a system, in collaboration with print workers, that aided rather than undermined traditional craft printing skills. In Germany, an engineer working within the Manufacturing Technologies programme of the German Research and Technology Ministry, Peter Brödner, emerged as the leading exponent of a collaborative project to establish workshop oriented programming (WOP) systems for decentralized craft-based machine programming, and decentralized production islands for computer assisted multi-skilled group work with both direct and indirect production activities carried out by craftspeople on the shopfloor (Brödner 1987, 1989, 1990). Although these three projects in particular were deliberately organized and influential as exemplary models, they do of course only represent a few of the multiplicity of initiatives that were carried out during these years, especially in Germany and Scandinavia.

These projects were the product of, and source of further inspiration for, a fairly extensive "humanization of work constituency" (Badham/Naschold, in this volume). This constituency is oriented to the redesign of the technological and social components of modern production systems to improve both production efficiency and the skill, responsibility and autonomy of the workforce. They are the product of a number of different developments but, in particular, they owe their intellectual origins to four main influences: firstly, the theoretical work and practices of the Tavistock Institute ("socio-technical theory") and the Lucas Aerospace/Greater London Enterprise Board activities ("socially useful production"); secondly, the post-war critique of the "inhumane" nature of automation emerging from control engineering and computer science[1]; thirdly, the movement for "alternative technology" and "socially responsible" growth that gathered momentum in the 1970s and found wider cultural acceptance and institutional embodiments in the 1980s[2]; and, fourthly, the quality of working life or humanization of work movement that became increasingly popular throughout the 1970s.

It was, however, the national traditions of work reform and national government programmes that provided the major direction and stimulation to the different projects. In the UK, the network of local enterprises fostered and co-ordinated by the Greater London Enterprise Board did much to promote the ideas of socially useful production associated with the promotion of human centred technology concepts developed by Mike Cooley while participating in the formation of the shop stewards plan at Lucas Aerospace (Cooley 1984). In West Germany, there have been a number of attempts to define and implement features of 'anthropocentric' production islands in an industrial relations context that includes a strong trade union structure, extensive legal support for workers rights, co-determination legislation and statutory consultation with works councils. Two of the key influences have been the programmes funded by the Federal Ministry of Research and Technology in the areas of manufacturing technologies *(Fertigungstechnik)* and the humanization of work (now work and technology). The 'Humanization of Work' programme was established in 1974 and by 1981 it had spent DM 400 million and budget plans were set to run at 100 million DM annually. The Manufacturing Technologies programme was less directly oriented to work humanization and subject to less formal trade union influence. However, it played a key role in the development of programming technologies and production island concepts that came to exert a strong influence on human centred ideas. Of major significance have also been the programmes established by the North Rhine-Westphalian government (Socially Oriented Technology Design) and the Bremen government (Work and Technology). In North Rhine-Westphalia, the activities of the Institute of Work and Technology, linked to a special research group on production islands at the University of Bochum, are playing an increasing role in the further development of anthropocentric concepts.

In Scandinavia, sociotechnical ideas from the UK were developed in Norwegian industrial democracy experiments from the mid-1960s (Thorsrud/Emery 1970; Elden 1979). In addition, a development agreement between the Norwegian employer and union organizations, the NAF (Norwegian Employers Confederation) and the LO (Norwegian Confederation of Trade Unions), allocated 5 million Crowns (US$ 700,000) annually for research and education programmes on the work organization and industrial democracy features of new technology. However, since the establishment of the Swedish Work Environment Fund *(Arbetsmiljofonden)* in 1972 (which allocates over US$ 40 million annually, funded by a tax on employers), Swedish projects have come to play a dominant role. Of particular significance for human centred systems design, have been the Development Programme for New Technology and Work Organization (1982-1987 US$ 8 million), and the Programme for Man-Computer Interaction (a joint programme with the Swedish National Board for Technical Development) (1987-1992 US$ 10 million). The former programme was established "to investigate the possibilities of developing job content and work organisation, as well as high productivity and competitive capacity, in conjunction with the introduction of new technology" (Oscarsson et al. 1988) and both advise and disseminate the information from the 24 projects. The latter programme is aimed "to promote the development of computer technology at work adapted to human needs and requirements by a systematic build-up of interdisciplinary knowledge. This is to be achieved by stimulating and supporting *collaboration between research and development in technology and work environment research*" (MDA 1990:3).

In addition, the Work Environment Fund provided funds for the Swedish Centre for Working Life *(Arbetslivscentrum)* which was established in 1977 as an independent tripartite body. Government legislation of codetermination and participation, provides the guidelines for the research activities of the centre which are directed by employer/union agreement. Part of the research philosophy of the centre is that in order to have a significant impact on new technology, unions have to present workable modifications and alternative designs. The emphasis is on local trade unionists working in collaboration with researchers (engineers, computer specialists, linguists and social scientists). The centre was directly involved in the UTOPIA project's activities to involve graphic workers, computer and social researchers working together to propose modifications to the design of new newspaper text and image technology.

In the second half of the 1980s, the activities undertaken in the different national projects have been formally written up and diffused in English in the form of books by Brödner, Ehn and Rosenbrock. At the same time the human centred activities have become more influential at the European level. This has been most prominent in the European Commission ESPRIT projects 534 and 1199, with the latter in particular combining representatives from the UK, Scandinavia and West Germany. Current projects of significance include ESPRIT 5603 and BRITE-EURAM 3302. The European Commission FAST programme has also taken up these ideas and encouraged their diffusion through the exertion of influence on European technology programmes such as ESPRIT and BRITE-EURAM, the establishment of a newsletter (AT&S), collaborative research projects on national conditions favourable to anthropocentric systems, the creation of an 'institute without walls' to assist European interaction, and a series of seminars and working papers in the area of human centred or anthropocentric technologies. This coordinating activity is strongly assisted by the work of the IFAC Social Effects of Automation Committee with its triennial conferences on Skill Based Automation and workshops in IFAC conferences, the establishment of a MODEM network linking national programme administrators in the technology, work and organization area, and the work of the SEAKE Centre in the UK in the publication of the journal *Artificial Intelligence and Society* and the Springer-Verlag Artificial Intelligence and Society book series.

This further development in the latter half of the 1980s has been strongly influenced by three additional factors: firstly, the apparent development within the computer and process control industries of both the capacity for decentralized production solutions and a greater degree of attention to "user-oriented" design practices; secondly, widespread debate and substantial agreement on the need for "new production concepts" or "broader" "comprehensive innovation concepts" to meet the "Japanese Challenge" and address the failures of the "automated" "unmanned" "computer integrated" factory of the future; and, thirdly, the closer integration of the European Community, and debate over the meaning of social citizenship and technology policy for the "new Europe". An influential part of these developments in Europe has been the debates surrounding the European ESPRIT CIM-OSA project to create a new set of technical/organizational standards for future systems development for the computer integrated manufacturing (CIM) "factory of the future", the role of the FAST programme in promoting a socially oriented view of technology policy that deliberately facilitates the integration of research networks around the design and implementation of "anthropocentric" production systems, and the rise and development of government programmes, research practitioners and consultancy firms oriented towards implementing "comprehensive innovation concepts" (Badham/Naschold, in this book).

6.2 The Socio-technical Trajectory Defined

In combination these developments have resulted in the creation of a European community of projects, programmes, institutions and networks that have a substantial interest in the promotion of human centred or anthropocentric systems. The major activities and influences upon this community are schematically outlined in Figure 1. Any discussion of the general approach adopted must be aware of the clear influence that different national contexts have had on the development of various projects and programmes. As socio-technical theory was differently emphasized in different countries such as the UK (liberal-pragmatic), Scandinavia (worker democracy), the USA (human relations), and the Netherlands (comprehensive-strategic goal oriented) (den Hertog/Schöder 1989; Ehn 1988) so the different human centred activities have received a different emphasis. This has involved, for example, a stress within Germany on production islands, within Scandinavia on democratic participatory design, and the UK on two themes of human/machine interface and socially useful production. Despite this divergence of national emphasis, however, certain common themes of a general approach can be observed.

A key focus is the direct promotion of new exemplary models and techniques of system design to enhance job enrichment and worker autonomy. Through the use of such dichotomies as "technocentric" and "anthropocentric" approaches to new manufacturing systems, this approach continues and develops upon the traditional socio-technical and organization theory contrasts between "mechanical" and "organic" systems, "low-trust" and "high-trust" cultures, "Theory X" and "Theory Y" management philosophies, etc. In this sense there are a number of similarities with the image of the skill-based socio-technical "integrated factory" provided by socio-technical writers in the United States. According to this view, the contemporary introduction of flexible and integrated technologies in an increasingly uncertain manufacturing environment results in a general increase in the need for worker's skills, responsibility and teamwork.

The human centred model is in three ways also a partial rejection or important development upon this traditional approach. Firstly, it emphasizes the socially shaped, negotiated and influential character of technological systems. It is stressed that job designers that fail to actively intervene in the initial shaping of technological systems are, in a sense, "rearranging deckchairs on the Titanic" (Rosenbrock 1990). Key decisions about work are built into technological systems, and given traditional Taylorist management philosophies and technocentric views of human operators as sources of "error" in automated systems, the *poten-*

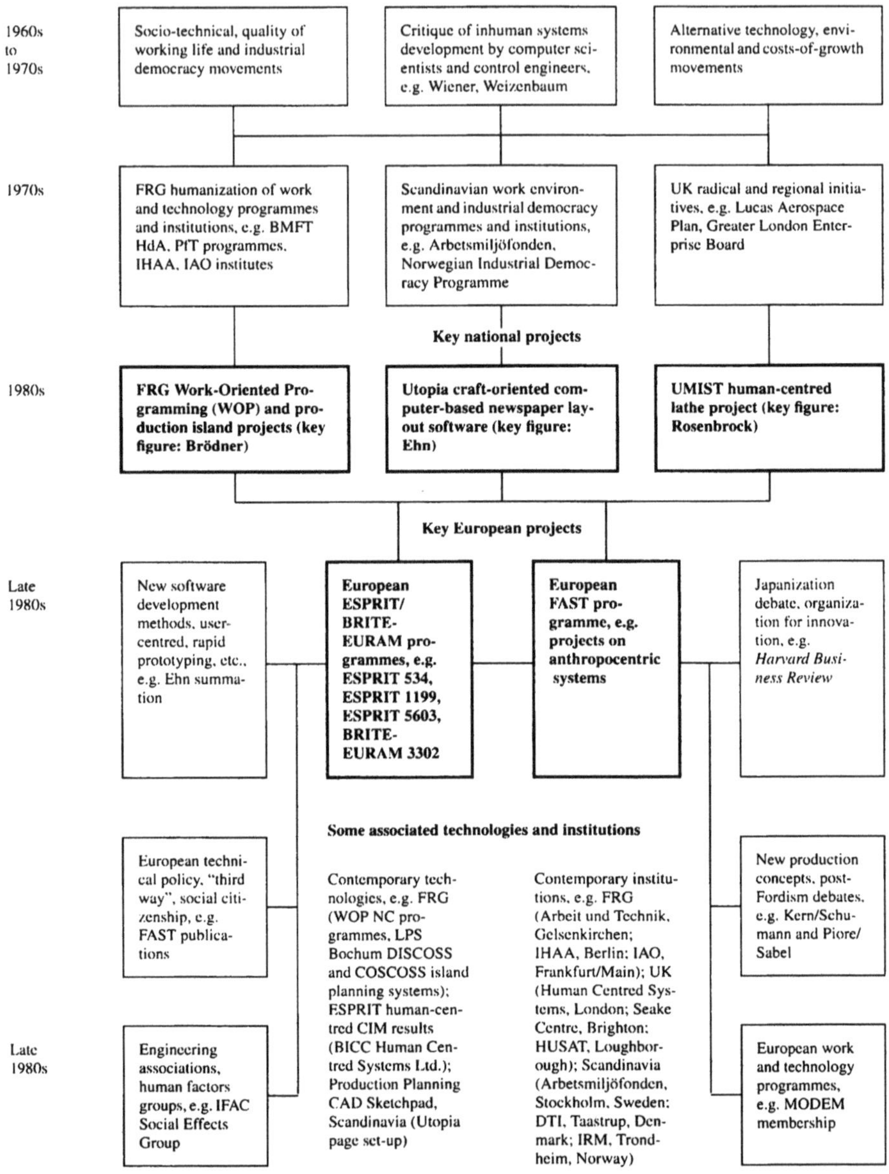

Figure 1: Schematic outline of human centred system design developments

tial of modern technology to facilitate enriched group work may be prevented. In Shoshana Zuboff's terms, systems will have "automatic" functions built into them and "informating" capabilities built out (Zuboff 1988). The human centred capabilities of new systems are thus a potential to be realized in truly interdisciplinary (Kidd 1990b) system design, not a consequence of fortuitous developments in information technology and market conditions. The development of the human centred CIM philosophy and the creation of exemplary models, practical design criteria and techniques, is consequently seen to be of crucial importance in influencing manufacturing technology paradigms. As Howard Rosenbrock commented, while "there is a strong tradition in the social science of 'socio-technical design', in which the technology and conditions of work are both studied together in the design stage", this is usually only related to "small changes at the fringes of technology". In contrast, human centred projects can be seen as attempting to apply socio-technical design principles at a deep level in technology.

Secondly, this approach is explicitly oriented to nations with higher levels of skills, established trade union structures, competitive advantages in higher-quality and more customized manufacturing products, and cultures and political systems based on individualism and democracy. These conditions, it is emphasized, provide *real* resources that can be deployed in the shaping of new production systems in a manner that builds upon and develops rather than undermines human skills. It is not assumed, therefore, that every industry in every country will (or can) follow the same route but rather that the above economic and political conditions broaden the range of options for some firms and nations. Those countries with, for example, higher trust production cultures and producing for more quality-oriented customized markets, will have substantial advantages (Sorge/Streeck 1988). The promotion of "human centred" approaches is explicitly based on this assumption, especially by the European Commission Forecasting and Assessment of Science and Technology (FAST) programme and European Strategic Programme on Research in Information Technology (ESPRIT) (Warner et al. 1990). The objective is therefore the design and implementation of competitive strategies and supportive technological and organizational structures that will adapt to, rather than undermine, these advantages.

Thirdly, as a corollary to the first two points, it is recognized that there are substantial constraints upon the development of the "human centred" path. Not only are the traditional Tayloristic management strategies and departmental structures highly resistant to change but the 'new production concepts' that are emerging pose potential dangers for human skills and the quality of working life. Under the umbrella of "flexibility" and "integration", new forms of work organization may be implemented that result in jobs that are increasingly stressful, and work processes (and workers) that are more rather than less monitored and controlled by central planning departments and upper level management. For the more radical advocates of "work oriented" or "labour oriented" systems design, this requires making sharp distinctions between restricted "socio-technical" perspectives and more radical "collective resource" approaches (Ehn 1988). What is commonly agreed is that one of the key challenges facing "human centred" CIM approaches is drawing upon existing technological, economic and political resources to create techno-organizational systems that avoid these dangers.

The design ideal of human centred approaches is generally characterized in opposition to "Taylorist" or "traditional engineering"/"technocratic" approaches that (i) identify efficiency with the maximum extraction, codification and automation of all production knowledge and skills and (ii) view the remaining human activities within engineering systems as sources of "error" or "disturbance". In opposition to this view, a design ideal is upheld that seeks to "subordinate" machines to human rather than vice versa, build upon and enhance existing skills and abilities, and place human purposiveness at the centre of system operations (Rosenbrock 1990).

In elaborating this general orientation, the "computer as craft tool" metaphor is used as an important design objective. This orientation is clearly expressed in the imperative "to design and use computer artifacts based on the practical skill of craftsmen, rather than just the calculus of engineers" (Ehn 1988:371) and assertions that "The perspective in system design has to be shifted from developing computer programmes imitating and replacing human expertise towards designing computers as a useful tool for human experts. Thus, not the computer but the computer aided craftsman might emerge as the new hero of production (Brödner 1989:45). The importance of these design ideals is not their originality, for such debates were widespread in discussions of the tool/machine question in the nineteenth century. They possess a different twofold significance: firstly, that they are now being operationalized in the design of new computer based production technologies, and secondly that they are intervening at a specific technological conjuncture in the development of CIM.

The operationalization of these concepts involves an emphasis on three design imperatives for human centred systems. Firstly, the status and operations of the system must be as far as possible made "transparent" to the user, i.e. data bases are accessible to users, system status information is available, information is organized and presented in a manner that is compatible with traditional forms of knowledge and work practices, etc. Secondly, the users of systems must be the main sources of "judgement", i.e. actions are at the user's discretion (even where these are supported, assessed and evaluated with the assistance of computers), machines are controlled by those most directly and immediately involved in their use, the choice between allocating functions to machines or humans is made by the users, etc. Thirdly, the maximum degree of "autonomy" must be given to users in controlling their own work and fulfilling the indirect support functions required for their tasks, i.e. maintenance, system selection and development, personnel recruitment, scheduling, incremental process improvement, etc. Human centred systems should therefore (i) accept the present skill of the user and allow it to develop, rather than incorporating this skill into the machine, and subsequently de-skilling the human; (ii) allow a greater degree of freedom for users to shape their own working behaviour and objectives, thereby increasing the extent to which the human controls the technology rather than vice versa; and (iii) unite the planning, execution and monitoring components of work, thereby challenging established divisions of labour (Corbett et al. 1991).

These principles have been explicitly embodied in the design of NC Workshop Oriented Programming (WOP) software in Germany, MDI NC lathe software developed at UMIST Manchester, and the computer aided design (CAD) sketchpad developed in ESPRIT project 1217 (1199). If we focus on one example, the WOP system was developed using research showing that there are economic advantages in allowing skilled workers to develop programmes, yet the experience and knowledge of these workers is at odds with traditional abstract programming methods and editing devices. The WOP system provides a unified user interface for new shop-floor programming methods suited for turning, milling, grinding and sheet metal cutting. The software allows the workers to draw upon their manufacturing knowledge, planning capabilities and situational knowledge. In contrast to earlier NC programmes, it is up to the worker to plan the programme's flow and decide on technological data. The system keeps recurring geometric elements and cutting cycles in a generalized pre-programmed form, and the worker uses his/her initiative in combining these modules and setting parameters. Further assistance is given by a tool data base and graphic real-time simulation of tool movements in a user interface that emphasizes the graphic (rather than abstract and symbolic) representation of objects and operations. The "user-friendliness" of the system is evidenced by the fact that workers can use the system unaided after a quarter of an hour, and that operators with no programming experience perform, after a five-day learning phase, as well as their experienced colleagues did at the beginning (Brödner 1989). In the case of the UMIST software, further developed in the ESPRIT 1217 graphic interactive CNC lathe project, an added advantage is that it can be turned off and left aside in order for the operator to work in a conventional way if this is appropriate (Gottschalch 1989).

As observed by Professor Rosenbrock, head of the UMIST project, this research appears at an important technological conjuncture. In the development of numerically controlled technology, shopfloor/manufacturing manual data input (MDI) systems and integrated computer aided design and manufacture (CAD/CAM) systems offer very different technological visions for the use and development of computer based manufacturing technology. Developments in microelectronics and computer hardware, accompanied by proven downtime and other costs of highly integrated systems, have created a techno-economic opportunity for retaining and developing traditional manufacturing skills and traditions. In this context, the importance of human centred system developments is that they are designed to reveal and promote the economic effectiveness of the MDI option, thereby enabling the retention and computer-assisted development of more traditional workshop-oriented production skills. Unlike Noble (1984), who emphasizes the defeat of "record playback" automation models as a result of a capitalist control imperative, Rosenbrock stresses the importance of focussing on the particular use and development of NC and the transcendence of traditional engineering design prejudices. It is this connection with these broader issues that has led the human centred approach to a computer as craft tool ideal that extends beyond the design of user-friendly software to assist shop floor operators. As one proponent of human centred design commented, "human centred is not user-friendly writ large".

The "tool" ideal has three other characteristics that support this difference. Firstly, the use of computers as tools supporting human judgement requires a deeper consideration of the allocation of functions between humans and systems than the design of easy-to-learn-and-use "interfaces" for a pre-given distribution of functions. Three essential characteristics are (i) *system architectures* to support decentralized design and production competencies; (ii) *hybrid automation* that ensures the appropriate allocation of functions between systems and operators at every level to enable and encourage the exercise of human judgement; and (iii) *user/system interfaces* that support rather than undermine production skills and tacit knowledge and enhance the "tool" like character of new systems. The current BRITE EURAM 3302 project is exploring this issue further, emphasizing the degree to which anthropocentric ideals require even more radical approaches to system design, involving further extending the understanding of anthropocentric systems from criteria of "*what* the system does" to "*how* the system operates". This is, in a sense, a far deeper level of 'transparency' than the design of user-friendly interfaces to support traditional craft skills, it involves not how the system *appears* at different levels but how it functions overall. It cannot, for example, involve the use of complex APT programming languages at one level and workshop-oriented programming (WOP) languages at another (Kidd 1990a). Secondly, in traditional craft production, production design emerged from a profound understanding of manufacturing tools and practices and involved a strong commitment to production quality and excellence. As a result, the computer as craft tool ideal orients systems designs towards more quality-oriented product markets and the manufacturing oriented MDI rather than design-oriented CAD/CAM options. Thirdly, craft tools were traditionally designed, maintained and purchased by craftspeople, and knowledge about the use of these tools occurred through master/apprentice experience based learning. This design ideal focusses on the broad range of tasks and activities fulfilled by craftspeople and, in the context of contemporary manufacturing, facilitates greater attention to group work and the decentralization of these varied 'indirect' responsibilities to the craft group. This ideal has been most developed in human centred or anthropocentric projects to design and support *Fertigungsinseln* or production islands.

Since the establishment by the Research and Technology Ministry of its production island project in 1980, there have been four major conferences held by the AWF Association in Eschborn, most recently in 1990, elaborating the principles and case study experiences of production islands. These proceedings represent the most extensive coverage of production island experiments. Some of the most significant projects are those established at Sulzer-Weise (Bruchsal), Felten & Guilleaume (Nordenham), and Kodak (Berlin), as well as others carried out by such firms as MAN Roland, Heidelberger Druck, Mercedes Benz, Krauss Maffei, Mannesman Demag and others. The institutes involved and the projects upon which they have worked have been fostered by a number of governmental programmes. Amongst the most significant are two programmes of the German Federal Ministry for Research and Technology (BMFT): the New Manufacturing Technologies Programme and the Humanization of Work (now 'Work and Technology') Programme; and state government *(Länder)* programmes such as the North Rhine-Westphalian 'Man and Technology: Socially Oriented Technological Development' (SOTECH) Programme and the Bremen city state 'Work and Technology' Programme; and other government research funds, such as the 5-10 year special research grant funding the establishment of the multi-institute Production Island Research Centre at the Rhein-Ruhr University, Bochum. As a result of this degree of support, the concept and reality of production islands has become a key feature of models of human centred CIM. The experiences and discussions about German projects have had a considerable effect on human centred CIM designs throughout Europe, represented at the European level for example in the semi-autonomous production islands produced by ESPRIT projects 534 and 1217.

Following from the technological research, for example, there are presently a wide variety of German computer based production planning systems to support production islands, in addition to the WOP programming software described above. These systems operate at various levels from strategic production planning (PPS/MRP systems and mainframe support), to medium-range co-ordinating planning (Shop Floor Control SFC systems and workstation/PC support), and short term local planning (Local Area Cell Controllers and PC support). Firstly, operators tasks are widened as far as possible. Secondly, computer aided planning facilities are located at shop-floor rather than planning-department level. Thirdly, as far as possible planning and scheduling functions are supported at production island rather than foreman/area control level.

Some of the German systems that fulfil these functions include the AHP Leitstand DOS system; INFOR CIM Leitstand DOS system; FI2 UNIX system; and Inteps, LPS DOS system. More recently, the capacities of the LPS system, developed as a production planning aid to production islands, has been supplemented through the development of a COCOSS system for planning and co-ordinating production islands. The University of Bremen partners in the ESPRIT 1217 project have produced an IKARUS programme that is not yet commercially available. From the same ESPRIT project, Appropriate Computer Integrated Manufacturing Technology (ACiT) PC based software is now sold by Human Centred Systems

Ltd. (BICC) in the UK for scheduling both within and between production islands. In one BRITE project co-ordinated by the Fraunhofer IPK in Germany, further additional support for decentralization has been provided through the creation of a knowledge-based shopfloor control system to assist group leaders or foremen in making complex production decisions (Schallock 1989).

From one point of view these initiatives could appropriately enough be seen as a further extension of job design and work humanization activities to take into greater account the technical components of socio-technical systems. From another point of view, however, they can also be perceived as the activities of a growing 'technological constituency' attempting to intervene significantly in the development of a technological trajectory that has not yet culminated in working systems and commercial products: the path towards computer integrated manufacturing (CIM). As such, they represent an important set of initiatives in a new type of technology policy – the direct intervention by government in the early introduction of social orientations (other than health and safety regulations) into the design and development of new forms of process technology.

As is comprehensively documented, the new technologies in design, manufacture and planning that go to make up CIM as yet affect only a very small minority of jobs, documented as 7.1% using programmable equipment in West Germany in 1988 (Ebel 1990:22-25). Moreover, the diffusion of the new CIM technologies in their more 'hi-tech' and 'global' forms of corporate integration is remarkably limited. This is apparent even within such sophisticated producers as the West German machine tool industry (Schultz-Wild 1990). This means that we are faced with a dynamic and changing situation not a fixed 'mature' technology with established and rigid operation procedures. The debates over CIM are, therefore, to a great extent, a discussion over what images or models should be adopted as a guide to the design and implementation of CIM. Moreover, radically different models abound of how best to use the new technologies, and serious discussions are carried out into the effectiveness of "Fordist", "Toyotist", "Volvoist" and even "Sonyist" models of production. This is accompanied by an equal diversity in the technical standards governing CIM technologies. Not only will the technical components of CIM in design, production and planning often not speak to each other, *there is no consensus about who should be speaking to whom and what they should be saying*. Integration standards are not well developed and are still a matter of inter-corporate and inter-national rivalry, the conflicts surrounding MAP, TOP and CIM-OSA standards are only some of those involved. The information requirements of highly integrated and formalist technical systems are also extremely complex to define and even more difficult to fulfil. Software unreliability, data inconsistency and limited control sensor ability all contribute to system vulnerability and breakdown. All that can comfortably be said is that CIM needs to be custom-built by and for specific enterprises, and that little is known about how this is to be most effectively achieved.

In this context, human centred or anthropocentric design activities can best be seen as the work of a growing 'technological constituency' proposing a particular model of CIM capable of addressing defined technological problems. Where, as is now commonly accepted, experiences with CIM implementation reveal low levels of productivity improvement, this 'critical problem' can be perceived differently as, for example, the result of inadequate levels of technological development and coordination, the fault of inadequate implementation processes, or design failures in a technocentric attempt to remove human skills from production systems. The solution to such "reverse salients", as they are described in history of technology research, can, consequently, be differently defined. In such situations, there are always what innovation researchers have called "socio-technical constituencies" promoting different forms of hardware, software, system architecture and organizational design as solutions (Molina 1989). These constituencies include R&D laboratories, corporate equipment vendors, engineering research and consultancy institutes, university research departments and applied consultancy centres, human factors and engineering associations, national and international standards and regulatory bodies, government innovation programmes, corporate system users and user groups, employee representatives and trade union associations, educational and training bodies and industrial regulation institutions etc. Moreover, there is substantial evidence to suggest that, in most cases, the activities of these constituencies are often carried out without a clear overview of the overall context of system development, and with mistaken views of the ultimate implications of the systems.

As observed in the ILO report on *Computer Integrated Manufacturing: The Social Dimension* (Ebel 1990), for example, there has been a strong US constituency promoting what some have dubbed the "moonshot" or "star wars" approach to CIM, i.e. the promotion of extremely capital intensive and high technology factory designs and exemplary models that attempt to utilize the latest developments in information technology and process control equipment to fulfil the long standing technocratic dream of the unmanned or automatic factory. Promoted by the American Department of Defense, such large corporate vendors as General Electric (MacDonald 1987; Noble 1984), and user multinationals such as General Motors and Boeing, this computer integrated and highly automated "factory of the future" concept has been developed as part of the attempt to restore American competitiveness in the face of low wage competition from newly industrialized countries and the "Japanese threat". This concept has been weakened by evidence that many of the exemplary factories of the future have become costly 'white elephants' unable to match Japanese productivity or meet their expected targets. In addition, growing experience with the introduction of information technology has revealed the central importance of human and organizational changes, both in ensuring the effective use of the technology and as a source of productivity increase in their own right.

In the face of evidence collected that between 40 and 70% of total improvements can be attributed to better logistics and organizational changes (Ebel 1990:47), one of the dominant themes now present in US business and organizational journals is the central significance of human and organizational support for efficient manufacturing rather than a reliance on what are now perceived to be excessively high levels of automation (Badham 1989). In the face of current experiences even in relatively low levels of CIM, there is now a not inconsiderable lobby of management consultants, engineering organizations, and university research centres, as well as vested interests within user firms and organizations, promoting the idea that the key to productivity is greater attention to the "human factor" and "organizational redesign". In this context, the human centre or anthropocentric design constituency represents an attempt to define these human and organizational factors in a specific direction – one that builds upon rather than undermines 'craft' traditions of worker autonomy, creativity and skill. The interventions of this constituency may occur at three strategic levels: in the development of global CIM concepts; in the creation of generic CIM component technologies; and in the development and introduction of CIM systems for specific firms. The constellations of human factors, social science, computer system, engineering and government groups and institutions supporting 'socially responsible' developments in these areas are worthy of promotion. This will only be so, however, if their activities are understood for what they are: the actions of a specific technological constituency to expand its role in CIM system design and development. It represents a constellation of vested interests and social concerns, groups and institutions, that are presently relatively fragmented and far weaker than other 'technocratic' constituencies. It is for this reason that the activities of coordinating groups such as the IFAC Social Effects of Automation Committee, the MODEM technology, work and organization network, the CAPIRN production culture research group, the FAST activities in stimulating the development of an anthropocentric research network, and the book series and journals established by publishers such as Springer (e.g. Artificial Intelligence and Society journal and book series) are so important.

The promotion by this constituency of 'European' solutions to the production problems associated with technocratic CIM approaches may or may not come to have a significant impact on the shape and diffusion of CIM. Its activities are best seen, however, as those of a constituency worthy of promotion rather than a group with unique insight into 'the requirements' of CIM or the vanguard of a revolutionary shift to a new 'humane' world of work. It is already observable that the earlier forerunners of the constituency in the late 1970s and 1980s such as Mike Cooley, Howard Rosenbrock and Peter Brödner, and the programmes they initiated or were part of, have now been supplemented by a new generation of researchers and programmes (Badham/Schallock 1990) The changes, debates and critiques that this involves should not be interpreted in terms of the collapse or fragmentation of the constituency but rather as inevitable developments within a constituency which needs to both give a coherence to its own activities and enroll outside interests. It is this constituency that can or should be influenced and promoted by labour representatives and others sympathetic to their aims.

Whether this group will continue to adhere to the concepts of human centred or anthropocentric systems, or replace these by more neutral terms such as "interdisciplinary system design" or "comprehensive innovation processes", will depend in part upon the dominance of different sections within the constituency and in part upon strategic negotiations with potential allies in government programmes, CIM system vendors, consultancy organizations, etc. From the point of view of a direct socially oriented innovation policy, however, it will be important to retain the explicitly political element of actual system development, in selecting for promotion systems that approximate more towards the "post-Taylorist" rather than "neo-Taylorist" options. Whether this is conducted under such labels as human centred and anthropocentric remains to be seen. These concepts do, however, suggest a design philosophy and general orientation that remains of central importance, in a sense because of, rather than despite, its inherently controversial nature.

Despite its presence as a technological constituency promoting a particular socially directed innovation path, however, in terms of numerical strength and practical achievements it remains relatively weak. As identified in the ILO report, initiatives can be observed within technology *agreements* established by trade unions, the development of new forms of system design by *socially responsible engineers and associations* and *government programmes promoting human factors considerations in CIM* system development. However, as discussed elsewhere, there is substantial evidence of the restricted character and limited effects of technology agreements, the isolation of many 'human factors' groups to a small and organizationally weak minority of engineers, and, despite important exceptions, the comparative insignificance of government activity in this area when compared to the funding of large-firm-dominated and technology-led forms of automation (Badham/Naschold, in this book; Badham/Schallock 1991b).

6.3 Key Issues of Human Centred System Promotion

The projects and programmes to promote human centred systems represent an important attempt at socially oriented innovation policy. Yet some reflections on the key problems and challenges facing these activities throw some light on the practical difficulties facing all such forms of innovation policy. Three issues will be briefly addressed here: the problems of *direct promotion* methods; the dilemmas of a *technology orientation*; and the adoption of a meaningful *social orientation*.

6.3.1 Direct Promotion

All forms of direct technology promotion are commonly recognized as facing two major problems: firstly, the relative insignificance of direct government expenditure compared to business expenditure and the effect of indirect government activities; and, secondly, the difficulty of ensuring that the recipients of government funds fulfil the intentions of the funding programme without the use of overly restrictive bureaucratic regulative measures.

With regard to the first problem, there has been much discussion within Germany and Sweden over the usefulness of promoting "exemplary models" of new technology and work humanization through government funding. In the case of the German Humanization of Work programme, there was substantial controversy generated early on in the programme over the large amount of funding given to Bosch as an exemplary model. More recently in Sweden, the LOM programme has been deliberately developed by the Work Environment Fund in opposition to the earlier Development Programme based on the philosophy of setting up "best-practice" models (Gustavsen 1987). As argued by critics of exemplary models, the allocation of moneys to companies large and sophisticated enough to become good models, and in sufficient quantities necessary to assure the influence of government programme criteria on the project, acts as a severe drain on programme resources. Moreover, it is questioned whether other companies wish to copy leading edge examples as considerable scepticism is expressed over the applicability of conditions in larger innovative firms to other enterprises. With specific reference to socially oriented models, it has been observed that firms may be less willing to copy organizational/social models than technological ones (Klein 1991b). In a current review of CIM implementation in Germany, for example, nearly all the firms brought in external consultants to assist with the technology, yet practically none hired externally for organizational assistance.[3] In opposition to the critical view, others stress the influence that models have in defining and transferring scientific paradigms, and point to the strong impact that economically successful models, such as that provided by Japanese manufacturing techniques, have on the production imagination. The whole question of the diffusion of such models remains controversial, with more 'realistic' proponents emphasizing the need for a detailed breakdown of direct economic benefits as a key diffusion factor, while others stress the priority of changing existing measures of productivity and economic benefits to incorporate a broader range of social as well as economic considerations.

With respect to the second problem, a key difficulty for direct promotion programmes is that it is the large firms and established research and consultancy institutions that have better links to policy-makers and administrators and have the resources and expertise available to apply for funding and carry out projects. In the process these companies and institutions have a strong impact on how the actual projects are conducted, and attempt to transform these projects in a manner that conforms with their own interests and expertise. There are, for example, a number of examples from the European programmes of German Fraunhofer Institutes interested in CIM influencing programmes such as Eureka and the FAMOS projects in the direction of their interests. Similarly, humanization of work projects carried out by engineering institutes in the areas of assembly and robotics have been known to define humanization primarily in terms of the automation of detailed jobs or the further development of robots, on the grounds of health and safety, that are then not specifically developed for the complex forms of welding etc. that create the ergonomic problems (FAMOS Project Members 1990). In order to counteract such reorienting of the programmes, potentially away from programme goals, increased time and effort must be spent on specifying project requirements, monitoring and evaluating progress, etc. Yet, these activities make it even more difficult for smaller firms and less substantial research institutes to obtain and effectively use the money available. Direct funding programmes must, therefore, be developed in a manner that addresses the difficult task of taking into account the effect of the institutional context within which the programme is embedded, and as part of a set of integrated measures to foster a favourable institutional and cultural climate for reform.

6.3.2 Technology Orientation

As commented above, one of the key considerations in socially directed innovation is to apply social criteria earlier on in the design of technology and work systems. Where, however, these activities are promoted by technology oriented public institutions or programmes – such as, for example, the European Commission ESPRIT and FAST programmes or the German Research and Technology Ministry, this may create an excessive concern with the reshaping of technology to the relative neglect of the organizational features of new production systems. As commented by leading researchers and trade unionists in this area, the amount of time and effort required to develop new technologies may be better spent on projects to assist the way in which these technologies are implemented and used (Hammarstrom 1990; Dankbaar 1990a, 1990b). Moreover, given the acknowledged difficulty in gaining effective collaboration between social scientists and engineers (Rosenbrock 1990; Corbett et al. 1991), the conduct of many such projects within engineering institutions is likely to lead to an overly dominant emphasis on the engineering and technical aspects of the problems, often in accordance with the particular engineering speciality that a particular institution wishes to develop and promote. A careful and balanced approach needs to be taken, therefore, to not only the importance of social intervention at different stages of the innovation cycle but also the realistic ability to make effective changes at each of these stages.

6.3.3 Social Orientation

As emphasized by Charles Edquist in this volume, the explicit introduction of political and ecological concerns into technology policy requires effective lobby groups in society with an interest in these areas. This phenomenon is more complex than the existence of organized interests, for there are a broad range of conditions that influence the formation of a popular sense of grievance and then act to enhance or prevent such grievances being transformed into political issues addressed at the centre of government. Involved in this process are cultural and institutional factors that influence public perceptions, interest group formation, and the general 'non-decision' making processes within public bureaucracies and political parties that select which issues are given consideration (Lukes 1975). Without the presence of favourable 'external' conditions and 'internal' support, government projects and programmes integrating social and ecological concerns into technology policy may be trivialized or interesting only as a legitimatory phenomenon to prevent protest. In the discussion of socially oriented innovation, for example, frequent references are made to the rising educational level and expectations of the workforce. These, it is argued, exert a pressure for appropriate work reforms additional to that created by new technology and changing markets (Grootings 1990; Brödner 1990). This theme, prevalent in post-industrial writings in the early 1980s (Scott 1990), is particularly emphasized by commentators on the pressures for work humanization in West Germany and Sweden.(Edquist/Glimell 1989; BMFT 1990). There has, however, been relatively little systematic research on the strength of this phenomenon in the context of the implementation of new technology, its relative influence on trade union and management perceptions, and its promotion by the education and training establishment. Some interesting suggestions have, however, been made by researchers on Sweden and West Germany concerning the influence of the supply of skills and government influenced employment levels on corporate approaches towards exploiting these skills and overcoming potential absenteeism problems (Sorge et al. 1983; Edquist/Lundvall 1989).

The existence of such conditions is of importance in assessing the significance of trends to replace terms such as "human centred design" or "work humanization" by "interdisciplinary design" or "integrated technology, work and organization" approaches (Kidd 1990b; den Hertog/Schöder 1989). Without the presence of social and political lobbies ensuring that these categories include effective consideration of working conditions desired by the workforce, they will all to easily be limited to restricted economic requirements and legitimatory phraseology. The more 'neutral' terms of "comprehensive" or "interdisciplinary" design of new systems is accompanied by an emphasis on the social gains achievable through the overlap between changes to enhance economic efficiency and work humanization. As detailed by Elam and Börjeson (1991), this has become a dominant "language of workplace reform" in Sweden (Elam/Börjeson 1991). However, as observed at the firm level by Janice Klein and at the industry level by Elam and Börjeson, where workers have previously had a substantial degree of work autonomy, there is greater resistance to organizational reforms that may restrict this autonomy – even when these are presented as "post-Taylorist" in many important respects (Elam/Börjeson 1991; Klein 1991a). Opposition is also observable amongst less skilled workers in countries such as Australia to reforms that may increase the gap between highly skilled and favoured "core" groups and less skilled "peripheral" ones (Badham 1991a). If socio-economic innovation processes are to be meaningfully socially oriented, they must be directed in a manner that takes into consideration these wants and desires. National conditions will clearly be influential in determining what these wants and desires are, whether they can be considered, and if there is a will to do so. These conditions range from the industrial structures and skill formation processes in the national economy, through the strength of collective bargaining associations and industrial relations system characteristics, to the governmental structures and the party in power. A realistic assessment of these conditions is an essential component of any attempt to set up effective direct promotion activities as part of a socially oriented innovation policy. Some important research in this area is being conducted by a number of European programmes and institutions including those of the European FAST programme, the CAPIRN network, and individual university-based programmes such as ADD-TECH (Linköping, Sweden) and MERIT (Maastricht, Netherlands).

7 Conclusion

In its report on *New Technologies in the 1990s: A Socio-Economic Strategy*, the OECD (1988a) argued for a better understanding of "the comprehensive innovation process" and the important technology policy role of "constructive technology assessment". This requires, as they observed, a radical change in the direction of technology policy. In their words,

> Government science and technology policy is at a crossroads. The policy trajectory which has survived over the last 30 years, with its almost unique focus on R&D, needs to have its boundary lines redrawn. These lines will be broader, less minimalistic, and less 'market failure' inspired than they appeared at the outset. They involve a new phase of experimentation with new institutional structures and arrangements (OECD 1988a:45).

In this process, there are important opportunities open for the introduction of more socially oriented innovation policies. Yet there are at the same time, as we have seen from the above discussion, substantial problems in defining and implementing such policies. As has been witnessed in discussions of ESPRIT and FAST programmes to introduce "human centred" or "anthropocentric" technologies, and in reviews of the German Humanization of Work programme, this has even led to opposition to the use of explicit terms such as "human", "socially oriented", etc. These terms, it is argued, arouse antagonism from many established interests. It is better, they argued, to participate in what has been aptly termed "the rhetoric of consensus politics". By this is meant conformity to the increasing acceptance of the central importance of socio-economic innovation and policies, and attempting to shape the definition of the "technical requirements" of new technology in a favourable direction, i.e. according to formally unvoiced socially oriented criteria.

In contrast, it has been and can be argued that we are presently at the crossroads in the public understanding of technology and innovation. If this is so, and the public has become increasingly awakened to the political nature and social significance of technologies, to promote a view that these problems can be 'managed' by 'policy experts' would be to seriously undermine the potential that the new technologies provide for increasingly open public discussion and consideration of technology policy issues. On these grounds the explicit raising of political and social issues by the use of such terms as "socially oriented technology policy" or "socially oriented innovation policy" is a strong argument in their favour, despite the necessarily controversial nature of any attempt to define their exact nature. Such terms may be easily condemned as "value laden", "ideological" or "idealistic" but these may be their strength and importance if the issues addressed correspond to broad ranging social concerns. If they do not, they are likely to fail in any case.

In the more particular task of addressing the key issues raised by socially oriented innovation policy, a recent OECD (1989) report on *Government Policies and the Diffusion of Microelectronics* provides some interesting guidelines. Four priority issues for what we have called socio-economic innovation policy are outlined. Firstly, the promotion of training and skill development. Secondly, improvements in the quality, availability and use of consulting services. Thirdly, widening of the technology base, to ensure ready access to technology from all sources. Fourthly, monitoring and assessment of the use and effects of new technologies.

These same issues are also key areas for socially oriented innovation policy initiatives in the manufacturing area. Criteria for skill based manufacturing could be clarified and applied in new training and skill development initiatives, as well as being incorporated into systematically organized consultancy approaches and institutions. Some of these criteria and initiatives can be seen in such varied contexts as the research report provided by the ISF Research Unit in Munich for the Karlsruhe CIM Centre and the "Contribution of Labour" component of the Australian National Industry Extension Service. The widening of the technology base can be further extended through further technological innovations promoted by humanization of work organizations and principles, and further attention paid to the way in which, for example, software development firms can be influenced to apply such principles. Due to the limited amount of public funds available, an attempt to influence the strategies of developer firms may be preferable to directly funded projects. Discussions around these issues are presently being undertaken in the *Arbeit und Technik* programme of the German BMFT.

Finally, as confirmed by the OECD report on government policies and the diffusion of microelectronics, further information is required on the nature and type of new technologies in use and the characteristics of users; the extent of their diffusion by industry, firm size and ownership; the obstacles in acquiring and using new technologies; and skills, training and employment impacts. This research agenda needs to be extended, however, to include (i) an examination of the dynamics of production technology design and development, the nature of present development trajectories, and criteria formulated at an adequate level of generality for assessing the social desirability of these developments, (ii) a classification of socially desirable inter-firm production systems, and the form these take in different industrial sectors, (iii) an examination of the extent to which these are diffused throughout different sectors, and the extent to which the structural characteristics of particular types of firms facilitate or hinder their development. The European FAST programme is presently conducting a number of studies in some of these areas in FAST Study No. 1 on Applied Technology Assessment. These issues cannot, however, be resolved in the abstract, nor can they be validly extended between cultures and across countries. An important part of this exercise, therefore, is an examination of the types of research institutions and policy making bodies suitable to carry out these studies and initiatives.

In conclusion, it is clear that if the concept of socially oriented technology policy is taken seriously it requires a radical reconsideration of existing policies and institutions, and in the case of socially oriented innovation policy faces institutional obstacles even more fundamental than those facing socio-economic innovation policy. However, if interesting and sporadic local initiatives or programmes are to be co-ordinated and given relevance such issues need to be seriously addressed. This could, and should, be perceived as the key technology policy concern of the 1990s.

Notes

[1] This trend is exemplified in the work of engineers and computer science critics such as Norbert Wiener, Ida Hoos, Mike Cooley, Howard Rosenbrock, Hubert Dreyfus and Joseph Weizenbaum. With particular focus on numerically controlled (NC) machines, computer aided design (CAD) equipment, and artificial intelligence (AI) programmes, this led to a general critique of the technocratic hybris that exaggerated the power and capabilities of the new machines and programmes, and an emphasis on necessity and desirability of preserving and supporting human "tacit skills", "operator experience", and human "purposiveness" as a central element in the new designs.

[2] The classification of these groups and initiatives as a "social movement" is contentious. As social movements are usually defined in contrast to traditional forms of politics

embodied in trade unions, government programmes, etc., the description of a set of human centred design initiatives that involve such forms as a social movement may be somewhat surprising. However, one of the central characteristics of the groups and initiatives involved is an explicit commitment to going beyond traditional trade union approaches and form of government action, and the espousal of many of the values associated with social movement activities, e.g. orientation to direct action and cultural innovation, the formation of network/grassroots organizations, changes in values and lifestyle, and participation in changes in civil society. Although, because of its strong links to more traditional trade union policies and structures, this movement has a far greater degree of ambiguity than many others. See the discussion in Scott (1990). In addition, as a movement attempting to directly reshape technology it involves a more or less fragmented network of researchers and practioners from different intellectual disciplines such as ergonomics, work science, industrial sociology, management science, computer science, and production engineering. Whether one can perceive a degree of coherence and integration amongst these sufficient to apply the label "movement" to their activities is a matter of not only debate but also action. What is certain, however, is that it represents a substantial establishment of vested interests and ideological predilections for the "humane" reshaping of technology and work.

[3] This is one of the preliminary unpublished findings of G. Doerr at the WZB, Berlin, in the study currently being undertaken of CIM implementation in German firms.

References

ACTU (1987) *Australia Reconstructed*, Canberra: Australian Government Publishing Service.

Annerstedt J., A. Jamison (eds.) (1988) *From Research Policy to Social Intelligence*, London: Macmillan.

Badham, R. (1989) The Factory of the Future: A Concept in Search of a Reality, Discussion Paper, Wollongong: Department of Science and Technology Studies, University of Wollongong.

Badham, R. (1991a) Human Centred CIM: Informating the Design/Manufacturing Interface, in: *Futures* December: 1047-1060.

Badham, R. (1991b) Computer Integrated Manufacturing: The Social Dimension, in: *International Labour Review* 3 (130): 373-392.

Badham, R. (1991c) Work, Organisation and Culture, in: *Artificial Intelligence and Society* 4 (5): 263-277.

Badham, R. (1991d) Anthropocentric Systems and the Australian Production Culture, in: *CAPIRN Report to FAST on International Production Cultures*, Bremen: Institut Technik und Bildung.

Badham, R. (1992) Skill Based Design: European Experiences and their International Relevance, in: *Prometheus* 2 (10): 239-259.

Badham, R. (1993) Smart Manufacturing Techniques: Beyond Electronic Drawing Boards and Turnkey Systems, in: *International Journal of Computer Integrated Manufacturing* (in press).

Badham R., J. Mathews (1989) The New Production Systems Debate, in: *Labour and Industry* 2 (2): 194-247.

Badham, R., B. Schallock (1990) The Human Centred System Design Constituency, Working Paper, Berlin: Fraunhofer IPK.

Badham R., B. Schallock (1991a) Human Factors in CIM Development: A Human Centred View from Europe, in: *International Journal on Human Factors in Manufacturing* 2 (1): 121-141.

Badham R., B. Schallock (1991b) The Human Shaping of CIM, in: R.A.Peacock (ed.) *Information Technology and People*, Milton Keynes: Open University/Peter Pereguinnes.

BMFT (1990) Research and Development Program "Work and Technology" (Project Management Team "Work and Technology" on behalf of the Federal Minister for Research and Technology), Bonn: BMFT.

Boyer, R. (1989) New Directions in Management Practices and Work Organisation, mimeo, prepared for the OECD conference "Technical Change as a Social Process: Society, Enterprises and the Individual", Helskinki.

Braun, E. (1984). *Wayward Technology*, London: Pinter.

Braverman, H. (1974) *Labor and Monopoly Capital*, New York: Monthly Review Press.

Brödner, P. (ed.) (1987) Strategic Options for 'New Production Systems': CHIM: Computer and Human Integrated Manufacturing, Brussels: FAST Occasional Papers No. 150.

Brödner, P. (1988) Options for CIM: 'Unmanned Factory' versus Skill-based Manufacturing, in: *Computer Integrated Manufacturing Systems* 2 (1): 67-84.

Brödner, P. (1989) In Search of the Computer Aided Craftsman, in: *Artifical Intelligence and Society* (3): 39-46.

Brödner, P. (1990) *The Shape of Future Technology*, London: Springer.

Burns T., H. Flam (1987) *The Shaping of Social Organizations*, London: Sage.

Child, J. (1987) Organizational Design for Advanced Manufacturing Technology, in: T. Wall, C. Clegg and N. Kemp (eds.) *The Human Side of Advanced Manufacturing Technology*, Chichester: Wiley, 145-163.

Collingridge, D. (1980) *The Social Control of Technology*, London: Pinter.

Cooley, M. (1984) *Architect or Bee?*, New York: Abacus.

Corbett, J.M., L. Rasmussen, F. Rauner (1991) *Crossing the Border*, London: Springer.

Couchman, P. (1990) The Rhetoric of Consensus Politics, Unpublished Ph.D Thesis, Wollongong: Department of Science and Technology Studies, University of Wollongong.

Daniel, W.W. (1987) *Workplace Industrial Relations and Technical Change*, London: Pinter.

Dankbaar, B. (1990a) Prospects of Anthropocentric Production Systems in the Netherlands, Draft Report for FAST, Maastricht: MERIT.

Dankbaar, B. (1990b) Interview, Berlin, January.

Dickson, D. (1984) *The New Politics of Science*, New York: Pantheon.

Ebel, K.-H.(1990) *Computer-Integrated Manufacturing: The Social Dimension*, Geneva: ILO.

Edquist, C., H. Glimell (1989) Swedish Frontiers of Change – A Guide to the Impact of New Technologies, Work Designs and Management Practices, report prepared for the Directorate for Social Affairs, Manpower and Education, OECD, mimeo, Linköping: Linköping University (ADD-TECH Programme).

Edquist C., B.-A. Lundvall (1989) Comparing Small Nordic Systems of Innovation, paper presented to the seminar on "National Systems Supporting Technical Progress", Maastricht.

Ehn, P. (1988) *Work Oriented Design of Computer Artefacts*, Stockholm: Arbetslivcentrum.

Elam, M., M. Börjeson (1991) Strategies of Flexibility or Rewording the World of Work in Sweden?, in: B. Jessop, H. Kastendiek, K. Nielsen and O. Pedersen (eds.) *The Politics of Flexibility*, London: Edgar and Allen, 167-198.
Elden, M. (1979) Three Generations of Work Democracy Experiments in Norway: Beyond Classical Socio-Technical Analysis, in: G. Cooper, E. Mumford (eds.) *The Quality of Working Life in Western and Eastern Europe*, London: Associated Business Press, 21-32.
Elliott, D.(1986) The GlC's Innovation and Employment Initiative – Second Report, in: *TPG Occasional Paper* 11: 1-3.
FAMOS Project Members (1990) Interview with Fraunhofer IPK Participants, Berlin: Fraunhofer IPK, January.
FAST (1991) Anthropocentric Technology and Systems (AT&S), in: *Newsletter* 1, Brussels: European Commission.
Fischer, F. (1988) Techno-Corporatism in the United States: A Critique of Contemporary Technocratic Theory and Methods, in: WZB Discussion Paper FS 11 88-105, Berlin: WZB, 1-25.
Fleck, J., J. Webster, R. Williams (1989) The Dynamics of IT Implementation: A reassessment of Paradigms and Trajectories of Development, in: *Edinburgh PICT Working Paper Series* 3: 1-24.
Gottschalch, H. (1989) Examples of Human-Centred Work Design in CIM Structures, IFAC International Federation of Automatic Control, Symposium on "Skill Based Automated Production", Vienna.
Grootings, P. (ed.) (1990) *New Forms of Work Organisation* Berlin, New York: de Gruyter.
Gustavsen, B. (1987) *Report on the LOM Programme*, Ontario: Ontario Centre for the Quality of Working Life.
Hammarstrom, O. (1990) Interview with Chief Research Officer, Swedish White Collar Unions, May, Stockholm.
Hayes, R., R. Jaikumar (1988) Manufacturing's Crisis: New Technologies, Obsolete Organizations, in: *Harvard Business Review* September-October: 77-85.
Hertog, J.F. den, P. Schöder (1989) *Social Research for Technological Change: Lessons from National Programmes in Europe and North America*, Maastricht: MERIT.
Hildebrandt, E. (1989) The Social Constitution of the Firm, paper presented to "CAPIRN Workshop", Santa Cruz, November.
Johnston R., P. Gummett (eds.) (1979) *Directing Technology*, London: Croom Helm.
Kidd, P. (1990a) Telephone Interview, December, Berlin: Fraunhofer IPK.
Kidd, P. (ed.) (1990b) Organisation, People and Technology in European Manufacturing: Interdisciplinary Research for the 1990s, Brussels: CEC/FAST Consultancy Report.
Klein, J. (1991a) A Re-examination of Autonomy in Light of New Manufacturing Practices, in: *Human Relations* X (43): 23-31.
Klein, L. (1991b) Contributions to CIM Gestaltungswissen: International Experiences Panel Meeting, Göteborg: IVF.
Lukes, S. (1975) *Power: A Radical View*, London: MacMillan.
MacDonald, L. (1987) Bargaining for Lynn's Factory of the Future: A Case Study, Working Paper 87-6, Lowell: University of Lowell.
Mathews, J. (1988) *A Culture of Power: Labour Movement Goals for the 1990s*, Sydney: Pluto Press.
Mathews, J. (1989a) *Tools of Change*, Sydney: Pluto.
Mathews, J. (1989b) *Age of Democracy*, Melbourne: Oxford University Press.

Mayor, F.(1989) Human Resources and the Cultural Dimension, in: L. Emmeru (ed.) *One World or Several?*, Paris: OECD, 176-191.

McLoughlin, I., J. Clark (1988) *Technological Change at Work*, Milton Keynes: Open University Press.

MDA (1990) *Description of Activities*, Stockholm: Arbetsmiljofonden.

Metall und Elektro Industrie Verband (1990) *Mensch und Arbeit*, no place: Metall und Elektro Industrie Verband.

Molina, A.H. (1989) The Transputer Constituency: Building up UK/European Capabilities in Information Technology, PICT Research Report Series No.1, Edinburgh: University of Edinburgh.

Naschold, F. (1980) Problems of a Society-oriented Research and Development Policy – The Humanization of Work Program at the Crossroads, Working paper series IIVG/dp/80-213, Berlin: WZB

Naschold, F. (1983) Technological Politics in the Federal Republic of Germany, IIVG Discussion Papers 7, Berlin: WZB.

Noble, D. (1984) *Forces of Production*, New York: Knopf.

O'Connor, J. (1973) *The Fiscal Crisis of the State*, London: St James' Press.

OECD (1988a) *New Technologies in the 1990s: A Socio-Economic Strategy*, Paris: OECD.

OECD (1988b) *Science and Technology Policy Outlook*, Paris: OECD.

OECD (1989) *Government Policies and the Diffusion of Microelectronics*, Paris: OECD.

OECD (1990) Conference on "Technical Change as a Social Process: Society, Enterprises and the Individual", Helskinki, December.

Oscarsson, B., O. Hammarstrom, G. Karlsson, R. Lindholm (1988) *A New World of Work, The Development Programme*, Stockholm: Swedish Work Environment Fund.

Piore, M., C. Sabel (1984) *The Second Industrial Divide*, New York: Basic Books.

Pock Too, N.G. (1987) Workers in Technology Transfer, in: Japan Institute of Labour (ed.) *The Role of Labor and Management in Technology Transfer*, proceedings of the "Asian Regional Conference on Industrial Relations", Tokyo.

Reich, R. (1983) *The Next American Frontier*, Harmondsworth: Penguin.

Rosenbrock, H. (1990) *Designing Human Centred Technology*, London: Springer.

Rothwell R., M. Dodgson (1989) Small Firms in Europe, in: *Science and Public Policy* February: 30-45.

Schallock, B. (1989) Skill Enhancing Shop Floor Control for SMEs, Conference Proceedings "New Manufacturing Imperatives", London: AXIOM Systems Design.

Schmandt, J. (1989) The Future of Science and Technology in the United States, WZB Occasional Paper FS 1189-108, Berlin: WZB.

Schultz-Wild, R. (1990) Application of IT and the Restructuring of Production Work in West Germany, paper presented to the "New Technologies, Institutional Changes and Working Life Seminar", Holmenkollen, Oslo.

Schumacher, E. (1974) *Small is Beautiful*, New York: Abacus.

Scott, A. (1990) *Ideology and the New Social Movements*, London: Unwin Hyman.

Self, P. (1975) *Econocrats and the Policy Process*, London: Macmillan.

Sorge, A., G. Hartmann, M. Warner, I. Nicholas (1983) *Microelectronics and Manpower in Manufacturing*, Aldershot: Gower.

Sorge, A., W. Streeck (1988) Industrial Relations and Technical Change: The Case for an Extended Perspective, in: R. Hyman, W. Streeck (eds.) *New Technology and Industrial Relations*, Oxford: Blackwell, 1-31.

Sweeney, P. (ed.) (1985) *Innovation Policy*, London: Pinter.
Thorsrud, E., F. Emery (1970) Industrial Democracy in Norway, in: *Industrial Relations* 2 (9): 35-51.
Walton, R., G. Susman (1987) People Policies for the New Machines, in: *Harvard Business Review* March-April: 98-106.
Warner, M., W. Wobbe, P. Brödner (eds.) (1990) *New Technology and Manufacturing Management: Strategic Choices for Flexible Production Systems*, Chichester: Wiley.
Weiner, E., A. Brown (1989) Human Factors: The Gap Between Humans and Machines, in: *The Futurist* May-June: 46-61.
Williams, B. (1988) *Review of the Discipline of Engineering*, vol. 16, Canberra: AGPS.
Zuboff, S. (1988) *In the Age of the Smart Machine: The Future of Work and Power*, New York: Heinemann.

Technology Policy: The Interaction between Governments and Markets

Charles Edquist

1 Introduction

Some of the questions which prompted me to write this chapter were[1]:

- How does state intervention interact with the market mechanism in the field of technological change?[2]
- Can technology policy have social and environmental objectives?
- How can research inform and advise policy-makers about technological change and technology policy?

Given the character of these questions, I can, of course, only scratch the surface of each issue. I will try to address the issues of why, when, where and how technology policies are or could be pursued.[3] I will attempt to do this by addressing the objectives of technology policy (section 4), the reasons for state intervention in the process of technological change (section 5) and the instruments used (section 6) to achieve the objectives. Before touching upon these issues I will, however, briefly discuss some features of technological change and technology policy (section 2) and how technology policy is actually formulated and implemented in practice (section 3).

Obviously, a chapter covering this vast ground can not be very detailed. My point of departure was that technology policies *are* designed and implemented and that this is often done on a rather weak analytical basis. Based on my experience in studying technology policies and evaluating technology policy programmes, I wanted to discuss how technology policies are actually designed in practice, as well as reflecting over alternative modes of doing it.

2 Some Characteristics of Technological Change and Technology Policy

In a long-term perspective – say one hundred years – it is intuitively evident to everyone that technical change has played an enormous role in shaping the presently advanced societies; especially in reshaping them from almost exclusively agricultural societies to dominantly industrial ones to service and information societies. Modes of transportation and everyday life have been similarly changed in fundamental ways. Christopher Freeman, one of the pioneers in the research field of the economics of technological change, writes: "technological capacity is the main source of the competitive strength of firms and nations" (Freeman 1987:1).

Hence, technical change is of great importance for things like productivity growth, competitiveness, welfare, working conditions, etc.[4] At the same time, it is a fact that the development, diffusion, adaptation and use of technologies are socially constructed in the sense that they are shaped by social economic and political factors. The market mechanism and private actors are crucial mechanisms shaping technical change in the current world. What is actually addressed when discussing technology policy in this chapter is how social and political forces can also shape technical change – as a complement to market forces.

Technology has certain rather peculiar characteristics. According to Ken Arrow – Nobel Prize winner in economics – technology can be regarded as information (Arrow 1962). And all information – such as a new method of production – should, from the welfare point of view, be available free of charge to everyone. This would lead to optimal utilization of the information (e.g. the technologies). However, if technologies are available for everyone free, no one has an incentive to invest in research and development (R&D), simply because no one could make money by developing new technologies.[5] Theoretically, it can be shown that the market mechanism leads to underinvestment in research (and development) in a market economy. This is one reason for state intervention in the process of technological change.

If technological change has such an enormous importance and if it cannot be left to market forces alone, then state intervention in the process of technological change may be very important. Therefore, in this chapter, I will deal precisely with the interaction between the market mechanism and public intervention in the process of technological change.[6] Part of this public intervention is often called "technology policy".

Before continuing, it may be useful to specify what I mean by technology policy. In the abstract it is, of course, *all* public intervention in the process of technical change.[7] More specifically technology policy is implemented by a number of public policy-making bodies that use specific instruments to influence the process of technical change.

It may also be useful to distinguish between direct or explicit and indirect or implicit technology policy. A *direct* technology policy is government intervention expressly intended to influence the process of technical change. Examples are R&D subsidies and publicly supported dissemination of information about new technologies. *Indirect* technology policy, on the other hand, includes policies that are not primarily designed to influence technical change, but which influence technical change to a greater or lesser extent anyway. Examples are trade policies, exchange rate policy, military security policy, industrial policy and fiscal policy. The consequences of indirect technology policies for technological change are often unintended as well as unforeseen by planners and decision-makers. Indirect technology policies may be inconsistent with, or even counterproductive to, the direct technology policies.[8]

However, it is the policy that is actually implemented that matters for the process of technological change – regardless of whether it has been explicitly formulated or not. In other words, direct and indirect policies should be studied together. Many policy actors make decisions which affect issues of technology policy. Technology policy that is actually implemented will by necessity, therefore, consist of many, very different and heterogeneous policy components.[9] This indicates that there are large problems involved in coordinating a consistent and compatible technology policy.

Currently technology policy is increasingly being considered to be an 'independent' policy field in many countries.[10] This means that public intervention in the process of technological change is becoming more integrated and coordinated. It also means that what I call indirect (implicit) technology policies are increasingly taken out into the light and are transformed into direct or explicit ones. Examples of this increasing independence are that technology policy is explicitly delineated and addressed in general, recurrent public documents (e.g. proposals to Parliament) and that national technology programmes are designed and implemented. When technology policy is becoming more 'independent' in these ways, inherent contradictions and conflicts between various objectives that exist within (the direct and the indirect) technology policy are often exposed – as discussed in section 4.

Technology policy is also a very important policy field in many countries, and in countries where it is less important, there are signs that this is changing. In Japan technology policy has been an enormously important policy field since the 1950s, although its significance has decreased in recent years (see section 3.2). The same is true for South Korea with a time lag of about a decade. In many European countries, technology policy is becoming increasingly important. This is also true for the European Community, where technology policy has also played a role of unifying the community (McKelvey 1989).[11]

It is very interesting to note that the importance of civilian technology policy also seems to be increasing in the USA. There are many indications in this direction. For example, in a report (OTA 1990), the Office of Technology Assessment in Washington has proposed the establishment of a "Civilian Technology Agency" to allocate government money to strategic technologies.[12] This is important, since "national security" has been the only reason, accepted by the administration, for public support to technical change in the USA.[13] Reasons behind this change are the ongoing demilitarization and the fact that Japan is taking over the lead in an increasing number of technology fields; radios, TVs, automobiles, semiconductors, etc. Countries like South Korea and Taiwan are also catching up fast.

In addition an integrated Europe may provide new challenges in the future. As a consequence, with Clinton as President the US administration has announced that it will pursue a civilian technology policy on a large scale.

The development of technology policy into a more independent and more important 'policy field' will probably continue in many OECD countries in the 1990s.

3 How Technology Policy is Formulated and Implemented in Practice

The most important goal of (civilian) technology policy is in practice increased productivity growth and competitiveness.[14] Let me, in this section, address a few possible mechanisms for technology policy-making, i.e. ways in which technology policies are – or could be – formulated. Such policies could be the outcome of or originate in theories, imitation, lobbyism or what I call "diagnostic" analyses. I will devote one subsection to each of these. I will not, however, present any systematic analysis of these mechanisms. What follows is rather a general and impressionistic discussion with illustrating examples from Sweden and Japan. In sections 4 – 6, I will address the objectives, reasons and instruments of technology policy in a somewhat more systematic manner.

3.1 The Usefulness of Theory as a Basis for Technology Policy

First I will discuss what assistance theory can provide to policy-making in the technology field. We saw, in section 2, that the automatic functioning of the market tends to lead to underinvestment in research and development. The reason is that firms have problems appropriating the results of their R&D; since technical advances can be copied by other firms they are reluctant to invest in R&D, in spite of the fact that new technologies would increase their efficiency (Arrow 1962).[15] This market-failure theory explains why firms underinvest in R&D. Economists and policy-makers often consider this to be a justification – or theoretical foundation – for government intervention in the process of technological change.

However, such policy implications which emerge from standard economic theory are actually not very helpful for policy-makers from a practical point of view. The policy implications of economic theories are too blunt to provide much guidance for government intervention in the process of technical change. These theories may simply indicate that there are sometimes reasons for state intervention in the field of research and development (R&D), particularly in regard to basic research. They do not indicate how much intervention[16] or in which particular research field (where and when) one should intervene. And they say almost nothing about how to intervene, i.e. which instruments to use and how.[17]

The main conclusion so far is that standard economic theory is not of much help when it comes to formulating and implementing specific technology policies.[18] Neither, however, have other social science disciplines developed theories which are directly useful for policy-making.

If theory shall be able to inform technology policy, we need theories about what determines and shapes technical change. Like all complex social processes, technical change has many determinants. And these are economic, social as well as political. What would be needed is therefore theories about the determinants of technical change which do not emanate from any particular academic discipline, but places technical change *as such* in focus. These theories should be interdisciplinary and not confined to, or restricted by, any one single discipline (Edquist 1988:section 4). We need to know what the *various* determinants are and how they influence various *stages* (invention, innovation, diffusion) of *different* classes of technical change.[19] Only then can theory be a guide for policy. However, such theories do not exist.[20] Therefore – for the time being – the specific formulation of policies are, and must be, largely based on something besides theory.

3.2 Imitation – A Common Mode of Technology Policy-making

If theory is not very helpful, how is technology policy formulated and implemented in practice? What are policy-makers actually doing – and why? Upon what do they base their decisions?

A very common means of formulating direct technology policy is imitation.[21] In most countries, policy-makers are simply doing similar things to what has been previously done in other countries (or in the same country). There is a visible copy-cat effect. A technology programme in an advanced nation gives birth to another somewhere else and all resemble one another. Everyone wants to be sailing on top of the wave of global technology. Obvious examples are the many national technology programmes in the fields of information technologies, new materials and biotechnology (Edquist 1989c:28).

This habit of imitation means that technology policy has become reactive; it is accommodated to, and becomes an uncritical importer of foreign technologies and policy instruments. Another important consequence of imitation is that the variations in structural conditions between countries are often not taken into account. In addition you are bound to be a lagger when imitating. If several countries intervene in the same way, late-comers will probably lose out. It is normally better to try to intervene in new and creative ways if you are a late-comer. And, above all, to adapt the mode of intervention to the national characteristics![22]

Some of us would like to see greater efforts to adapt technology policy to the specific socio-economic characteristics and objectives of different countries.

By doing so, technology policy could – to some extent – influence the direction of technical change and the way technologies are used.[23] In this way, national characteristics can – in a dynamic sense – even be the basis of advantages which can give a certain country, and its firms, a competitive edge.

However, there must be someone to imitate! Someone has to be a creative innovator. Various countries have played this role in particular time periods. The experience of relatively backward Japan in the 1950s was more important in this respect – on the civilian side – than most other countries. Therefore, I will briefly address the historical experience of Japan.[24]

Japan owes much of its economic success in the postwar period to the very active role played by the state in many sectors, including state intervention in the form of industrial and technology policies. According to Allen (1981) the Japanese government took advice from bureaucrats and administrators whose economics had been learnt by experience – rather than from professional economists. Some of these advisors were engineers who had been drawn by the war into the management of public affairs. "They were the last people to allow themselves to be guided by the half-light of economic theory. Their instinct was to find a solution for Japan's postwar difficulties on the *supply* side, in enhanced technical efficiency and innovations in production. They thought in *dynamic* terms. Their policies were designed to furnish the drive and to raise the finance for an economy that might be *created* rather than simply to make the best use of the resources it then *possessed*" (Allen 1981:68; my emphasis).

Japan's factor endowment at that time was characterized by a large supply of unskilled labour, an extreme scarcity of capital, and out-of-date technology. If Japan had specialized in line with its resource endowment, it would have been common sense – i.e. conventional economics wisdom – to concentrate on labour-intensive sectors such as textiles, clothing and simple metal products. Instead capital- and technology-intensive products were chosen; like steel, shipbuilding, and chemical fertilizers, and later radios, television sets, watches, petrochemicals, cars and machine tools (Allen 1981:70).

This was obviously in conflict with standard economic theory.[25] A long-term dynamic rationality was allowed to dominate over a short-term static rationality. The objective was to create new resources and comparative advantages, rather than to build upon existing ones.

How was then the selection of industries for development made? Two criteria had to be satisfied before a product was selected.

- The first one was that the product must be one for which the income-elasticity of demand in the world as a whole was likely to be high. When choosing among possible candidates, the one with the highest income elasticity was preferred.
- The second criterion was that of comparative technical progress. Even if the costs of manufacturing a product were high at the initial stage, it would be preferred to others if there was a probability of exceptionally rapid technical

progress which would be reflected in steeply falling costs (Allen 1981:71, who refers to Shinohara 1980). These criteria reveal those products for which global demand grows rapidly, i.e. what can be called growth products.[26]

A strategic technology and industrial policy, intended to stimulate growth and competitiveness, must be able to contribute to the creation of conditions favourable to specialization in growth sectors and growth products. Therefore, these criteria are certainly interesting to other countries trying to do the same thing.[27]

A Vice-Minister of MITI expresses a similar view:

> The Ministry of International Trade and Industry decided to establish in Japan industries which require intensive employment of capital and technology, industries that in consideration of competitive cost of production should be the most inappropriate for Japan, industries such as steel, oil refining, petrochemicals, automobiles, aircraft, industrial machinery of all sorts, and electronics including electronic computers. From a short-run static viewpoint, encouragement of such industries would seem to conflict with economic rationalism. But, from a long-range viewpoint, these are precisely the industries where income elasticity of demand is high, technological progress is rapid, and labour productivity rises fast (quoted in Fransman 1985:55-56).

Christopher Freeman draws the following conclusion:

> The main point about MITI's role ... is that it has persistently pursued long-term strategic objectives to stimulate new developments in relation to technology and industrial structure over the past 40 years. It has been pragmatic about its methods, depending upon the particular stage of development, but it has always striven, in collaboration with the scientific and technological community and with industry, to develop a well researched 'vision' of the future (Freeman 1987:38-39).

The strength and importance of MITI and other Japanese ministries is also stressed by Clyde Prestowitz (1988 100-121).

On the whole, there has been no profound conflicts between the bureaucracy and private industry in Japan. They have been aiming at the same target – rapid industrial growth. The chief media for policy-making have been the Industrial Committees of MITI, where civil servants and industrialists met to decide on future strategies and on the means required for reaching the goals agreed upon. According to Allen, the means most commonly employed by government during the postwar period were lending policies, protection (high tariffs or quantitative import controls), subsidies, tax concessions (e.g. for R&D expenditures), exchange controls to promote exports, encouragement of dumping (Allen 1981:73-75). To these means should be added the Japanese hostility towards foreign direct investment as a means of acquiring technology from abroad. Other mechanisms of transfer of technology were – for a long time – strongly preferred; mechanisms which implied a lower degree of foreign control over the Japanese economy.

Some additional lessons and conclusions of the Japanese experience are the following. Japanese technology policy has, as a whole, been fairly *supply* oriented, i.e. dominated by a technology push approach (as opposed to demand pull).[28] To a large extent, the Japanese policy has also been *product* technology oriented; the composition of products produced in Japan was, as we have seen, changed through state-induced structural adaptation. This structural change was, of course, accompanied by the introduction of advanced *process* technologies and new forms of organization in the economy.[29] Until recently, Japanese technology policy was concerned mainly with the *diffusion* (including acquisition, imports, absorption and assimilation) of product technologies invented and innovated in foreign countries.[30] Japanese technology policy was *extremely pragmatic* and not based on economic theory nor theory emanating from any other scientific discipline. The policies in Japan were characterized by a high degree of industrial targetting and strategic 'picking the winners', e.g. by allocating foreign exchange to certain firms, industries and products. The policies were extremely interventionist and the economy strongly regulated by the state – even at the firm and product level.[31] It is clear that the combination of state intervention and the market forces in Japan was much more efficient than it would have been to rely only on the market forces and private actors. At the same time, it has been convincingly shown that the proportion of government financing of private R&D carried out in industry has been very low as compared to other large OECD countries (Odagiri/Goto 1990:table 1). In other words, the government role of advisor, coordinator and catalyst has been more important than its role as financer.

3.3 Lobbying as a Mechanism behind Technology Policy

Another mechanism behind the formulation and implementation of technology policies is lobbying. Private firms, state agencies and other organizations often act with the objective to influence technology policies in order to get them designed and implemented in their own interest. This is a common mechanism, often combined with imitation.

One example is the National Microelectronics Programme (NMP) in Sweden. The decision in Parliament concerning this was taken in 1983 and the programme ran for the next five years. In a detailed analysis of the conception of NMP, Hans Glimell has shown (Glimell 1988) that this programme, to a large extent, emanated from military needs. At that time – just like the present – Sweden was highly dependent upon the USA with regard to the supply of electronic components. Fifth generation computer technology, based upon rapid VLSIs, were developed. In the USA, the VHSIC (Very High Speed Integrated Circuits) programme was running. At the same time, the USA had indicated that a more restrictive policy with regard to exports could materialize.

The NMP emerged out of a project within FMV (The Swedish Defence Material Administration). The possibilities of establishing domestic production of advanced integrated circuits intended for JAS, the then planned lightweight, multirole military aircraft had been investigated. However, this was not very successful.

To put further pressure on the Swedish component producers, relations were established between FMV and STU (The Swedish Board for Technical Change). After preparation and lobbying, the decision to establish the programme was taken by Parliament in December 1983.[32]

A similar combination of imitation and lobbying was behind the succeeding programme: The National Information Technology Programme (1988-1992). This IT programme was system-oriented rather than component-oriented (like NMP). The agencies behind were in this case Swedish Telecom, STU and FMV, with Swedish Telecom playing the role FMV had in NMP (Glimell 1989).[33]

In both the national programmes mentioned, the large Swedish firms also played an important role in the implementation of the programmes and they used most of the public money involved.[34] Of course, these firms and the state agencies involved greatly influenced the specific formation of the projects constituting the programmes. They were important pressure groups, lobbying in their own interest.

External actors have also played a very important role in the normal or continuing activities of STU (The Swedish Board for Technical Change). STU has dozens of reference groups designed to give advice to STU in the formulation and implementation of its various programmes. It is, of course, natural that representatives of firms and other actors in the reference groups play the role of promoting their own interests. They can therefore be expected to be a conservative force cementing old structures rather than one flexibly initiating activities in emerging technology fields – which are normally not represented by strong lobbyists.[35] In the next section, I will try to show this by summarizing an evaluation (Edquist 1989a) of state support to technology renewal in the engineering industry – which I have carried out for the Swedish Ministry of Industries.

3.4 "Diagnostic Analyses" – an Alternative Basis for Technology Policy-making

I will first summarize the evaluation of state support to technology renewal in the Swedish engineering industry. To be brief, I will only present the main conclusions and concentrate upon the activities of STU (The Swedish Board for Technical Change).[36]

- The Swedish engineering industry accounts for about 40% of industrial production, 45% of value added and 53% of Swedish industrial exports (Edquist 1989a:13-14).
- Swedish engineering industry is very advanced as regards automation of production. Sweden is the only 'old' industrial country that keeps pace with Japan as regards the diffusion of new *production* technology.[37] There are more computer controlled machine tools, industrial robots, CAD systems and flexible manufacturing systems per employee in the engineering industry in Sweden

than in the USA and the rest of Western Europe. Countries like the UK and the USA are far behind.[38] (Edquist/Jacobsson 1988:104; Edquist 1989a:59, 81, Tables 4 and 6; Edquist/Lundvall 1991:Table 10)
- When it comes to *product* technologies Sweden is much less advanced. If high technology products are defined as products with a high R&D expenditure in relation to the value of production, the high technology content of Swedish engineering production is lower than in most other OECD countries. The proportion of production emanating from R&D-intensive subsectors in the engineering industry was in 1986 38.3% for all OECD countries on the average, and only 24.8% in Sweden. As a ratio of exports the corresponding figures were 39.1 and 28.1, respectively (Edquist 1989a:62, 83, Tables 7 and 8.).[39]
- The total STU support to production technologies in the engineering industry was 67 million SEK in 1987/88. The expenditure of the industry as a whole was 941 million SEK, which gives us a relation of 7.1% (Edquist 1989a:50).
- The total STU support to the development of products for the engineering industry was at most 60 million SEK. The industry expenditure was here 8,587 million SEK, which gives us a relation of 0.7% (Edquist 1989a:50).
- Hence STU spends 10.1 *times* more on production technologies than on product technologies in relation to the total R&D expenditures of the engineering *industry*.

The overall conclusion is of course that STU priorities must be questioned. STU is spending a lot of resources in a field where Sweden is among the most advanced countries in the world while spending very little in a field which must be considered a serious long-term problem for the Swedish economy.[40] In the former field, the market functions in a satisfactory manner. In the latter, it does not. No one seems to have questioned STU activities, and the engineering programme has been going on in a similar fashion for years – to a cost of hundreds of millions of Swedish crowns. There is, of course, a steering committee for the STU engineering programme which includes actors with a stake in the engineering industry. Their vested interests seem to have been to continue in a similar manner over the years.

My method – used in Edquist (1989a) – in trying to show that the STU money allocated to the engineering industry has been used for the less important purposes may be called a "problem-identifying" or "symptom-describing" analysis. With this I mean an empirical and comparative analysis of the level of technology diffusion. The symptom-describing analysis must be *empirical* because there is no way to identify these kinds of problems specifically enough – for purposes of policy-making – on the basis of a theoretical analysis alone.[41] A problem-identifying analysis must be *comparative* since one can not determine whether a level is high or low in a country without comparing it with other countries. Such an analysis requires that a large amount of data is collected – and normally, it is not readily available. Further these data must be made comparable between countries, i.e. it has to be normalized in one way or the other. This means that a symptom-describing analysis is quite labour intensive and time-consuming as shown by Edquist (1989a). However, it is perhaps the only efficient way to find out when there is reason to consider state intervention and in which situations there are no reasons to do so.[42]

Ideally, however, we should also know *why* the technology level in a certain country is lower than in other countries before a policy of intervention is formulated and implemented. A description of the symptoms is not enough, since it only indicates when and where to intervene. Therefore, causal analyses, i.e. explanations of development and diffusion of technologies are also an important precondition for the formulation and implementation of effective government policies. Such analyses may indicate *how* to intervene, e.g. which instruments to use; causal explanation is thereby important to ensure that intervention is efficient.[43] The combination of a symptom-describing analysis and a causal explanation of the problems may be called a "diagnostic" analysis. This may be an important basis for an efficient therapy or treatment.

If differential technology levels cannot be causally explained, technology policy is out of necessity relatively ad hoc in nature: much is left to chance and to leaps in the dark – or at least in the fog. Hence, it is – again – strongly justified to stress the need not only for empirical and comparative, but also for explanatory and theoretical work. However, the inability to explain is not a reason to abstain from intervention in the process of technological change. Because problems identified may sometimes have very severe consequences – for the economy, for the environment or for the social conditions – trial-and-error intervention may be necessary.

4 Objectives of Technology Policy

As indicated in section 3, the primary objective of the civilian technology policies of most countries are often increased productivity growth and competitiveness.[44] The policies are governed by a rather narrow economistic orientation. This has so far on the whole implicitly been assumed to be the case also in this chapter.

An important question is whether this necessarily has to be the case. Can noneconomic objectives be the rationale for technology policy? More specifically: can, for instance, technology policy have environmental or social objectives, e.g. less pollution or a less risky working life.

The answer is, of course: "yes, in principle". It is perfectly possible to have non-economic objectives for technology policy. The main argument for this is that it has been the case for decades in one very large and important part of the society. I am, of course, thinking of defence-related intervention in the process of technological change. This intervention has absorbed a major part of the resources spent on technology policy in many countries. And it has certainly not been governed by narrow economic objectives like economic growth and competitiveness, but by national security, military strength and independence. Of course, this intervention has been supported by strong lobby groups in the relevant countries. And these lobby groups have economic interests. And the military-related intervention has had spin-off consequences for civilian technologies and for growth and competitiveness. But the main *objective* of the state intervention have been non-economic. At the same time this intervention has often been widely politically accepted by the public in the relevant countries.

An obvious conclusion is that state support to develop and diffuse socially or environmentally sound technologies can be an objective of technology policy if this has political support. It is certainly common that the market produces unacceptable results for the environment. The same is true, for example, of health hazards and social conditions in industry. There are therefore reasons to consider state intervention in the process of technological change to the extent that technologies are involved as determinants of these problems.

However, this does not happen automatically. First of all, it must be *known* that the technologies used have unacceptable environmental or social consequences. It was not long ago that we were ignorant about very dangerous social and environmental consequences of certain technologies. An obvious conclusion is that *technology assessment* should be practiced to a much larger extent than what is today the case. At present most product and production technologies can be used before we know very much about the consequences thereof. Historically, asbestos and DDT were widely used for decades before the general public knew that they were dangerous for humans and/or for the environment.

However, firms using or producing the technologies are often much better informed about the risks than the general public. It is therefore absolutely crucial that technology assessment is done in a way that makes the results available to the public and that this information is widely disseminated.

In one sector, technology assessment is practiced to a much larger extent than in others: the pharmaceutical industry. When a drug based on a new chemical entity is developed, its use has to be approved in every country before being put on the market. Each country has an agency with the purpose of approving new drugs. The objective is of course to make sure that the drug does not have dangerous side-effects. This takes years and can, of course, be very costly in terms of opportunities foregone (patients may die in vain because of the delay) and in terms of costs for experiments and analyses.

The drug approval system seems to function fairly well and is accepted by the industry as well as by politicians and the general public. Perhaps the technology assessment regulations in the pharmaceutical industry could be a model for the evaluation of new technologies in other sectors as well – at least as regards new chemicals? It could lead to a drastic reduction in the ignorance about the negative social and environmental consequences of these new technologies. It is, of course, advantageous if technology assessment can also be as 'proactive' and 'constructive' as possible – as discussed in other chapters included in this book by Badham and van Boxsel.

The case of nuclear power in Sweden is an example where non-economic factors have partly been governing technology policy. After the oil crisis in 1973, energy policy started to emerge as an independent policy field. In the 1970s the public became increasingly concerned over the risks attached to the operation of nuclear plants. The Harrisburg accident increased this concern and in 1980 a referendum was held. The outcome was that no new nuclear plants shall be built in Sweden and that existing plants shall be taken out of operation. Later on, Parliament decided that all nuclear power shall be taken out of operation at the latest by 2010.[45]

A lesson to be learned from the Swedish nuclear case is that one way in which non-economic aspects can influence technology policy is when the issues become matters of intensive public debate and enter the general political agenda.[46] If technology issues enter the public agenda, then political, social, ethical and environmental issues will have a greater chance of influencing technology policies – in addition to economic issues.

Furthermore, technology policy may mean that the technology issues enter the democratic sphere. Technology policy is, in principle, democratically controlled in a pluralistic democracy. This may be advantageous from the standpoint of accountability and participation.

However, if non-economic objectives become more important for technology policies, then contradictions and conflicts between various objectives will certainly emerge. For example, there will be conflicts (or trade-offs) between economic objectives and social or environmental ones. Economic growth may, for example, lead to increasing social and environmental problems – more than balancing the benefits of growth. Such conflicts will probably become more common in the future. Or perhaps it is more correct to say that they will become more visible and apparent.

There are examples where socially oriented technology policies have been proposed and/or attempted. For example, the so-called Brooks Report (OECD 1971) represented a view where social goals and qualitative aspects of growth were stressed as much as quantitative ones in the field of science and technology policy. The report also stressed state intervention to complement the operation of the market (OECD 1971:89). However, these tendencies have mainly been undercurrents and have not become dominant state technology policies in the OECD countries. For example, we have seen – in section 3.2 – that the Japanese technology policy was strongly oriented towards quantitative industrial growth with a supply-push perspective.[47]

The main conclusion so far is that the objectives of technology policy must be *politically* determined. Social aims may be determined through a political process. Then objectives of technology policy may be derived from these wider social goals – or the objectives of technology policy may be politically determined in a direct way. In the first case the – politically determined – social objective may be higher productivity growth or less pollution. If there is a relation between these objectives and technology, technology policy may be a means of achieving the goal. In the second case, it may, for example, be politically determined that a certain technology level is an objective as such.

5 Reasons for State Intervention in the Process of Technological Change

I have implicitly dealt with reasons for state intervention in the previous sections. Sometimes I have also done so explicitly, but without being systematic about it. Let me therefore, in this section, try to summarize the arguments or reasons for state intervention in the process of technological change in a market (or mixed) economy. It is here taken for granted that the social objectives and the objectives of technology policy – whichever they are – have been *politically* determined, as discussed in the previous section.

The state should consider intervention in the process of technological change if the market and private actors fail to spontaneously achieve the objectives wanted.[48] This is the main reason for state intervention. There are no reasons to consider intervention if the market fulfills the objectives. On the other hand, there *are* reasons to consider intervention when this does not occur. Hence, the role of policies is to complement markets; not to substitute for them.[49]

In other words there must be a *problem* which is not automatically solved by the market forces for state intervention to be considered.[50] Normally such problems can not – as argued in section 3.1 – be identified only with the help of theory (emanating from economics or from other disciplines).[51] The best way to identify these problems seems to be very pragmatic (and primarily empirical) about it and carry out what I have called a diagnostic analysis, i.e. a combination of a description of the symptoms and problems with – if possible – a causal explanation of them. Diagnostic analysis is what has been practiced in Japan (and South Korea) in a very pragmatic and undogmatic manner during the post World War II period (see section 3.2).

The fact that the problems warranting the consideration of state intervention cannot be identified only or mainly with the help of theory means that the argumentation here is very different from the one pursued within standard economics. The theory of market failure in economics is formulated and discussed within the framework of a strict and formal model: the theory of general economic equilibrium in an abstract economy. As we saw in section 3.1, it produces policy implications which are too blunt to be useful for technology policy-makers. This is related to the fact that the policy implications are considered to be valid for all economies; differences between specific economies are not considered. There are also issues which cannot be satisfactorily handled within the framework of the model, e.g. changes in product technology (and product quality).

However logical and strict the abstract theory of standard economics is, it must, in the end, be confronted with empirical reality. The conclusions reached in the abstract model world must be shown to be valid also in the real world. This is not an easy task and is rather seldomly attempted. Hence, the use of market failure theory also leads to problems of translating the conclusions reached in the model world into policy conclusions valid in the real world.

It would have been advantageous, and easier, to argue within the framework of a strict model of some kind, using some explicit criterion – e.g. Pareto optimality – as a guiding principle for technology policy formulation. However, for reasons mentioned in section 3.1. and above, this would not provide much guidance for policy-makers when considering intervention in the process of technological change.[52]

Therefore, it is necessary to take a step backwards as regards rigour and formality compared to standard economics. When I talk about "problems" which are not spontaneously solved by the market, this is not done within the framework of a formal model, but on an empirical basis and in a pragmatic way. This is intentional. The reason is that this mode of analysis is more useful as a basis for policymaking in the field of technological change. It is basically an empirical question to determine whether politically determined goals are fulfilled or not by the automatic operation of the market.[53] Because of the underdeveloped state of the art with regard to theory (of technological change and technology policy), there is no alternative to a pragmatic basis for technology policy.

However, even if there is a problem which is not automatically solved by the market forces, state intervention in the process of technological change may still not be justified. The state must also have the capability to solve or mitigate the problem. Otherwise the intervention should not materialize, since the result would be political failures. There may be two reasons why the state does not have this capability. Firstly, the problem may be unsolvable from the political level in an absolute sense. Then state intervention would, of course, be futile. Secondly, the state may need to develop its capability in order to be able to solve the problems. A detailed analysis may, for example be needed before intervention can take place. Then it may be necessary to carry out a diagnostic analysis. New policy instruments or implementing institutions may also be needed. Then these have to be developed or created before intervention shall – or can – take place.

Accordingly, there must be an identified problem and a state capability to solve it, in order for state intervention to be called for.

To conclude, both the market and the government have a role to play in the process of technological change. It is not a question of a choice between them. It is rather a question of finding an appropriate combination of the two in the specific field of technological change. A relevant question is which activities in this process are most efficiently handled by the market and which by the government. The answer to this question may certainly vary between countries as well as over time.

6 Instruments of Technology Policy

Once the objectives of technology policy are specified and it is clear that intervention is motivated, it is necessary for the relevant actor(s) to find out *how* the objectives can be achieved. Therefore there is reason to discuss various technology policy instruments further – and somewhat more systematic than in earlier sections. This will be done in a rather general way below.

Let me start by distinguishing between instruments on the supply side and on the demand side. If we combine this distinction with one between development and diffusion of technology we get the Figure 1 below. Let me briefly discuss the four boxes in the order in which they are numbered.[54]

	Technology policy instruments on	
	the supply side	the demand side
Development of technologies	(1)	(4)
Diffusion of technologies	(2)	(3)

Figure 1: Technology policy instruments

(1) A technology policy instrument on the supply side geared towards development of technologies is public financing of research. These subsidies can be channelled to basic research at the universitites or they can benefit private actors such as firms. This is what I have previously in this chapter called "R&D policy".

(2) When it comes to the diffusion of technologies, one instrument on the supply side is the (Swedish) system of technical attachés. These offices are placed in the major technologically advanced countries and their role is to report back to Sweden about the existence and use of new technologies. The objective is to help widen the spectrum of technologies available to Swedish actors. This is a supply-side instrument since it is intended to increase the supply of technologies which are available for absorption and diffusion to domestic firms and other actors. Other ways to increase the domestic availability of globally existing technologies is to establish data bases, take part in international educational exchanges and contribute to making technological and research cooperation between domestic and foreign actors as smooth as possible.

Of all new basic technologies available in a small country like Sweden, the absolute majority have their origin abroad.[55] In this sense *diffusion* of technologies – including absorption or assimilation – is many times more important for Sweden than domestic development of technologies. However, technology policy geared towards diffusion does not thereby necessarily have to be as many times more important than policies to influence technology development. It may be the case that diffusion works satisfactorily by itself and that state intervention thereby is not warranted. This probably varies between different fields and categories of technology.[56]

(3) A technology policy instrument on the demand side which influences diffusion of technologies (including choice and use) is the procurement by the public sector, i.e. public demand. For example, the public sector can, through its own demand, influence the choice between different kinds of energy sources and thereby which energy conversion technologies are used. The choice between different kinds of public transport systems can have this effect; for example, choosing trams or underground instead of diesel busses. The state also has an indirect influence over private demand through laws, taxes and subsidies. A prohibition of direct electricity heating or higher petrol taxes obviously affects the choice between alternative energy technologies.

(4) The most important technology policy instrument on the demand side intended to influence the development of technologies is public technology procurement. In this way not only the choice between existing technology alternatives can be influenced, but also which new alternatives are developed. This is important in the production of defence (and space) material. The state simply orders weapon systems which do not exist – for delivery far in the future.[57] Public technology procurement is a way for the state to use its own (future) demand as a technology policy instrument.

On the civilian side, technology procurement has also led to considerable results for the competitiveness of Swedish industry. Some examples are the relations between Swedish Telecom and Ericsson (telecommunication switching equipment), between the Swedish State Railways and ASEA/ABB (railway equipment) and between The Swedish Electricity Agency (Vattenfall) and ASEA/ABB (equipment to generate and transmit electricity). Public technology procurement in other countries has also had very important consequences for technology development.

A similar policy instrument would be if the state decided to stop buying petrol or diesel driven vehicles from a specific future date. This can be a part of a public technology procurement programme if it is combined with an order for vehicles with some other kind of engine. The state could also use its influence over private demand by deciding that petrol and diesel driven vehicles cannot be used in certain (urban) areas from a future date. The latter is being discussed in Los Angeles, where regulations with regard to maximum degree of pollution considered, would outlaw present petroleum-based engines. Given the huge market in Southern California, this kind of 'negative technology procurement' is a very potent technology policy instrument. The incentive for motor-vehicle producers to develop alternative kinds of engines is extremely strong, particularly if other metropoles can be expected to follow the example of Los Angeles.[58]

An important conclusion which emerges from the discussion of the diagram is that instruments on the demand side are very important for technology policy-making. In addition they seem to be very efficient. Therefore, if only a part of the total swedish public procurement of about 60 billion dollars annually were simultaneously used for technology policy purposes, the effects on the development and diffusion of technologies could be considerable.

There are also signs that public procurement is increasingly beginning to be consciously used for technology policy purposes in Sweden. One example is the procurement of refrigerators, which was coordinated by the Swedish Energy Administration (STEV) in 1990. A competition was announced where the price was an order for a couple of hundred refrigerators – with larger orders likely in the future. The requirement was that the amount of freon and electricity necessary for the production and use of the refrigerator should be considerably smaller as compared to existing designs. In a very short time, one firm had designed a new refrigerator with the required characteristics. This is, by the way, also an example which illustrates that the objectives of technology policy do not need to be only economic, but can also be concerned with protecting the ozone layer and limiting energy use.

As mentioned earlier, however, explicit technology policy is often – in practice – considered to be equal to (supply-oriented) R&D policy in a narrow sense, i.e. square (1) in the Figure.[59] This means that a supply-push perspective is heavily dominant over a demand-pull perspective on technological change and technology policy. If this dominance is broken, the possibility of integrating social and environmental dimensions in technology policy increases.

Such a broader view of technology policy – including supply and demand and development as well as diffusion – also increases the possibilities of small countries like Sweden to pursue a successful policy for technical change; the prospects of a small country to pursue a successful R&D policy are very limited since the possibilities to influence technical change in a global sense are very limited. The possibilities to influence diffusion (absorption, assimilation) and use of technology are much greater (Edquist 1989b).

However, on the whole the very rapid process of internationalization which is currently going on makes it increasingly difficult to carry out technology policy – as well as national policies in other spheres – particularly for small countries. The Swedish home-based multinational firms are becoming increasingly footloose. ASEA's merger with Brown-Bovery from Switzerland in 1989 has recently been followed by mergers involving, for example, SAAB Automobile (with General Motors), Volvo (with Renault) and STORA (with Feldmühle). In addition to the fact that the nationality of several of the largest 'Swedish' firms can be questioned, the process of internationalization is also proceeding in other fields. For example, a very important change in the 1980s was the development of a very free flow of capital across boarders.[60] This means that the possibility of pursuing an independent monetary policy in a country like Sweden is severely restricted. Further, the Swedish flow of direct investments abroad is enormous. For example, during the first half of 1990, Swedish corporations bought firms in Western Europe for a total of 72.2 billion SEK. This is to be compared to the acquisitions of US firms of 32.7 billion and of Japanese of 10.5 billion (Edquist/Lundvall 1991:35)

The process of internationalization changes, drastically, the basic workings of the so-called Swedish model. The old model was based upon the fact that the Social-Democratic state and the big firms were *mutually* dependent. If this dependency only goes one way in the future, the basis for corporatism will erode and the position of the state (as well as of the unions), in relation to capital, will be weakened.

One consequence of the process of internationalization is that it becomes more difficult to pursue a national technology policy. The possibilities of the Swedish state to influence firms by means of R&D subsidies (supply side) decrease. Also concerning, for example, technology procurement (demand side), the possibilities of national influence decrease. Between ASEA and the Swedish state there was a mutual relation where both parties had advantages from cooperating and were important to each other. For ABB, the Swedish state is much less important.[61] These transnationals do not need Sweden any longer, but the Swedish economy needs them. It seems like the difficulties for Swedish technology policy increase most with regard to the supply oriented instruments. Here, the state resources decrease over time; i.e. in relation to the resources controlled by the growing firms. At the demand side, there are still considerable possibilities to influence technological change. The reasons for this are that technology diffusion and assimilation take place in Sweden, that the choice between technology alternatives occur in Sweden and that the technology is used in Sweden.

Notes

[1] This article has been written within the context of the ADD-TECH (The Anatomy of the Development and Diffusion of New Technologies) Research Programme at the Department of Technology and Social Change, University of Linköping, Sweden. The department is an interdisciplinary and problem-oriented institute carrying out research and offering a Ph.D programme (but no undergraduate programme). The relations between technology and socioeconomic change form the central concerns of research. ADD-TECH is one of several research programmes at the department. It deals with the economic, social and political determinants and consequences of the development and diffusion of new technologies in the production of goods and services.

Various parts of a previous version of this article have been presented and discussed at three conferences:

- at "II Encuentro Hispano-Sueco sobre Democracia Industrial", organized by *Fundación Sistema* in Madrid, March 29-30, 1990,
- at "Sozial orientierte Technologiepolitik", organized by the *Institut für Höhere Studien* in Vienna, April 3-4, 1990, and
- at "New Technologies, Institutional Changes and Working Life: Research Issues for the 1990's" organized by *The Norwegian Council for Information Technology Policy* and *Norwegian Computer Society* in Oslo, June 20-21, 1990. The article has also been presented at seminars at the Department of Economics, Stanford University and at Crown College, University of California, Santa Cruz.

I am very grateful for comments received from participants at these conferences and seminars. I have also greatly benefitted from comments by Mark Elam, Hans Glimell, Maureen McKelvey – all members of the ADD-TECH Research Programme. The comments from Ted Bradshaw, Harvey Brooks, Björn Elsässer, Lennart Erixon, Christopher Freeman, Johan Hedrén, Bengt Jönsson, Kent Lindkvist, Fredrik Westander and two anonymous commentators have also been very useful.

[2] The terms "technological change" and "technical change" will here be used interchangeably. It must also be mentioned that I deal only with technical change and technology policy proper in this chapter. The analysis of organizational and managerial innovations as well as education and training – and related policies – are not addressed here. This does not imply that I consider them unimportant! On the contrary, they may be more important than technological changes proper for working conditions, productivity, competitiveness, employment and life as a whole. They are also often very intimately related to technological changes and a precondition for the implementation of technological changes. In Edquist (1992), I analyze organizational innovations, i.e. the emergence and diffusion of new organizational forms in working life and in other spheres, and the implications thereof for productivity and employment. Organizational and managerial matters as well as education and training are also dealt with in other chapters included in this book.

[3] The issue of who – or which institution – is formulating and implementing technology policies will be dealt with only in passing. Institutional aspects of technical change are placed more in focus in Edquist/Lundvall (1991), where the Danish and Swedish National Systems of Innovation are analyzed.

[4] Technology is, of course immensely important also for warfare, but this paper will deal mainly with civilian technologies.

[5] In a market economy, property rights are created on the basis of inventive activity. To the extent that this is successful, there is a underutilization of the information. There are basically three means through which firms can appropriate returns to their technical innovations:

- through the patent systems,
- through secrecy, and
- through exploiting a head start.

[6] I will concentrate on intervention in the production sector, as opposed to, for example, infrastructural intervention.

[7] The word "all" is emphasized because technology policy is often used interchangeably with R&D policy. The arguments for broadening the perspective will be presented in section 6.

[8] This is actually an indication that this policy field is an 'underdeveloped' one – an issue to be dicussed later.

[9] These are not even described in an integrated and holistic manner, for most countries and various parts of the policy as a whole are normally formulated and implemented by different agencies. This means that the right hand may not know what the left one is doing.

[10] This happened – as a comparison – with energy policy in Sweden in the 1970s.

[11] There are, however, exceptions to this rule. In Sweden and some other countries, the degree of state intervention in the process of technological change is quite limited – and not increasing. But this has not always been the case. In the late 1960s and early 1970s, there was a drive for a more active industrial and technology policy, which included the creation of the Ministry of Industries and of STU (The Swedish Board for Technical Change) (Glimell 1986:108-112). Later in the 1970s this developed – during the right-wing government 1976-82 – into excessive state support to ailing industries (shipyard, textile, etc.). The negative attitude of the Social Democratic government (until 1991) to offensive future-oriented intervention, for example, to change the structure of industry in a hi-tech direction, may be a consequence of these negative experiences. The new centre-right government seems to be even less interested in pursuing an offensive technology policy. There is, however, a tremendous difference between state support to activities that will die anyway and to those that are in the process of being born! If the new ones are R&D-intensive, they can be expected to give rise to external economies and to create competitive advantages in a dynamic manner. In another paper (Edquist/McKelvey 1992), we have also shown that productivity growth is higher for production of R&D-intensive goods than for other goods.

[12] This agency would be a civilian counterpart to the Defence Advanced Research Projects Agency (DARPA), which supports technology development considered to be critical for defence.

[13] However, the head of DARPA was fired in 1990, and the reason was said to be that he talked too much about civilian technology policy. Accordingly, there was still resistance within the administration to civilian technology policy at this time.

[14] However, other objectives are certainly possible, as we shall see in section 4.

[15] In the case of imperfect market competition there may, on the other hand be *over*-investment in R&D because of duplication of efforts – as Dasgupta and Stiglitz (1980) have shown. From a somewhat different angle, Richard Nelson (1987) argues that technical change is an inherently inefficient process characterized by uncertainty, duplication and waste.

[16] This has been expressed in the following way: "The theory of market failure suggests that whatever is being spent by firms on industrial R&D is not enough; it says nothing about what is enough" (MacDonald 1986:271).

[17] One could even argue that the market failure theory – summarized above – has contributed to a very narrow-minded kind of state intervention in the process of technological change: direct (or explicit) technology policy has tended to be the same as R&D policy. Another force leading in the same direction is the so-called

linear model of technological change – which stresses linear causality from basic research to applied research to development work, innovation and diffusion. (See MacDonald 1986 for similar arguments.) Inherent in this model is that resources should be allocated to basic research – in order to produce economic growth at the very end of the chain. Other parts of the innovation process tend to be neglected. Setting technology policy equal to R&D policy means that a supply-push perspective is heavily dominant over a demand-pull perspective. As we shall see in section 6 there are many other important dimensions of technology policy.

[18] As we saw, it can even be argued to have led to a narrow-minded view of technology policy.

[19] At the present immature stage of theory development, it would be too ambitious to try to formulate holistic theories of technological change. It is wiser to try to formulate partial theories for different groups of technologies (process technologies, product technologies, interactive technologies, etc.) and for each stage in the process of technical change separately.

[20] This is a strong argument for supporting theoretical work in the field of determinants of technological change even from the policy point of view.

[21] Indirect (or implicit) technology policies are, of course, formulated with some other objective in mind, and the effect on technological change is simply a spin-off.

[22] However, if the ambition is not to overtake, but only to catch up, imitation may be a viable mode. This is because of the fact that it may be advantageous to let someone else make the mistakes. In addition, imitation can be successful if the context in the imitating country happens to be more favourable to a positive outcome than in the country imitated. However, this is then more by chance than by design.

[23] For small countries, it is naturally easier to influence the use of technology than technological development per se, simply because most technologies are developed abroad. Whether this is successful or not is another matter. This issue will be further discussed in section 6.

[24] I concentrate on the historical experience of Japan, and the latest decade or so will not be covered at all.

[25] Consequently, the policy was not formulated by economists, but by bureaucrats and engineers at the Ministry of Trade and Industry (MITI). There was even a conflict between the professional economists at the Central Bank and the bureaucrats in MITI. In the early 1950s the Central Bank was an obstacle to the process outlined above, by denying loans and foreign exchange licenses. However MITI won the struggle during the course of the 1950s.

[26] Referring to a Japanese called Amaya, Prestowitz writes: "A key objective in any economy ... is to create an industry that produces technologically sophisticated products with high income elasticity ... and a rapid growth rate ... . That objective, according to Amaya and most other Japanese observers, cannot be achieved without government intervention" (Prestowitz 1988:128).

[27] This is so in spite of the fact that it is unclear whether these criteria were formulated ex ante or if they have later been derived from the actual selection of products and industries – which was then perhaps done in an intuitive and heuristic manner. On this latter issue, Allen writes: "it is debatable how far those principles were consciously applied *ab initio*, and how far they have been subsequently inferred from the decisions and conduct of the industrialists and the bureaucracy" (Allen 1981:71).

[28] This is, for example argued by Okimoto (1989:106). The distinction between technology push and demand pull will be further adressed in section 6. However, to avoid misinterpretations, it must be mentioned already here that policy instruments intended to influence the processes of absorption and diffusion at foreign-developed technologies can be considered to be supply-side measures from a Japanese point of view. Although this constitutes a *demand* for technologies developed in other countries, the policy objective is to increase the *supply* of technologies available in Japan.

[29] Production technologies or process technologies are used in the process of production, to produce goods and services. Product technologies, on the other hand, are or are included in products (goods and services). These terms are defined more precisely in, for example, Edquist (1992). It would be interesting to investigate in detail to what extent the introduction of new product technologies was given priority in the Japanese policy or whether process technology and work organization were more important objects of state intervention. (This is, of course, something different from their relative importance in the *economy*.)

[30] Now, of course, Japan has taken the global lead with regard to certain technologies – which means going into *development* of technologies, e.g. basic research, invention and innovation.

[31] Of course, the degree of intervention and regulation has diminished during the 1980s.

[32] The NMP has been evaluated in Wennerberg (1989).

[33] The initial stages of the IT programme have been evaluated in Arnold and Guy (1989).

[34] This was not possible in a later programme which was starting to be implemented in the Autumn of 1990. It was intended to support technology renewal in the engineering industry and only small and medium-sized firms were eligible to receive support. However, this programme was terminated in the Autumn of 1991, after the formation of the centre-right government in Sweden.

[35] As will be indicated in section 3.4, it seems natural for Swedish firms to use their R&D resources for improvement of their traditional products and processes. And they are reluctant to go for new products. However, if they have a strong position in various state agencies, they also tend to influence state technology policies in the same 'conservative' direction. Thereby state policies may duplicate or substitute for private efforts, rather than serve as complement to them. (As will be argued in section 5, this is not a good reason for state intervention.)

[36] Facts and figures are found in Edquist (1989a) (in Swedish) and, to some extent in Edquist and Lundvall (1991) (in English).

[37] Production and product technologies were defined in footnote 29 in section 3.2. Here it could be added that product technology is a question of *what* is produced. An empirical measure of "diffusion of product technology" is simply which types of products are produced – for example the%age of R&D intensive products. Production technology is a question of *how* things are produced. Some product technologies are obviously simultaneously – or rather, in a later stage – production technologies; when a machine is *produced* it is a product technology and when it is *used*, then it is a process technology. A product technology which is not intended for final consumption is, in fact, also a process technology. (See also Edquist/McKelvey 1992:9 and 13.)

[38] Eleven factors which may partially explain this picture were discussed in (Edquist 1989c:10-11). It is very unlikely that public technology policy has been an important determinant behind the high degree of diffusion in Sweden as regards production technologies in the engineering industry.

[39] Much the same picture is given if we look at the proportion of production and exports of all manufacturing products emanating from high-R&D-products. In Edquist and McKelvey (1992), we provide a much more detailed analysis of the Rand intensity of output (product technology) in the Swedish engineering industry as well as in industry as a whole. In an attempt to causally explain the low level of product technology in Sweden the crucial thing is to understand why firms in other countries do go for new (R&D-intensive) products, but Swedish firms do not. This is also attempted in Edquist and McKelvey (1992).

[40] There are several reasons why it is important for a country to be present in the R&D-intensive sectors and product groups. Firstly, there is a correlation between R&D-intensive products and growth products. Secondly, if OECD countries do not have a considerable proportion of their production in R&D-intensive (high-tech, knowledge intensive) industries, they will soon find themselves subject to fierce price competition from developing countries and NIC countries. Thirdly, R&D-intensive sectors are associated with external economies, see Edquist and McKelvey (1992) for an elaboration of this.

[41] However, a certain amount of conceptual work is a precondition for the empirical analysis. It should also be mentioned that a "problem" identified can be the result of the operation of the market mechanism as well as a result of previous state intervention.

[42] An analogy with the field of economic policies may be illuminating. Policy-makers in the field of fiscal and monetary policies need indicators about the balance of payments, employment, inflation, etc. in order to be able to know when to intervene. A basis of empirical indicators is needed also in the field of technology policy. However, such indicators are much less developed in the field of technology policies than with regard to economic policies.

[43] The causal analysis may also reveal that state intervention can not solve the problem identified through the symptom-describing analysis. In such a case, the correct decision is of course non-intervention.

[44] Technology policy is, however, certainly not the only means to achieve this.

[45] However, the referendum decision has increasingly started to be questioned; and there is in the early 1990s an intense debate over the issue.

[46] Of course technology policy-making is also a part of the political process when it is not a matter discussed by the general public.

[47] In the West, the pattern was somewhat different, at least in the 1970s. On the Swedish scene STU had three tasks or roles in the first years of the 1970s: an industrial role, a research-supporting role and a sectorially oriented one. The third one was new and meant that STU was supposed to actively cooperate with state agencies and with sectorial R&D agencies in certain sectors of society, like transportation, health care, environmental issues, etc. The intention was to direct technical R&D into certain priority areas and at the same time induce demand which could lead to advanced technologies and new products. During several years towards the mid-1970s, different strategies coexisted within STU. However, the more socially oriented one was never

dominant and gradually faded away around the mid-1970s (Glimell 1986:109-112, 1989:95).

[48] It is important to keep in mind that the objectives are not necessarily economic. They may also be of a social, environmental, ethical or military character.

[49] The reason for the unfulfillment of the social objectives may also be previous state intervention – in which case this political failure must be balanced.

[50] That a problem has to be carefully identified before intervention occurs is, as we saw in section 3.4, not a banal proposition.

[51] It would, for example, certainly not have been possible to identify the fact that the low-R&D character of Swedish *product* technology is a severe problem exclusively on the basis of theoretical analysis. Neither would it have been possible to say that the low-tech character of *process* technology in the engineering industry is a problem in the USA and the UK (see section 3.4.).

[52] It may also be mentioned that the theory of market failure presupposes that the objective of technology policy is maximum (economic) welfare or Pareto efficiency, i.e. the policy objective is implicit in the theory.

[53] This means that I address "problems" which the market does not solve automatically and "state intervention" at the same pragmatic common-sense level. This solves a problem of imbalance from which standard economics suffers. Economists talk about "market failures" within the framework of a formal model but "government failures" in an *ad hoc way,* unrelated to any general theory. This makes systematic comparisons difficult. Charles Wolf has recently tried to solve the problem in the way opposite to what I proposed above – by attempting to develop a general theory of government (non-market) failure (Wolf 1988). However, he failed to do so in a coherent and consistent manner (Sharkey 1990).

[54] The following discussion is based upon Edquist (1989b).

[55] It is assumed in this article that minor innovations and adaptations are part and parcel of diffusion. Hence marginal innovations are distinguished from radical innovations and the development of generic technologies.

[56] Referring back to the case of the Swedish engineering industry we saw in section 3.4 that the diffusion of new production technologies is very rapid while the diffusion of R&D-intensive product technologies is very problematic.

[57] The technical development of systems within the defence firms – in Sweden, for example, SAAB (fighter planes) and Bofors (guns, missiles) – are often also directly subsidized by the state.

[58] In addition, the direct cost for the public sector of using such an instrument is zero.

[59] This bias can, at least partly, be traced back to the market failure theory and the linear view of technological change (see section 3.1).

[60] For example, the Swedish restrictions with regard to international currency flows were gradually lifted in the late 1980s.

[61] To provide support in the field of technology development to the large – increasingly footloose – 'Swedish' firms may also be done in vain, since they may move the exploitation – and thereby economic benefits – of the R&D results outside Sweden. In this sense, state support to R&D infrastructures or to small and medium-sized firms seems to be better, since they are still more closely integrated in the national economy. However, large multinational firms get a lot of state support in other countries as well as from the European level. This may induce the Swedish state to do the same in

order to give these firms incentives to keep their R&D activities in Sweden. (One country which has been involved in direct negotiations with firms on such issues is Canada.)

References

Allen, G.C. (1981) Industrial Policy and Innovation in Japan, in: C. Carter (ed.) *Industrial Policy and Innovation*, London: Heinemann, 68-87.

Andersson, F. (1990) *The International Diffusion of New Chemical Entities – A Cross-national Study of the Determinants of Differences in Drug Lag*, Linköping: Linköping Studies in Arts and Science 51.

Arnold, G., K. Guy (1989) Evaluation of the IT4 Programme – Phase I, Report to the Delegation of the Swedish Government, Stockholm, 8 September 1989.

Arrow, K. (1962) Economic Welfare and the Allocation of Resources for Invention, in: *The Rate and Direction of Inventive Activity*, Princeton: Princeton University Press, 164-181.

Dasgupta, P, J. Stiglitz (1980) Industrial Structure and the Nature of Innovative Activity, in: *The Economic Journal* 90: 266-293.

Edquist, C. (1988) *Technology Policy – Conceptual Remarks and European Experiences*, Paper presented at the Conference on "Technology Policy in the Americas", Stanford University, California.

Edquist, C. (1989a) Utvärdering av statligt stöd till verkstadsindustriell teknik (Evaluation of State Support to Engineering Technology in Sweden), Tema T. Report 22, Linköping: Department of Technology and Social Change, University of Linköping.

Edquist, C. (1989b) Teknisk förändring och teknikpolitik, in: *Energin, makten och framtiden – samhällsvetenskapliga perspektiv på teknisk förändring* R16, Stockholm: Statens Energiverk.

Edquist, C. (1989c) The Realm of Freedom in Modern Times? New Technology in Theory and Practice, Inaugural Lecture, Tema T Report 18, Linköping: Department of Technology and Social Change, University of Linköping.

Edquist, C. (1992) Technological and Organizational Innovations, Working Paper No. 233, Geneva: ILO.

Edquist, C., H. Glimell (1989) *Swedish Frontiers of Change – A Guide to the Impact of New Technologies, Work Designs and Management Practices*, Report prepared to the Directorate for Social Affairs, Manpower, and Education, Paris: OECD. (To be published in OECD (1993) *Technological Change as a Social Process: Society, Enterprises and the Individual*, Paris: OECD.)

Edquist, C., S. Jacobsson (1988) *Flexible Automation – The Global Diffusion of New Technologies in the Engineering Industry*, Oxford: Blackwell.

Edquist, C., B-Å. Lundvall (1991) Comparing the Danish and Swedish Systems of Innovation, Tema T. Working Paper 77, Linköping: Department of Technology and Social Change, University of Linköping. (Published in: R.R. Nelson (ed.) (1993) *National Systems of Innovation: A Comparative Study*, Oxford: Oxford University Press, 265-298.)

Edquist, C., M. McKelvey (1992) The Diffusion of New Product Technologies and Productivity Growth in Swedish Industry, Consortium on Competitiveness & Cooperation (CCC) Working Paper No. 91-15, Berkeley: Center for Research in Management, University of California. (In Swedish it has been published in *Forskning, Teknikspridning och*

Produktivitet, Expertrapport nummer 10 till Produktivitetsdelegationen. The Swedish title was "Högteknologiska produkter och produktivitet i svensk industri".)

Fransman, M. (1985) Conceptualizing Technical Change in the Third World in the 1980s: An Interpretive Survey, in: *Journal of Development Studies* October: 34-76.

Freeman, C. (1987) *Technology Policy and Economic Performance – Lessons from Japan*, London: Pinter.

Glimell, H. (1986) Hur kan teknisk forskning påverkas av statlig forskningspolitik? – Sjuttiotalets strävanden att öka sektors- och samhällsintressenas inflytande över den tekniska utvecklingen, in *Tekniska Museets årsbok 1986-Daedalus* Vol. 34, No. 1, 17-32.

Glimell, H. (1988) Den svenska vägen till trängre kretsar – framväxten av det nationella mikroelektronikprogrammet, in: H. Glimell (ed.) *Industriförnyelse i Norden – (80-talets programsatsningar på mikroelektronik*, Roskilde: Forlaget Samfundsökonomi og Planläggning, 74-93.

Glimell, H. (1989) Återerövra datapolitiken – En rapport om staten och informationsteknologin under fyra decennier, Tema T Rapport 20, Linköping: Graduate School of Technology and Social Change, University of Linköping.

MacDonald, S. (1986) Theoretically Sound: Practically Useless? Government Grants for Industrial R&D in Australia, in: *Research Policy* 15: 20-37.

McKelvey, M. (1989) *European Economic Community: Research Policy, Political Ideology and Decision-making Process*, Master's thesis, Lund: Research Policy Institute, University of Lund.

Nelson, R.R. (1987) *Understanding Technical Change as an Evolutionary Process*, Amsterdam: Elsevier.

Nelson, R.R. (ed.) (1993) *National Systems of Innovation: A comparative Study*, Oxford: Oxford University Press.

Odagiri, H., A. Goto (1990) National Systems Supporting Technical Advance in Industry: Japan, paper presented at the conference on "National Systems Supporting Technical Advance in Industry" at Stanford University. (Published in R.R.Nelson (ed.) (1993), 76-114.)

OECD (1971) *Science, Growth and Society – a New Perspective*, Report of the Secretary-General's Ad Hoc Group of New Concepts of Science Policy, Paris: OECD.

Okimoto, D.I. (1989) *Between MITI and the Market – Japanese Industrial Policy for High Technology*, Stanford: Stanford University Press.

OTA (Office of Technology Assessment) 1990: Making Things Better – Competing in Manufacturing, Washington, D.C.: Congress of the United States, Office of Technology Assessment, February 1990.

Prestowitz, C.V. jr (1988) *Trading Places – How America Allowed Japan to Take the Lead*, Tokyo: Tuttle.

Sharkey, W.W. (1990) Book Review (of Markets or Governments: Choosing between Imperfect Alternatives by C. Wolf), in: *Managerial and Decision Economics* (11): 207-210.

Shinohara, M. (1980) *Industrial Growth, Trade and Dynamic Patterns in the Japanese Economy*, Tokyo: University of Tokyo Press.

Wennerberg, S. (1989) Nationella Mikroelektronikprogrammet (NMP) och Industriell Utveckling (Utvärderingsrapport), unpublished paper.

Wolf, C. jr. (1988) *Markets or Governments: Choosing between Imperfect Alternatives*, Cambridge, MA: MIT.

Promote or Regulate: The Dilemma of Innovation Policy

Ernest Braun

1 Introduction

Technological innovation is regarded as one of the main driving forces behind economic growth, and as one of the main determinants of competitiveness of manufacturing firms and of nations. As governments wish to encourage both growth and competitiveness, they regard it as their duty to do all they can to foster technological innovation. Yet the days when unrestrained growth was viewed as unconditionally desirable are long gone. Although many still believe that there is much scope for further growth, not many believe that growth in profligate material consumption can go unbridled for ever. Thus one horn of the dilemma of innovation policy is the unease created by the suspicion that growth, though desirable for many reasons, is not a long-term option for humankind.

Technology has long been recognized not only as the motor of economic growth and the very essence of human society, but also as a source of danger. Technology can be the cause of hazards to life and limb, to health, and to the natural environment. It has long been recognized that the development and use of technology cannot be left unregulated and must, like so much human activity, be subject to rules and regulations. On the other hand, it is obvious that rules hem in the free development of technology and thus might slow it down. The second horn of the dilemma of innovation policy is thus the need to create rules which govern the development and use of technology, in the full knowledge that this limits freedom and can slow down technological innovation.

Technology policy has to tread the middle path and must find suitable compromises between enhancing technological innovation and controlling the dangers created by rapid expansion in the use of technology.

2 Technology Policy and Technological Innovation

2.1 The Scope of Technology Policy

In the broadest sense, technology policy includes all policy measures which, in some way, affect the development and use of technology. As technology permeates all spheres of human life, this broadest definition is somewhat vacuous – it includes everything and thus defines nothing. We must narrow the definition of technology policy to include only those policies which deliberately set out to affect the development and use of technology. We regard policies as intentions to influence some development, while policy measures are taken to carry out these intentions. Technology policies and policy measures are all those specifically designed to influence technology for some goal or purpose. Some policies do, of course, have mixed objectives. We include such policies, if at least one of their objectives is to influence technology.

The scope of the word policy itself needs to be qualified here to refer mainly to public policies, i.e. policy measures taken by government or quasi-government bodies at all levels. Thus policies pursued by industrial firms will be referred to only exceptionally.

We use a rather narrow definition of technology, i.e. technology are artefacts designed and manufactured for specific human purposes and the software required for their operation. Thus, a telecommunication system is here regarded as a technology and includes all its multifarious software; while a social security system is not a technology, though it uses technologies, e.g. computers, in the pursuit of its activities. Our definition concentrates on artefacts, not on institutions or organization.

There are two main types of technology: that used for the production of goods and services (production technology); and that used directly as items of consumption (consumption technology). The division is by no means neat, as, for example, an electric drill may be used either by a do it yourself (D.I.Y.) enthusiast or a worker in a manufacturing plant. In one case, it is an item of capital equipment, in the other case an item of final consumption.

Generally speaking, technology policy in the public domain pursues two broad objectives:

- It attempts to support, enhance and accelerate the development and use of technology, with the ultimate goal of strengthening the economy. In some cases, the strengthening of the political and military power of the state may be a further, or even sole, aim of technology policy.

  There are strong links between Research and Development (R&D), as one of the main sources of technological innovation, and technology. Hence, we expect a strong up-stream linkage between R&D policy and technology policy. On the other hand, there are strong down-stream links between technology pol-

icy and industrial policy, as industry is the producer of technological products and the user of production technologies.
- Technology policy attempts to regulate the use and development of technology in such ways as to eliminate, or minimize, risks posed by technology to health, safety, the social fabric, and the natural environment. Technology policy also regulates the use of technology in those cases where its unregulated use would cause chaos and render the technology inoperable, e.g. in transport systems.

The discussion of possible conflicts or synergies between the two fundamental aims of technology policy forms the substance of this chapter.

The ways to influence technology by policy measures are many and varied (Braun 1984). In the main, possible direct influences are:

- financial, e.g. direct subsidies or tax incentives for producing technological innovations or for pioneering use of new technologies; tax allowances for investment in production machinery, subsidies for training courses, etc. Similarly, financial incentives or disincentives may be used to achieve reductions in levels of pollution, e.g. by charging for effluent or giving grants for introducing filters;
- legislative, e.g. regulations, such as on fire safety or on emission levels; patent laws; requirements, such as for testing of drugs or vehicles; or prohibitions;
- public procurement, both civil and military, e.g. purchases of computers for departments of government or the procurement of weapons systems. The provision of a technical infrastructure might be included here, as governments see to it, directly or indirectly, that roads, railways, canals, communications systems, etc. are built, operated, maintained and modernized. This requires a great deal of procurement either by public bodies or by private institutions under public supervision;
- information, including measures such as organizational arrangements for industrial collaboration, the operation of public R&D institutions, standards institutions, technological training and education, technical libraries, data banks and advisory services.

The above categories are not intended to convey the impression of watertight compartments. Any policy of any substance is likely to consist of many related measures, and the categories apply to single policy measures, rather than to policies. Indeed, many measures cannot stand on their own, e.g. legal requirements may necessitate organizational and institutional change.

In summary, we find that technology policy is intended to influence technology and cannot be easily isolated either from R&D policy or from industrial policy. Even some aspects of education policy, and many concerns of infrastructural and military policies interact with technology policy. The two main thrusts of technology policy are the support of technology and the regulation of technology. The former stems mainly from the desire to strengthen the national economy, while the latter is mainly necessary in order to reduce health and environmental hazards caused by the use of technology.

The main thrust of supportive policies is aimed at enhancing technological innovation. The reasons for this are twofold: first, technological innovation is believed to lead to economic growth and is regarded as necessary to avoid economic stagnation and maintain international industrial competitiveness (Freeman et al. 1982; Stoneman 1983; Coombs et al. 1987). Secondly, it is very much easier to influence new technologies by policy measures (Collingridge 1980). Old established technologies rarely require state intervention, except when they run into trouble and need to innovate, or when they come to be regarded as hazardous and need to be regulated. Technological innovation is the prime target of supportive technology policies, much as young plants are the prime concern of the gardener's supportive and regulatory measures.

2.2 Definition of Technological Innovation

We define technological innovation as a new or substantially improved product or process used or purchased on a commercial basis.

This definition requires some explanation. First, we do deliberately distinguish between innovation and invention. An invention is merely an idea, a drawing, possibly a prototype. The invention is probably, though not necessarily, the subject of a patent. At the stage of the invention no product on a commercially usable scale exists, though possibly the prototype, or prototypes, may be used experimentally. The great majority of inventions never lead to commercially used products or processes, and thus play no significant economic role. The road from invention to innovation is indeed an arduous one. It requires a great deal of effort, money and tenacity, and often ends in failure.

We distinguish essentially between three types of innovation. The most clear-cut is the product innovation. A new or substantially altered product, i.e. a product with some entirely new features and not merely some re-styling or re-design, which is newly offered for sale on the market, is defined as a product innovation. Strictly speaking, the product remains an innovation only during the very first stages of being marketed, though the concepts of new, newly on the market, and even market itself or commercial scale, are all slightly blurred.

The second type of innovation is the process innovation. In the strictest and simplest sense it means a new way of producing something, used for the first time in commercial production, i.e. not merely on an experimental scale. If, for example, insulin is produced by methods of genetic engineering instead of by extraction from pigs' pancreas, we have a clear case of a process innovation. The product, in this particular case, remains essentially unchanged. Unfortunately for those who love clearcut definitions, very often process and product innovations go hand in hand, a new product often requiring a new process for its manufacture.

The third type of innovation, commonly subsumed under the term process innovation, is here termed manufacturing innovation (Bessant/Grunt 1985). This is defined as the introduction of significant changes in the manufacturing process new to a firm. Thus we recognize the innovative activities of manufacturing industry in improving their manufacturing methods, even if what they introduce is neither a world novelty, nor an entirely new process. Take, for example, the introduction, in the early days of automation, of CNC machinery or robots into an engineering firm which had never used such machines before. Although the machines were not entirely new to the world, they were of fairly recent origin and required a good deal of innovative activity within the plant using such equipment for the first time. Another example might be the change from batch production to continuous production in, say, a brewery. The process of brewing remains the same and the idea is not entirely novel; nevertheless it requires a major reorganization, retraining, and rethinking on the part of management and workforce. The reason we are dealing with this type of innovation at such length is that, although it is often regarded simply as the diffusion of process innovations, it is subject to a great deal of support and attention by policy-makers. It is manufacturing innovation, or diffusion of process innovations, which, when all is said and done, is the major determinant of the use of manufacturing technology and of industrial competitiveness.

Indeed, it is always the final stage of the innovation process, the process of diffusion of the innovation through the economy, which determines the ultimate fate of the innovation, and thus its economic significance (Landau/Rosenberg 1986).

2.3 The Role of Technological Innovation in the Economy

The two main roles of technological innovation are as follows: first, to counteract market saturation, and thus stagnation, by offering ever new products, and to cause economic growth by offering an ever expanding range of goods. Secondly, technological innovation is a main weapon in the armoury for competitive battles. Only those manufacturers whose production technology is constantly renewed to ensure competitiveness, and whose goods are up to the latest standards, will thrive, or even survive.

The market for many types of technological products tends to reach saturation and the product can then essentially be sold as replacements for worn-out devices only. If, however, a substantially improved product comes on the market to replace the one for which the market is saturated, a new major wave of purchases can begin. A classical example is the saturation of the market for black and white TV sets and the new wave of purchases for colour TV sets. The industry now hopes that its fortunes will revive again with the likely introduction of high-definition TV, and thus a new cycle of investment will start. In some products, it is sufficient to make small stylistic changes to re-kindle flagging markets. This holds mainly for clothing and other items strongly subject to fashion. Obviously, manufacturers always attempt to get away with such small changes and with so-called built-in obsolescence; but in many cases, the only remedy to flagging markets is true innovation.

Technological innovation extends the range of goods available on the market, and thus, if successful, increases total demand. As long as the economy functions sufficiently well to ensure that customers willing to buy the new goods are also able to do so, i.e. that purchasing power expands in step with an expanding range of goods, economic growth will ensue. This argument is independent of whether we believe that new products merely satisfy existing needs, or whether we believe that new needs are created by new products, or whether we think that needs and the offer of their satisfaction can, in most cases, not be meaningfully separated. The only condition that must be fulfilled is that the new product must find buyers, i.e. it must succeed on the market. Otherwise, the investment embodied in the innovation will have to be written off and the total economic effect will be negative.

Major successful innovations tend to come in clusters: one innovation leading to another and thus to a whole technological system. Microelectronics is a very good example of this. The integrated circuit required a great deal of new production technology to be produced on a large scale at a reasonable price and this requirement led to a whole range of new production equipment. The integrated circuit found a very large scale application in computers of all kinds, as it improved their performance, and particularly their price-performance ratio, quite dramatically. The success of the computer led to many associated innovations, as for example a range of printers, computer numerically controlled machine tools, new telephone exchanges, and thousands of other innovations of greater or lesser importance.

Clearly the economic significance of this cluster of related innovations cannot be overstated. Indeed economists believe that such major technological changes cause long-term growth in economic activity, while down-turns are caused by a lack of such technological impetus (Mensch 1977). Even without considering the so-called long waves of economic activity (Kondratiev waves), it is clear that innovations of such importance cause major economic change. Whether this change is always beneficial is quite another matter, which will be discussed later. Generally speaking, new technologies tend, at least initially, to provide economic growth, though some innovations may have strong labour-saving effects and thus cause unemployment. Whether unemployment will prevail, depends on the balance between increased economic activity caused by technology and other factors, and the degree of labour saving, again caused by technology and other economic factors (Freeman et al. 1982; Rothwell/Zegveld 1981). Competitiveness of firms, i.e. the degree to which they can maintain, or increase, their market share and their profitability, is also determined by a large number of factors, and some of them may well be outside the control of the firm. An inescapable necessity, however, is that the firm should use appropriate manufacturing technology, in order to ensure high productivity and adequate quality of its products.

Economists used to speak a great deal about using the correct mix of factors of production, i.e. inputs of labour and capital or, sometimes, other factors, such as energy. The correct mix is determined by factor prices. In high-technology production, however, such considerations have become secondary. High-technology products, such as integrated circuits, cannot be manufactured without the appropriate automated machinery. No amount of labour can substitute for capital in these cases. Even in traditional construction work, the choice is not as flexible as it used to be, mainly for social reasons. It is inconceivable that in a major engineering project, as for example the construction of the channel tunnel, huge gangs of labourers with pick axes could replace the sophisticated monster machines cutting their way through the ocean bed. Appropriate production technology may be determined by technological or economic considerations, but may also be strongly influenced by social factors. Regulation may well be an important social factor. It is to be hoped that in the future the choice of production technology will be strongly influenced by environmental considerations, though these might well come in via regulations.

Production technology, whatever the influences upon its choice, is seen as a major determinant of industrial competitiveness. For if a firm produces its goods by an inefficient technology, they will become too expensive. If an ineffectual technology is used, the goods will fail in the market because of their lack of quality, or will be uneconomically produced because of a low yield of products able to pass the quality hurdle. The fate of individual firms may well concern only a small number of people, but the collective fate of manufacturing industry, or that of very large firms, affects the lives of whole nations and is therefore very much the concern of governments.

3 The Perceived Need for Innovation Policy

Despite the above argument, the reader may well ask why technological innovation should be the concern of government. As it is an economic activity, why can it not be left to private economic actors? There is no unequivocal answer to this question, and the matter is subject to political controversy. Interestingly, even governments totally committed to extolling the virtues of private enterprise, do not refrain from intervening in the process of technological innovation, be it by support or by regulation. The present chapter attempts to answer the question: why? We consider several aspects of the answer, in arbitrary order (see also Roessner 1988).

3.1 Reduction of Risk for Private Enterprise

Innovation involves risk. Treading unknown paths, encountering unexpected technical difficulties, meeting unforeseen market obstacles; these are but some of the hazards faced by the innovator. As governments are vitally interested that as much innovation as is reasonably possible should be undertaken, steps are taken to smooth the path of innovators and to reduce their risks.

The risks must be seen in relation to the size of the firm and the degree of novelty of the technology. Clearly, a very large and affluent firm can afford greater risks than a small business. Obviously, a moderate improvement of a product sold on a known market does not involve the same degree of risk as that encountered by embarking upon a radically new technology, to be sold on as yet non-existent markets. It is not, however just the novelty of either technology or market, but also the sheer size and complexity of an undertaking which determines the cost, and thus the risk involved.

If the potential economic or other benefit to the community is deemed to be great, and the risk involved for the innovating firm to be greater than it can reasonably be expected to bear, then public bodies feel justified in taking measures to share, or even bear, the risk. Sharing can be done by government alone, or else by collaboration with other firms, and even with other governments. The development of Concorde, the Anglo-French supersonic airliner, is a good case in point. The British and French governments got together and arranged for the collaboration between British and French aircraft manufacturers, while the governments footed a large part of the bill for this technically highly successful, but commercially disastrous, radical technological innovation.

The risk to the innovator may thus be shared with government or, with or without the help of government, with other commercial firms. However, reducing the risk to the innovator may take many forms – financial underwriting or sharing of the risk is not the only means of making it bearable. Other measures attempt to increase the chances of success and thus aim at reducing, rather than sharing, the risk. Details of possible measures will be discussed later, but essentially support can be given to make technical, commercial, or managerial information and advice more readily available and thus enhance the chance of a positive outcome. Risk flows from the very nature of the innovation process, but competent and well-informed handling can reduce the risk by increasing the chances of success.

Much the same arguments can be applied to process innovation, and even to manufacturing innovation. The literature dealing with the diffusion of innovation distinguishes between pioneering users, early users, and laggards in the application of new production technologies. The pioneering users clearly run large risks when buying untried radically new machinery, as many early users of robots found to their cost. If government believes that overall the new technology is likely to be beneficial, and likely to improve the competitive position of domestic industry on world markets, then it might assist pioneering users of the new technology.

Assistance can be financial, e.g. by buying or subsidizing the machinery for pioneering users, or it can be advisory, e.g. helping users to make the best of the new technology and to overcome early malfunctions.

Interestingly, virtually all governments in advanced countries with market economies (these are the ones we are speaking of here, and this statement does not exclude other governments), do support innovation for the reasons discussed above. In fact support for innovation has become an integral part of government attempts to keep their countries internationally competitive. Competition in support policies has been added to industrial competition in the battle for shares in international trade in manufactured goods.

3.2 Public Requirements for New Technologies

There is a perceived public requirement for progress. Whatever else is meant by this ambiguous word, technological progress is certainly included. Somehow, governments have become burdened with the responsibility for ensuring that the advanced country should remain advanced and, if possible, be at least as, or more, advanced than other countries. Being advanced means, among other things, to invent, produce, and use the latest technologies. Being wealthy has become associated with being technologically advanced. Surprisingly, social inequality, high crime rates, poor educational facilities and stark poverty do not seem to detract from the conception of being advanced. If a few of your countrymen can go to the moon, it does not seem to matter if many others are homeless and perish in the streets.

The quest for technological innovation is one of the main strands in the quest for progress. If governments shoulder the responsibility for progress, they must see to it that technological innovation proceeds as fast and as successfully as possible. Whether the public does or does not require particular technologies shall be discussed later; what is certain is that the public requires a constant stream of new technologies in order to foster economic growth as much as in order to be progressive.

The above critical viewpoint undoubtedly has some validity, though a more positive point of view is more generally accepted and this will now be explored. There are three types of applications for which technologies are required by the public, and in all of these innovation is, rightly or wrongly, constantly demanded. The three applications are: (i) public technological infrastructure, such as transport and communications systems or sewerage; (ii) defence and security, such as equipment for the armed forces, the police, or fire brigades; and (iii) technologies to relieve societal problems, such as urban congestion, or environmental pollution.

If a country is to remain internationally competitive, and is neither to put its own industry and commerce at a disadvantage nor discourage industrial investment by foreigners, it must have an up-to-date infrastructure. The infrastructure is either run directly by the state or by state-owned companies, or is, at least, controlled by the state. In any case, new technologies for the infrastructure are a public requirement and governments must see to it, one way or another, that the infrastructure is up to the required modern standards. Innovation in this sphere is therefore a concern of government, and hence some form of policy for innovation is inescapable.

That innovation in defence and public security is a responsibility of government goes without saying. Even the most market-oriented and non-interventionist of governments accept responsibility in the areas of defence and public security, and thus accept the task of equipping the security forces with adequate technical means. Unfortunately, technical means do not remain adequate for long and have to be brought up to the latest standards continuously. In no sphere of human activity is technological innovation pursued at such a forced pace as in defence; and agreement on a slower pace can only be achieved with great difficulty, as it must be obtained internationally through disarmament agreements.

On a more cheerful note, technology can make genuine contributions to the solution of societal – particularly environmental – problems. No matter whether the problems were created by the use of technology in the first place; we cannot dispense with technology in seeking solutions. As the environment cannot look after itself and looking after it often runs counter to individual immediate interests, particularly those of industrial firms, it must be a public concern if it is not to deteriorate beyond all hope of repair. It is, of course, enlightened self-interest which should make the environment everybody's concern, but only public bodies can channel enlightened self-interest into codes of conduct and, more importantly, into technologically supported action. Thus government is ultimately responsible for obtaining technologies required to repair and safeguard the environment. As most of these technologies do not exist, it is government which needs to see to it that they are brought into existence and into use.

We have discussed mostly the natural environment, but the same considerations apply to problems of urban congestion, to crime, and to any other societal problem one cares to mention. Obviously, none of these problems can be solved by technology alone – there are no technological fixes. Social, political and educational action is indispensable and social innovation is as much a requirement as technological innovation. It just so happens that the present book deals with technology; and hence, it is technology we concentrate on.

3.3 Directing Private Innovation toward Public Needs

Most innovative activity in market-oriented societies is carried out in private industrial firms. The fact that some of them may be wholly or partly owned by the state makes little or no difference, as nowadays even such firms are usually run strictly on commercial principles. As discussed earlier, straight self-interest is not able to look after certain matters of importance to the whole of society. The natural environment is the most blatant case in point; though traffic congestion and many other issues are not far behind. If, however, innovation in private firms is carried out with immediate self-interest as the sole motivation, it is unlikely that these types of problem shall be solved. It was argued above that governments – or rather the public – need certain new technologies. In addition to public requirements, it is necessary to try and give new directions to the huge creative forces in private industry by tilting the balance of market forces somewhat in the direction of social needs, instead of leaving them to cater only for strictly self-interested needs.

Adequate and environmentally acceptable equipment for the technological infrastructure is not the only public need for technology. It is at least equally important that privately used technology, whether for consumption or for production, should become benign. It is therefore necessary to tilt market forces in such directions as to ensure that private demand will move toward non-disruptive and sustainable technologies. Damage to the environment and to the social fabric can only be repaired and avoided if both private and public demand shift in socially required directions. Technology policy must, very slightly and sensitively, intervene in the market to achieve this change of emphasis.

It has been argued elsewhere that demand, or need, cannot be neatly separated from supply (Braun 1984). There is no fixed shopping list for technologies which would-be innovators could consult to find all unsatisfied needs. Indeed, innovators merely guess that their innovation might find a resonance, might kindle potential wishes into actual demand. For societies and individuals whose basic needs are fulfilled, market demand is a process of selection from available offerings. Additional needs are satisfied by additional products purchased by, and creating, additional wealth. Any attempt by government to steer the offerings on the market into socially acceptable directions cannot therefore be interpreted as depriving people of their right to have essential needs satisfied.

The need to steer technology for private consumption arises out of four considerations: (i) to ensure the health and the safety of users of a technology, whether in production or in consumption; (ii) to safeguard the natural environment; (iii) to ensure compatibility and mutual non-interference when a technology is used by many consumers; and (iv) to promote technologies which enhance the quality of life.

The first consideration, i.e. safety and health, is necessary because users of technical products cannot rely on the market alone to guarantee the harmless and safe nature of products offered for sale. Indeed even public bodies find it difficult to ensure absolute harmlessness, as very often previously unsuspected hazards come to light after many years of apparently safe use of a product. Quite apart from that, the market would discover the lack of safety of a product only in a haphazard manner and necessarily in arrears, while the legislator at least attempts to provide safeguards in advance of the introduction of a product onto the market.

Although most of the above is achieved by regulation and will be discussed later, innovation policy does also have a role to play. It is possible to build social criteria into the innovation process and, therefore, into any policies for the support of innovation. To make this possible, however, it is necessary to make early assessments of what the likely impacts, problems, and results of introducing the innovation might be. This type of early assessment, known as Technology Assessment, is a pre-condition for the successful incorporation of social criteria into measures for the support of innovation. It is possible, for example, to support an innovation which is likely to increase the safety of cars, or one that is likely to give public transport a boost, while not supporting innovations which are likely to have the opposite effects. Similarly, it is possible to promote innovations for crime prevention, reduction of risks from fire, improvements in treatment and prevention of disease, while withholding support from frivolous and potentially hazardous innovations. We are discussing here the support or lack of support for certain innovative activities, while leaving questions of regulation, prohibition or limitation till later.

All the above examples do, of course, assume a political will in a certain direction – if the political will tends toward different goals, the examples have to be modified accordingly. There is nothing prescriptive in Technology Assessment, which merely assesses likely consequences of new technologies and analyzes possible policy options. The policies adopted toward such innovations are an act of political will, and are guided by values and social objectives of the policy makers. Technology Assessment is a science, albeit a social science with all the imprecision this implies, whereas decisions on policies, and the administration of such policies, are acts of governance.

The arguments advanced in support of policies designed to enhance the safety of technological innovations can be applied with equal force to directing innovations toward environmentally benign technologies. Indeed, many critics will argue that this mild stance is insufficient, and that we require a complete change of paradigm. No longer should we seek economic growth, no longer should we seek the gratification of ever more trivial desires; instead, we should seek to save the globe from what they regard as certain destruction. Without necessarily wishing to disagree with these sentiments, they fall outside the scope of this chapter. We are not leaving the prevailing paradigm of the desirability of growth and the further raising of material living standards, but merely adding some dimensions of socially desirable criteria for the stimulation and control of technological innovation. We are discussing the possibility of a subtle shift in direction, in the hope that a small change will ultimately lead to the desired goal, rather than advocating a radical – possibly desirable but probably unrealistic – change of paradigm.

The compatibility and mutual non-interference of technologies requires some explanation. In his remarkable book Hirsch (1977:22) discusses "individual satisfaction in a specific activity being obstructed by the similar activity of others". The obvious examples are road traffic and tourism, though the latter is largely outside the scope of technology policy. In addition to the need to regulate every type and aspect of traffic because of the very obvious potential interference of each participant with every other participant, there may be possibilities to ease the problem by technical means. Technological progress in this direction is clearly preferable to innovations which make the problem ever more unmanageable. This is not to argue that the problem can be solved by technical means alone, but merely to suggest that technical solutions may help. Other technologies are equally, though less disastrously, prone to interference. Radio communications are an obvious example, and the introduction of cellular radio and fibre-optics communications illustrate how technology can ease the problem.

The final item in our list, improvements in the quality of life, refer not merely to urban congestion yet again, but mainly to improvements in working conditions, or in the life of the disadvantaged. It has been argued, for example, that CNC machines can be introduced in ways which maintain the skills of their operatives, or in ways which reduce the operators to the state of morons (Wilkinson 1983). If technological innovation can weigh in on one side or the other, the 'good' side should clearly be encouraged. Obviously, decisions on the use of technology are made in the political microcosm of the factory, and the same technology can be used in many different ways. Nevertheless, some useful positive features can be built into the machinery without economic detriment, and innovation policy can be formulated so as to encourage such developments.

Another example might be the care of the elderly. Technology can do little or nothing to change attitudes of families to their elderly relatives, nor can it change policies on residential care or on the provision of sheltered housing, home helps, and so forth. Technology can, however, provide some aids to help the elderly, or the disabled, to maintain a degree of autarky. If so, promising innovations aiming at improving the chance of independence of people might be better candidates for public support than many of a rival technology.

In summary, it is suggested that innovative activities might be assessed for their social utility and those that appear to have the potential to improve the quality of life – as defined by the policy maker – should be given priority in support.

There is a final – to some the most obvious and primary – point to make about support for private innovative activity. Governments may be worried that the sum total of innovation is insufficient to maintain economic growth and support innovation for this reason alone. However, even if the achievement of economic growth is the prime purpose of support for technological innovation, some additional criteria, such as those discussed above, may be usefully applied without detriment to the prime purpose.

We live in a highly competitive world, and governments are involved in international competition in innovation policies. Alas, to avoid such competition may prove as hard, or harder, than slowing down, or eliminating, the arms race. Progress has been made on the arms race, and this may be an auspicious moment to attempt some reduction in the overall levels of support for technological innovation. One day it may become apparent that the world would be a better place if technological innovation were slower. It may also turn out that much of government support is futile and ineffectual, anyway. The day that these things are recognized and acted upon internationally is far away, but the day when some agreement on reduced levels of support is attempted may be closer at hand.

4 Ways of Promoting Innovation

We have discussed a number of reasons why governments might wish to support entrepreneurs attempting to produce technological innovations. We must now look at the means available for this purpose.

4.1 A Taxonomy of Possible Support Measures

The first dimension in a taxonomy of support policies for technological innovation is the targeting of policy measures. The first distinction we must make in the targeting of support measures is that between direct and indirect support for technological innovation. The second distinction is between general and specific support. The third distinction is the stage at which support is given, which can be at any stage from early research, to advanced development, to early production, and to the first stages of diffusion (Braun 1984:39-71).

The second dimension is given by the domain of policy into which the support measures fall. Here we distinguish between; fiscal/financial, information/education, organizational/political, procurement/infrastructure, legal/regulatory, science/research, industrial/commercial (Braun 1984:125-130). Actual policy measures fall within one of these policy domains (though the list is not exhaustive), and are targeted in one of the ways described. Table 1 illustrates the above.

The targeting of support for technological innovation requires some discussion. Direct or indirect method of support: it is possible to support technological innovation by indirect measures, such as creating a good technological infrastructure, a good educational system, and an entrepreneurial climate. Tax incentives for investment might also foster innovation in an indirect way, and so might the ready availability of risk capital.

Many measures from the above list can, on the other hand, be applied very directly to technological innovation. In education, for example, specific courses for managers of innovation and for entrepreneurs can be arranged. In creating an entrepreneurial climate emphasis can be put on technological innovation, for example by rewarding innovators and inventors with honours and with publicity, rather than rewarding entrepreneurs of a purely commercial, and possibly

Table 1: Targeting and domains of policies for the support of technological innovation. A few examples of policy measures which fit into the matrix

Target / Policy Domain	direct/ indirect	general/ specific	research/ development/ diffusion
fiscal/financial	company taxation	innovation finance	R&D finance
information/ education	technical libraries	training courses	data banks
organizational/ political	international cooperation		R&D
procurement/ infrastructure		railway investment	purchase of prototypes
legal/regulatory	patent laws	emission standards	
science & research		national labs	support for environmental R&D
industrial/ commercial	trade agreements		

shady, type. Tax incentives can be very specifically directed toward using the latest production technology, or toward supporting R&D, patent activities, scientific equipment, etc. Even risk capital can be made available specifically for technological innovation, rather than for any kind of risky commercial enterprise.

General or specific support: General support aims at fostering innovation in general. It stems from the belief that technological innovation is a good thing and cannot proceed at a sufficient pace without public support. Some degree of selectivity can be introduced. Thus there might be a programme of support for innovations in pollution control technology. This would give support to all reasonably promising innovation projects in this area, with a varying degree of selectivity. If the selectivity becomes very strict, i.e. only very few applicants for support actually receive it, then the support is no longer general but becomes specific for selected projects. Specific support may or may not be aimed at a particular area of technology; it may simply aim at highly promising and deserving projects, irrespective of the kind of technology involved. On the other hand, it is perfectly feasible to apply any policy criterion, beyond the criterion of timeliness and promise, to selective support. The transition from general to specific support is continuous, from little or no selectivity to a high degree of selection.

Technological innovation is a multi-stage process. The literature lists at least three distinct stages, often more, and emphasizes that the boundaries between the stages are blurred and not all innovations necessarily go through the same phases. The three stages of interest here are: early research and development, further development and early production, early marketing and diffusion. For simplicity, we have called them "research", "development", and "diffusion". At what stage, or stages, of the innovation process support should be given by the state is a matter of some debate.

The important issues in this debate are (i) What is the legitimate role of the state, i.e. should it support purely commercial enterprise or should its role be confined to the support of public requirements and to research as a cultural exercise? (ii) If the state is to support technological innovation, what is the stage at which help is most needed and can be most effective? The former point has been discussed already to some extent, the answer to the latter question depends very much on specific circumstances of individual innovations. Because the research and development stages of innovation are the ones most commonly supported, the next section will be devoted to state support for R&D.

4.2 Research and Development

R&D has become a major enterprise. Ever since the Second World War, which proved the prowess of R&D for all to see, technological innovation has become more and more systematically based on R&D. While in the past innovative activity in many industries was craft-based and often not systematically organized, the post war years have changed all that. Innovation has become part of corporate strategy, and R&D has become the spearhead of innovative activity. Industrial firms in some sectors, particularly in pharmaceuticals, chemicals, electrical engineering, electronics, computers, and aerospace have become so research-intensive that a substantial proportion of their turnover is invested into R&D. In 1975, the British aerospace industry spent 29% of its output on R&D, though only 5.2% was self-financed; the rest was financed by government (Pavitt 1980:57). In 1985, the ratio of company-funded R&D to net sales for all Japanese manufacturing industries was 2.7%, compared to 2.8% in the United States (NSF 1988:11). Japanese industry financed 69% of all R&D; while in the US, industry financed only 49% of all R&D. It must be said, however, that only 1% of Japanese R&D goes toward defence, while about one third of US R&D is devoted to that purpose (NSF 1988:9).

As Table 2 shows, private and public R&D have become so large that total expenditure on it is of the order of several percent of gross domestic product (GDP) in all advanced countries.

Table 2: Total R&D expenditures for 1987 as % of GDP for several countries, and the shares of government and industry finance

	R&D as % of GDP	government (%)	industry (%)
USA	2.72	50.8	47.1
Japan*	2.59	19.4	73.7
FRG	2.71	35.8	62.8
France	2.28	52.9	41.0
GB*	2.42	39.8	48.2
Austria	1.31	48.5	48.8

* 1986

Source: Passweg 1990

Governments spend a very large amount of money on R&D and it is interesting to see their own reasoning in justification of such expenditure. Several themes recur in virtually all official or semi-official statements in support of public R&D expenditure. They include the need to maintain and enhance international competitiveness, the duty of the state to support pure research as a public responsibility, and the need for publicly funded research into solutions of urgent societal problems, such as environmental pollution. A few extracts from official statements of various governments and government advisory bodies may illustrate the points made. They give the general flavour of the justification for government support for R&D, though there are obviously considerable variations in R&D policies of different governments.

In the Technology Policy Survey 1989-1990, presented to the Netherlands Parliament, the following statements form part of the introduction:

Economic growth is generated in particular by the broad application of new technologies in a wide range of products, services and processes. ... Innovation is increasingly recognised in industry as a strategic factor which is essential for international competitiveness. ... Between 1984 and 1988, nominal corporate research and development (R&D) expenditure rose by some 50%. ... The technology policy outlined in this Survey aims to (help) create a climate in which technological progress can be translated into economic growth (Ministry of Economic Affairs 1990:5-6). (Note: the levels of corporate and government R&D spending as % of GDP in the Netherlands are roughly comparable to those in France.)

The Swedish Government Research Bill states:

In many respects it is justifiable to say that the scientific search for new knowledge is a societal concern, indeed a task set by society. It is regarded as natural and necessary for companies to invest in the development of new products and production processes. ... It is just as natural and necessary for problems in society to be made the subject of research. ... The natural and social sciences are indispensable in modern society; they deepen our perspective on life and help us to find rational solutions to human problems. ... In the final decade of the twentieth century and at the beginning of the twenty-first, the countries of the world are faced with the task of safeguarding ecologically sustainable development. ... Environmentally oriented research in all fields has a key part to play in this context (Cabinet Office 1990:8-10).

The Federal German government defines the aims of its science, research, and technology policy as contributing to the following general aims:

1. The expansion and enhancement of scientific knowledge and thus to the further structural and organisational development of the federal German research system. ... 2. To humane working and living conditions full of opportunities, to a sustainable use of natural resources and hence to positive developments in the natural environment. 3. To improved chances for technological innovation and thus to increased efficiency and competitiveness of the economy. 4. To a strengthening of international cooperation in research and technology (*Bundesbericht Forschung* 1988:18).

The Federal German government stresses that the state should mainly devote its resources to basic research and to research into long-term opportunities and societal problems, while devoting less effort to the support of economically motivated R&D performed in industry.

As stated earlier, technological innovation is a multi-stage process and the stage of an innovation at which support is given differentiates strongly between different attitudes to state support. It is argued by some that support should only be given at a pre-competitive R&D stage. This means at a stage when the feasibility and general features of a technology are being developed and no individual final designs are in sight. It is at this stage, some argue, that firms can cooperate on major new technologies without compromising future competitiveness. One example of such R&D support is the Research into Advanced Communications for Europe (RACE) programme of the Commission of the European Communities (CEC). This is an attempt to bring virtually all European manufacturers and operators of telecommunications systems together to develop an advanced system of integrated broadband communications (IBC). The programme requires uneasy compromises between commercial secrecy and open research, between competition and cooperation, and between political will at the centre and in individual countries.

The RACE programme is an example of support at the R&D stage, but with a very definite technological, political and economic outcome in mind. Indeed, the programme is the result of the political will to develop and install a European broadband communication system. The aims are (i) enhancement of the European telecommunications industry, and (ii) general improvement in the European communications infrastructure and thus in economic performance.

It is somewhat unusual to have such a very clear picture of what is wanted at the R&D stage. Usually this stage is more exploratory, leaving many avenues open, full of surprise and of risk. Many feel that this is the stage at which state help is needed most and, more significantly, is most legitimately given. Those who feel that competition and market forces should be left undisturbed, argue that only R&D with unknown outcomes is a legitimate subject for state support, because at that stage it is not clear who, if anybody, will reap commercial rewards from the investment. As soon as the R&D comes close to a commercially promising innovation, further investment should be left to those who hope to reap profits from it.

At the R&D stage state support can be given in the most general and unspecific way. The state often runs its own research laboratories and results from these are either openly available, or can be purchased by firms. The terms of availability depend on the degree to which the state wishes to use such laboratories for the support of innovation. At the R&D stage a whole host of measures is available: they can consist of information support, such as help from government laboratories or data banks. Organizational help can be given to foster university-industry collaboration, or financial assistance or tax incentives can be given for industrial R&D laboratories.

It is equally possible, however, to be very specific in R&D support and to give this only to selected programmes, selected scientific areas, or selected laboratories or individuals. The selection can be based on perceived societal needs, or simply on scientific excellence, or both. Most governments support a measure of pure research with no technological applications in mind, simply as a cultural enterprise, an attempt to find out more about the structure of the natural world or of society. In the absence of private patrons, governments have felt obliged to take over the patronage of culture in general, be it in the arts or the sciences. Sometimes pure science may bring unexpected practical rewards and indeed the distinction between pure and applied science is not at all clear-cut. Hence, the support for pure science is fuelled by the hope that something unexpected may turn up and new discoveries may open new technological avenues (see e.g. Braun/MacDonald 1982).

Selection of R&D topics by societal needs is certainly very appealing and few would dispute the need for more R&D into environmental problems or into some aspects of health or disease. However, support has to be geared to opportunities and to quality: there is no point in pouring in money in excess of the needs of available promising ideas and creative researchers. Another danger of gearing support too closely to perceived societal needs is the necessity to give serendipity a chance. Freely roaming and foraging by creative minds may turn up the unexpected. An answer to a problem in one area of human activity may come from activity in an entirely different area. Cross-fertilization is one of the modes of progress of ideas, and this must not be stifled by too rigid a framework of support.

Education, especially the higher education sector that is of interest to us here, has also become a major enterprise in support of technological innovation. As education policy is outside the scope of this chapter, let it suffice to say that the training of engineers and scientists at all levels, as well as the training of skilled workers and continuing education in the use of new technology, play a vital supporting role in technological innovation.

4.3 Direct Support for Innovation

Direct support for technological innovation can be given at all stages of the innovation process and can be as general, or as specific, as governments may wish. The basic mode of direct support consists of payment of the full, or an agreed proportion of, the cost of the selected part of an innovation process. The supporting organization may also help in a variety of ways, e.g. by arranging partnerships or making available information otherwise unavailable.

The Federal German government, for example, has recently launched a major new interministerial programme in support of information technology. Support is envisaged right across the board for this key technology, i.e. for all its branches and at all stages of R&D and diffusion. This is an interesting example of the conscious intermeshing of economic and technology policies. The measures range from re-regulation of telecommunications and new legislation on intellectual property, right down to financing major research institutions (*Zukunftskonzept Informationstechnik* 1989).

There is a large number of various support schemes for various technologies in most countries. Some, as the above German scheme, comprehensively supporting a large and widespread innovation, others in support of a very special one-off project. Generally, governments announce schemes in support of certain technologies and invite applications for grants. Sometimes the work to be done is spelt out in great detail; in other schemes, it is left to the imagination of the innovator. Though most governments do have priority areas for support, some scope is usually left for initiatives from innovators to apply in non-priority areas, because it is recognized that new ideas, indeed new ranges of technologies, may offer opportunities not foreseen by the originators of support programmes.

It is a frequent feature of such schemes that they attempt to encourage cooperation and partnerships either between industrial firms, or between firms and universities, or firms and government laboratories, or any combination of the above. The reason for these attempts is the recognition that cooperation, though potentially highly beneficial, is not without its own costs and is not easily achieved between either rivals or unequal partners. An example of such a policy is the LINK scheme of the UK Department of Trade and Industry. One of the aims of the scheme is described as: "To help industry exploit developments in science and make scientists more aware of industry's needs by strengthening the links between industry, higher education, the Research Councils and other research establishments. … LINK will build on, and extend over a much wider field, the experience already gained in collaborative programmes to the efficiency, innovative capacity and competitiveness of industry and commerce" (LINK Secretariat 1988).

4.4 Procurement

Procurement, i.e. purchases by public bodies and, very significantly, the military, is an old-established method of supporting innovative technologies in areas deemed significant to government. It is particularly a method much favoured in the United States, as it does not have the bad odour of public involvement in private technological enterprises, and yet achieves support of innovation and, at the same time, has an economic regulatory effect.

It has often been said that the microelectronics revolution came about in the USA because of military support for this technology. Although it can be disputed that military support brought about this revolution, there can be no doubt that it smoothed its path. "In the fifties and early sixties the Military was a main market for semiconductor products and often the only one for new and expensive components. New firms often found the military market a particular boon. ... That situation has now changed and the computer industry and consumer markets are of much more importance" (Braun/MacDonald 1982:142). In this case, procurement played an important role in the innovation, and is playing a lesser role in the more mature technology.

That many a development in civil engineering, or in railway equipment, and the like, owes its existence to public procurement goes without saying. What is less obvious, and indeed remains a moot point, is the degree to which normal civilian products owe their life to innovation undertaken either for the armed forces or for space exploration. It is probably safe to say that space exploration would either have not taken place at all, or at a much slower rate, had not the military forced the development of both rockets and satellites. Though the only consumer product that may safely be called a by-product of space exploration is said to be the non-stick frying pan, yet there is no denying that satellite communications are equally a by-product of space investments. Thus the argument that military procurement brings no civilian benefits, though appealing, is not entirely tenable. The argument that it would be more cost-effective to go for civilian developments without the detour of military development is, of course, unassailable.

Quite a few products, for example the hovercraft, have mixed sponsorship. They received a great deal of direct help in their development, and they also achieved a significant military market in their early life, thus contributing to the viability of the innovation. No doubt the direct aid was given with one eye firmly fixed on military possibilities, and thus direct aid cannot be neatly separated from procurement aid.

5 Need for Regulation

Whether we like it or not, modern society, or indeed any society, cannot function without certain rules. Unfortunately, the use of modern technologies has not decreased this need. To the contrary, the use of technology itself requires a vast array of rules. It has frequently been argued that regulation retards technological progress; and it has almost equally frequently been argued that the demands made by regulations are equivalent to market demands and indeed present new opportunities for technological innovations. We shall leave both possibilities open, as it is highly likely that both mechanisms do indeed operate. Which dominates, depends entirely on the technology and on the regulation.

Be this as it may, the regulation of technology is inescapable, and we shall now briefly explore why.

5.1 Adequacy and Compatibility of Products

The consumer of technological products needs to be protected from a variety of dangers against which the old rule of "buyer beware" is no adequate safeguard. Most products of modern technology are too intricate and complex to be subject to anything like an exhaustive and revealing inspection by lay consumers. Thus certain rules are needed to guarantee to the consumer that products are indeed fit for their purpose, i.e. that their performance is up to the standards given in their description. It is not good enough to rely entirely on the market in this respect. The market forms its judgements but slowly and retrospectively, and the cost of learning about the quality or fitness for purpose of products by individual consumers would be very high. Indeed very few consumers would risk buying new products, other than from a small number of companies with high reputations. In the absence of some consumer safeguards, innovation by small and unknown firms would be virtually halted, while innovation by large companies would be slowed down. Fitness for purpose implies minimum standards of quality. It also implies safety, which will be discussed later, and a quality we might term non-interference. An example of this are rules requiring electrical machinery that is liable to interfere with radio and television, to be appropriately shielded. Another example is the requirement for headlights of motor cars to fulfil certain standards, including dippability. Motor cars have to fulfil quite a number of requirements: about rear lights, direction indicators, number plates and numerous other items, all designed to bring the motor car under some sort of social control, however inadequate, as fit to be used on public highways.

Another example of regulation for non-interference is furnished by broadcasting, and generally the use of electromagnetic waves for the transmission of information. If unbridled competition were allowed to take place, nobody could use electromagnetic waves for the transmission of information, because each message would be likely to interfere with many other messages, thus leading to a breakdown of the system. Only by allocating specific wavebands, and making very clear rules, has it become possible to use electromagnetic waves for broadcasting of radio and television, for public telecommunications, and for communications of the emergency services or radio-controlled taxis.

Telecommunications, even without terrestrial or satellite electromagnetic transmission, has traditionally been subject to regulation for a number of compelling reasons. First and foremost, to ensure that every subscriber could reach every other subscriber; secondly to ensure that privacy is safeguarded for conversations and data transmission, and thirdly for reasons of social equity. Some of the regulation can and has been achieved by voluntary arrangements within the telecommunications industry, though not without some involvement by government. Other social objectives, such as social equity, requiring that every subscriber, irrespective of location or intensity of use of the facility, should have the right of access to a telephone at a reasonable and affordable price, were achieved largely by government regulation.

5.2 Safety and Health

As consumers cannot protect themselves against possible hazards posed by the use of certain technologies, they need to be protected by society, i.e. by government or its agents. Recent developments in fillings for upholstered furniture provide a good example for the need to regulate the safety of products. How are consumers to know whether the upholstery of their new expensive furniture is easily ignited and then gives off lethal fumes? Indeed, even furniture manufacturers did not initially know, and it took a number of fatal accidents before regulations were formulated to protect consumers from this particular hazard. Similarly, consumers cannot tell whether their electrical appliances are likely to electrocute them and must rely for their protection on the compliance of the manufacturer with regulations. Some products cannot be made entirely safe without losing their effect, e.g. lavatory cleansing materials and the like. In this case, there are regulations to make the product as safe as is consistent with its efficacy, and for labelling the materials adequately to warn of residual dangers.

Similar regulations are provided in the working environment. Every country has some act laying down the ground rules, and every country has something like a factory inspectorate which enforces the rules. Essentially, these rules make sure that machinery used in production, or even in offices, is as safe as it can be and that production processes do not pose avoidable hazards.

All the above examples fall into the large class of technological products that require regulations for reasons of safety. Indeed, this class is one of the first which gave rise to government intervention in technology. The boilers of early steamers on the Mississippi apparently were wont to explode, until the US government found it necessary to promulgate rules for the safety of steam boilers.

It is remarkable that one of the most dangerous products of technology, the motor car, is allowed by society to continue to wreak havoc among the population. Although a large number of rules and regulations attempt to make the motor car somewhat safer, the number of deaths and injuries continues at a very high level without drastic actions being taken.

Regulations can only be made when they are politically acceptable, i.e. when the users of a technology wish to be protected without too high a cost to themselves, either in monetary terms or in terms of unacceptable restrictions on the use of the technology.

A class of regulations somewhat similar to those related to safety, are those designed to safeguard the health of the users of products. An example is the use of lead as a component of paint or as an additive to petrol. Small quantities of lead in paint are unlikely to lead to fatal accidents, but there is a certain danger that the health of consumers, particularly that of small children, may be impaired. Thus the use of lead in paint has been severely restricted. Similarly, lead in the exhaust of motor cars is considered a health hazard, and attempts to eliminate the use of lead as an additive to petrol are in progress. For the same reason, the use of asbestos has been restricted, because the inhalation of certain types of asbestos fibres poses a severe risk to health. The latest risk, being slowly recognized at the present time, is that from aluminium as an additive to water. No doubt it will take quite some battles before regulations on the use of aluminium are tightened up in the light of new scientific findings, but eventually such regulations are very likely to be seen as a necessity.

A major class of products which is strictly regulated for reasons of hazards to health are pharmaceuticals. As drugs used to cure disease are potential causes of disease, an elaborate machinery to test the safety of drugs and a system of licensing them has been established. Some say that the system is inadequate, as now and again a previously undetected hazard comes to light after a drug had been in use for some time. Others say that the system is too severe and has crippled the innovative potential of pharmaceutical companies because it takes too long and costs too much to get a new drug approved.

5.3 Environmental Protection

The third class of technologies which require regulation are those that pose a danger to the environment. Initially, this danger was regarded more of a nuisance than a danger and the first perceived need was to regulate emissions from factories simply to reduce the nuisance to the neighbourhood. Thus the first British Lighting and Washing Act (1833), regulating effluents, and the Alkali Act (1863), regulating emissions, were created; and other countries followed with similar legislation in due course.

As knowledge about dangers to the environment increased, so regulations became more comprehensive and more severe, although there are very great lags between knowledge and regulation, and very great discrepancies between actual severity and that demanded by environmental activists.

Even the most ardent believer in the benefits of unhampered market forces generally admits that the environment cannot be adequately protected by these forces. The reason is simple: any costs that a producer can shift away from himself and turn into so-called externalities are beneficial to the consumer in terms of reduced prices, and/or to the producer in terms of increased profits. One way of reducing costs is to save on filtering and cleaning processes and shift as much waste as is permissible into the environment.

Another danger to the environment is its excessive exploitation as a source of raw materials and means of production. This is precisely what is happening in the rain forests of the world. It is also happening to agricultural land, which is being farmed so intensively as to pollute the soil, the rivers and lakes, and the ground-water.

Without wishing to recapitulate the well-known catalogue of destruction being wreaked upon the environment, we wish to establish that unregulated market mechanisms are inherently incapable of safeguarding the environment. This has been true historically, and is equally true today.

Regulation is one possible – and indispensable – way of safeguarding the environment. Strict rules on emissions and effluents must be formulated, enforced, and continuously updated. The disposal of waste must be regulated and made safe, while every effort must be made to reduce the amount of waste produced. This can be done by regulating the use of packaging and making rules on re-cycling of materials.

Regulation is necessary, but not sufficient. It is equally necessary to create incentives for the reduction of waste in general, and dangerous waste in particular. The incentives can be purely financial, but they can also be institutional. A good infrastructure for recycling and for the safe disposal of waste should be high on the agenda of public administrations. It may be possible to provide market incentives for environmentally-benign technologies, and it may also be necessary to provide subsidies to accelerate the diffusion of such technologies.

Support should be given for technological innovations which decrease waste, improve recycling, improve waste disposal, reduce the consumption of energy and raw materials, reduce emissions, etc.

It is impossible and unproductive to separate the two faces of technology policy entirely. Regulation and support are the two sides of the coin of technology policy, and if the coin is to be worth anything, the two faces must be well coordinated. It is not a matter of carrot or stick, but a judicious combination of the two.

6 Modes of Regulation

Regulation is controversial. There are those who argue that government interference has gone too far and should be decreased or virtually eliminated. The arguments are in part about freedom, but a larger part revolves round the issue of the cost of regulations and their benefits.

The first argument cannot be resolved, as it touches upon fundamental beliefs largely unaffected by rational argument. The only universal agreement that can be reached is that suggested by John Kenneth Galbraith: "there is a large range of such activity where the larger interest – the more general liberty – requires control (and it will certainly always be imperfect control) over the lesser and more specific interest". In another passage Galbraith makes the unassailable, but not very helpful, point that "a distinction between wise regulation and unwise regulation" must be made (Galbraith in Siegan 1980:8-9).

Regarding the costs of compliance with regulation, a recent summary of existing studies comes to the conclusion that "existing studies shed little light on the dispute between those who hold that regulation benefits the regulated and those who hold that regulation is of necessity a burden on those who are regulated" (Peacock 1984:39).

In the absence of hard evidence to the contrary, and with the vital proviso that regulation must be wise, it seems that the arguments produced earlier in support of a regulative role of technology policy, taken together with historical evidence of the disastrous effects of laissez faire in matters technological, provide a watertight case in favour of a modicum of regulation. The crucial questions of when, how, and how much cannot be answered in a generally helpful way and have to be resolved from case to case and issue to issue in the light of everything that has been said before.

Regulation rests, as it were, upon three pillars:

- Laws, statutes, and by-laws, i.e. legislation passed by competent legislative authorities. The laws themselves can specify regulatory detail, but generally they enable executive authorities to issue specific orders and to make detailed arrangements.
- Orders specifying details of regulations within the given legal framework. The details provide the flesh on the bones of legislation and specify what should or should not be done in specific cases. The orders are issued by the authorities, e.g. ministries or local authorities, or other statutory bodies, specified as competent in the relevant laws.
- The final pillar consists of the various enforcement agencies. As all powers of the state, final enforcement and arbitration rest with the courts, but the all-important day-to-day enforcement and detailed supervision rest with regulatory agencies. The institutions created for regulatory supervision are at least as important as the legal framework in which they operate.

Detailed arrangements clearly differ from country to country and from subject matter to subject matter. No detailed account can be given in a short text of this nature; only outline principles and a few examples. The matter is further complicated by the fact that some regulatory functions are fulfilled by supra-national organizations, particularly the Commission of the European Communities.

What we described as laws, statutes and by-laws consists, in the United Kingdom, mainly of primary legislation embodied in Acts of Parliament, which may be public, and thus applicable quite generally, or private, and thus relating to a specific issue. At the local level, local authorities can enact by-laws, which are, however, considered as subordinate legislation and thus are, unlike primary legislation, open to legal challenges.

What we described as orders specifying details of regulation can, in the United Kingdom, take a variety of forms and are invariably regarded as secondary legislation. One form are Orders in Council, another are orders, regulations, rules or schemes issued by a Minister of the Crown. Some statutory bodies may also issue orders, though these need to be endorsed by a minister.

A variety of enforcement agencies are set up under either primary or secondary legislation. Many of these agencies operate under approved Codes of Practice, which do not have the force of law, but constitute an important instrument setting down methods by which regulation is enacted in practice.

Examples of such agencies in the UK are the Health and Safety Executive, which operates under the Health and Safety at Work etc. Act. The executive controls the work of an inspectorate, which has comprehensive powers. Two further examples are the Office of Telecommunications (OFTEL), which regulates the deregulated telecommunications industry; and the Office of Water Services (OFWAT), which regulates the privatized water and sewage services.

In many cases of regulation of the use of technology, for example in the case of road traffic, there is no specific statutory body enforcing legislation and enforcement is in the hands of the general law-enforcement agency, i.e. the police.

7 Summary and Conclusions

Technology policy consists of a mix of measures in support of technology, and particularly in support of technological innovation, and measures regulating technology.

Support is given mainly with the general aim of strengthening the competitive position of industry in world markets, and of causing economic growth by satisfying ever new consumer demands. Another form of support of technology consists of the provision of a modern technological infrastructure, such as transport and communications, and of supplying public services, including the armed forces, with adequate equipment.

Regulation of technology is not an aim in itself but is necessary for a variety of reasons. The main reasons are the need to safeguard safety and health, and to protect the natural environment. It is argued that a degree of regulation is unavoidable and, if wisely chosen and judiciously balanced, can have highly positive effects, while lack of regulation can cause considerable damage.

The support of technological innovation by the state is a somewhat more doubtful enterprise. It may be argued that if the economy is arranged in such a way as to ensure that the main benefits from technological innovation accrue to private industry, then private industry should also bear the costs and risks associated with the innovation. It seems damaging that international competition in technology has been supplemented by international competition in support measures for technological innovation. On the other hand, it is argued that the state ought to try and support such technologies that are of obvious benefit to the public at large, first and foremost among them environmentally-benign technologies. As with regulation, no very general rules can be stated – there is no substitute for handling the support of technology with sensitivity and good judgement.

References

Bessant, J., M. Grunt (1985) *Management and Manufacturing Innovation in the United Kingdom and West Germany*, Hampshire: Gower.

Braun, E. (1984) *Wayward Technology*, London: Pinter.

Braun, E., S. MacDonald (1982) *Revolution in Miniature*, Cambridge, GB: Cambridge University Press.

Bundesbericht Forschung (1988) Bonn: Der Bundesminister für Forschung und Technologie.

Cabinet Office (1990) *Government Research Bill* 1989/90:90, Stockholm: no publisher stated.

Collingridge, D. (1980) *The Control of Technology*, London: Pinter.

Coombs, R., P. Saviotti, V. Walsh (1987) *Economics and Technological Change*, London: Macmillan.

Freeman, C. (1982) *The Economics of Industrial Innovation*, London: Pinter.

Freeman, C., J. Clark, L. Soete (1982) *Unemployment and Technical Innovation*, London: Pinter.

Hirsch, F. (1977) *The Social Limits to Growth*, London: Routledge & Kegan Paul.

Landau, R., N. Rosenberg (eds.) 1986) *The Positive Sum Strategy: Harnessing Technology for Economic Growth*, Washington: National Academy Press.

LINK Secretariat (1988) *LINK Mechanisms and Guidelines*, London: Department of Trade and Industry.

Mensch, G. (1977) *Das technologische Patt*, Frankfurt/Main: Fischer.

Ministry of Economic Affairs (1989) *Technology Policy Survey 1989-1990*, The Hague: Government Printing and Publishing Office.

NSF (National Science Foundation) (1988) The Science and Technology Resources of Japan, Special Report NSF 88-318, Washington, DC: NSF.

Passweg, M. (1989) Forschung und Entwicklung. Trends in der OECD und in Österreich, in: *Wirtschaft und Gesellschaft* No. 4: 511-536.

Pavitt, K. (ed.) (1980) Technical Innovation and British Economic Performance, London: Macmillan.

Peacock, A. (ed.) (1984) *The Regulation Game*, Oxford: Blackwell.

Roessner, J.D. (ed.) (1988) *Government Innovation Policy*, London: Macmillan.

Rothwell, R., W. Zegveld (1981) Technical Change and Employment, London: Pinter.

Siegan, B.H. (ed.) (1980) *Government Regulation and the Economy*, Lexington: Lexington Books.

Stoneman, P. (1987) *The Economic Analysis of Technology Policy*, Oxford: Clarendon Press.

Wilkinson, B. (1983) *The Shopfloor Politics of New Technology*, London: Heinemann.

Zukunftskonzept Informationstechnik (1989) Bonn: Der Bundesminister für Forschung und Technologie und Der Bundesminister für Wirtschaft.

New Technology Policy Concepts: Some Reflections on Technology and Work Humanization in West Germany

Richard Badham and Frieder Naschold

1 Introduction

As knowledge of the varied social influences on technological innovation increases, so are technology policy concepts are needed that come to terms with this complex reality. This is not least apparent in the area of process innovation in the workplace. Simplistic models of "R&D push", "product cycles" or "market demand" are giving way to a more sophisticated recognition of the national institutional and cultural factors influencing the rate and direction of technical change. As different nations expand the role of technology policy in the development and diffusion of new production technologies, these conditions are attracting greater attention. Research and policy orientations are starting to take into account the broader institutional influences on the competitive advantage of nations and address the effect of national systems of innovation, industrial clusters, regional networks and patterns of skill formation.

One component of these changes is the emergence of a new policy orientation towards promoting increased productivity. Influenced directly by the enormous successes of Japanese manufacturing techniques, more attention (as represented, for example, by the Sundquist Report) is now being paid in technology policy circles to the organizational as well as technological components of successful innovation. In European countries such as West Germany and Sweden, it also involves a significant re-orientation of national work humanization activities. In Sweden from the late 1970s and in West Germany from the middle 1980s, a 'new language of workplace reform' emerged. In Sweden, it took the form of a growing consensus between peak management and union bodies on the need to establish a 'post-Taylorist' work form with an emphasis on "capital rationalization rather than division of labour" as the key to both efficiency and participation (Elam/Börjeson 1991:11). In West Germany, it was apparent in the trend within the 'Humanization of Work' programme away from an anti-rationalization humanization stance, to the acceptance of "comprehensive innovation concepts" that accepted a degree of variance in the rationalization drive; and it emphasized the promotion of development pathways or paths of rationalization that combined increased productivity with improvements in the quality of working life.

This trend in technology policy has been clearly apparent throughout Europe in the late 1980s. As one extensive review of European programmes in the area of technology, work and organization points out, it was during this period that "a trend becomes gradually discernible: a development in the direction of a design and policy-oriented integral approach" based on a tight coupling of "effectiveness and technological innovation with socio-organizational renewal" (den Hertog/Schöder 1989:88). The development of this approach is clearly controversial. For some critics, it constitutes a "flexibility orthodoxy", "a language of social integration of the 1980s: how to live with insecurity and unemployment, and bow to market forces"(Pollert 1990:98; see also Pollert 1991). For others, it is perceived merely as the latest stage of rather ineffective socio-technical strategies in work design, or the search for more sophisticated forms of 'speed up' through quality of working life initiatives. In the eyes of its supporters, however, it represents both an absolute requirement if European firms are to remain, or achieve, world competitiveness in productivity, and the only realistic form in which humanization of work initiatives can be pursued.

If these New Technology Policy Concepts (NTPCs) are taken seriously, however, three major challenges have to be met: firstly, overcoming many of the traditional limitations of work design reforms and programmes; secondly, addressing the competitive strength of new Japanese manufacturing techniques; and, thirdly, establishing the conditions necessary for new forms of systemic and interdepartmental technology policy. After briefly reviewing the nature of NTPCs and the nature of these challenges, this chapter focusses on the experiences of the Humanization of Work programme in West Germany in order to throw some light on the practical difficulties facing programmes in this area, and on the types of political as well as socio-technical issues that have to be addressed if *comprehensive* innovation concepts are to live up to their name and incorporate an effective broader approach to innovation promotion.

2 New Technology Policy Concepts

A common theme of NTPCs is the need for the consideration of the "social" aspects of process innovation in a "proactive" manner as a key factor in the "promotion" of effective technological change. Commitment is given to promoting "post-Taylorist", "post-mechanistic", or "post-technocratic" approaches to organizational change that involve: greater collaboration between system designers and users in the design phase, greater workforce consultation in the implementation phase, and increased attention to operator-workforce initiative in ongoing incremental innovation and continual learning organizational structures.

A set of assumptions and priorities underlie this emphasis, i.e. a commitment to the priority of enhanced international competition and the development of advanced technological systems to meet this end – while identifying advanced systems with flexible skill enhancing techno-organizational structures; a significantly greater emphasis upon the importance of developing appropriate forms of training and work organization as a crucial component of innovation promotion; and a stress upon the significance of system operators or users in the design, implementation and further development of usable and efficient hardware/software systems. In this sense, a 'social' orientation is firmly embedded within the commitment to direct technology promotion for enhanced international competitiveness. This techno-economic concept lies at the core of NTPCs just as much as the idea of improved technology regulation to avoid costly citizen opposition "down the line" represented a key feature in economic justifications in the 1970s.

Such a "comprehensive" approach to innovation requires technology ministries to become involved in areas of human factors in system design, work organization, education and training, and organizational design – as a necessary component of their commitment to innovation promotion. These are all issues that have in the main either not been explicitly addressed in government policy or have been dealt with by other ministries. In the area of process innovation, therefore, they represent the initial emergence of new technology policy concepts that transcend traditional dualisms of industry oriented technology "promotion" versus socially oriented technology "regulation", or technology policy and educational policy. In the process, it is asserted that both productivity-oriented technical considerations and quality-of-worklife-oriented social considerations are addressed by these new initiatives. This can be clearly seen in many of the latest statements of the German Ministry for Research and Technology (BMFT 1990a, 1990b). Thus, in April 1990, the BMFT Work and Technology programme *(Arbeit und Technik, AuT)* statement confirmed this orientation in its focus upon a "comprehensive innovation" concept that "serves the objective to attain a higher level of production and services and likewise to design and organize the world of work in accordance with human needs and social equity. In the process, the development of human resources and qualification, comprehensive industrial safety and protection of health must be considered on equal terms with technological issues" (BMFT 1990b:3). The official general report of the BMFT on the modification of the Humanization of Work programme *(Humanisierung der Arbeit, HdA)* and its significance for the ministry's overall activities is that there is now less emphasis on the "subsequent improvement" of technology, and a new orientation towards "humane and user-friendly design", "involvement of the users", "adaptation of the organizational environment", etc., on the basis that they are "acknowledged principles of technological innovation" and "essential for our country to be able to survive in international competition" (BMFT 1990b:27).[1] As mentioned above, however, these concepts have to face up to a number of challenges.

3 Traditional Limitations of Work Design Reforms and Programmes

European countries have long been known for their range of humanization of work programmes. Drawing upon socio-technical concepts and ideas developed during the 1950s – most forcefully at the Tavistock Institute in the UK – many of these programmes emerged from the quality of working life debates in the 1970s and found most prominent expression in the Norwegian industrial democracy experiments, the West German Humanization of Work programme and the projects and programmes funded by the Swedish Work Environment Fund (Badham 1991b). Serious doubt has, however, been thrown on the theories and strategies that have informed and been employed by these programmes. As is revealed in comprehensive evaluations of job design experiments and the establishment of new forms of work organization, not only do these often affect a relatively small proportion of the workforce but their past success in achieving their stated goals of increased participation, enhanced productivity and extension beyond 'local' experiments is highly questionable (Kelly 1982; Grootings 1990). Moreover, systematic assessments of humanization of work projects have suggested that many fail to continue beyond the initial pre-phase, rarely involve major attempts to transform technology, and have not emerged as "exemplary models" successful in diffusing new production "paradigms" (Leydesdorff/van den Besselaar 1987:3-4; Schallock 1991).[2]

In part, this may be attributed to the limited view of the *object* of work humanization reforms, i.e. an overly exclusive focus on direct production work and management labour control strategies, rather than a broader assessment of and intervention in the range of organizational conditions, management objectives, inter-firm relations and industrial support structures that either facilitate or hinder work humanization reforms. This is a factor increasingly recognized by some traditional proponents of labour process theory, new production concepts, and socio-technical initiatives (Sorge/Streeck 1988; Knights et al. 1985; Kelly 1982).

In addition, however, these problems have been linked to the *methods* of work humanization reforms. For example, while considerable critical attention has been paid within technology policy analysis to the usefulness of direct promotion measures, many work humanization initiatives have continued to rely upon the direct promotion of "exemplary" models as a key to effective diffusion. Yet, as observed in critical evaluations of the Development Programme of the Swedish Work Environment Fund, these models are often not copied and a more complex understanding of diffusion processes is required (Gustavsen 1987).

European proponents of NTPCs are concerned to address these issues through more effective introduction of work humanization concerns into technological design and implementation, and the integration of organizational redesign into the innovation process. With an increasing stress on "interdisciplinary system design" (rather than simply "human centred" or "anthropocentric" design) and "comprehensive innovation processes" (rather than work humanization), it is asserted that these represent important attempts to design, implement and diffuse competitive *socio*-technical systems as a key component of effective innovation and enhanced productivity (den Hertog/Schöder 1989; Kidd 1990).The key role attributed to technology policies that promote such comprehensive innovation processes in an exercise in "*constructive* technology assessment" indicates the arrival of new technology policy concepts in the stimulation of process innovation (OECD 1988).

These concepts draw some plausibility from their effective critique of, firstly, the productivity problems facing US-influenced images of the hi-tech "unmanned" factory of the future that neglects the central role of organizational factors in innovation and, secondly, the lack of effectiveness of past idealistic humanization projects that fail to successfully address efficiency considerations and economic diffusion in their pursuit of humanization goals (Ebel 1990:22-25; Warner et al. 1990). Where they have so far been less successful, however, is in, firstly, convincingly showing how European conditions can be successfully exploited to compete effectively with the Japanese challenge and, secondly, detailing how this is to be promoted through national political processes that constrain and direct technology policy initiatives.

4 The Japanese Challenge

One of the areas in which NTPCs have been less successful is in addressing those apparent competitive strength of Japanese models of production that *have* adequately integrated organizational conditions into the innovation process and *do* combine efficiency considerations with, at least, developed personnel policies. What basic organizational principles or key concepts of the Toyota Production System (TPS) (Fruin/Nishiguchi 1990) or 'Japanese' model are advocated by proponents of interdisciplinary systems design and comprehensive innovation processes? To what extent do European work humanization ideals and production conditions require a modification of such models, and to what extent is this likely or possible in the face of increased internationalization and competition?

While in many respects 'post-Taylorist', the Japanese 'Toyotist' models of production innovation provide an alternative model to many of the more 'craft-oriented' work reforms in Europe. The European programmes, strongly influenced by craft unions drawing their membership from manufacturing industry and production personnel, have been strongly biassed towards the utilization and development of traditional craft skills and freedoms. This orientation conflicts in a number of ways with Toyotist or Japanese forms of work organization, in particular in the amount of discretion given to shopfloor operators, the levels of pacing and work intensification, and the degree of co-determination in corporate planning (Klein 1991; Jürgens 1990). In the face of the 'Japanese challenge', however, the competitive viability of the craft-oriented model is seriously thrown into doubt. As revealed in a variety of reports, this model has been remarkably successful in achieving major gains in production flexibility, quality *and* cost reduction. As illustrated in the recent MIT report *The Machine that Changed the World* (Womack et al. 1990) the introduction of this "lean production model" in the automobile industry has achieved productivity benefits that far surpass the achievements of such European producers as Volkswagen.

The view of "total quality management" (TQM) coming out of Japanese experience and promoted by the Japan Productivity Centre, involves such key principles as: a demand-driven production system more strongly linked to the market; production just-in-time with rapid throughput and low inventories; emphasis on product quality and built-in quality in the production process; step-by-step quality assurance by skilled production workers; organization for incremental innovation through worker suggestions and quality circle structures; production by small self-managing multi-function teams; training and payment systems based on accurate productivity measures and reward for skills; cooperative internal firm relations; and collaborative inter-firm sub-contracting networks (Mathews 1990a). As an embodiment of new managerial principles, its general characteristics are well illustrated by Robert Boyer (1989) in his list of 12 key principles of a post-Fordist model (see Table 1).

As recognized by the 'systemation' or 'systemofacture' analysts, however, there may be a number of variations within this model. The technical and organizational trend towards computer integrated manufacturing (CIM) may take a number of different forms. As Carlotta Perez observed in her classic analysis of systemation, "the new paradigm combines trends towards centralization and decentralization, towards more control and more autonomy, so the variety of combinations is likely to be quite wide, not only in the present transitional period, but probably into the future upswing" (Perez 1983:453). With specific reference to the Toyota Production System, a detailed study by Fruin and Nishiguchi significantly concludes that while "the 'success' of the TPS is evident, much studied and discussed in Japan and elsewhere" it is paradoxical that it is "*even without a clearcut definition and understanding of the TPS in Japan*, (that) the TPS has become an example, inspiration, and vision of what is possible in the world of production organization and management" (Fruin/Nishiguchi 1990:18). They cite the acknowledged founder of the TPS, Mr. Taiichi Ohno, who observes that "We know that the TPS works but we don't know why it works so well" (Fruin/Nishiguchi 1990:19). It is just not clear to observers or even the founders of the model themselves what the key unchangeable features are of the model, and what is largely superfluous, particularly Japanese, or even inefficient. Thus it has been commented:

Just how much Japan's distinctive forms of company organisation and industrial relations should be credited for the astounding competitive power of Japan's manufacturing economy

Table 1: From Fordism to a new model: A synoptic presentation

Fordist principles	The challenges of the 1970s and 1980s	The principles of a new model
F1: Rationalization of labour is the main target, mechanization is the means	C1: Under-utilization of equipment, large inventories of work in process	P1: Global optimisation of the whole productive flows
F2: First design and then manufacture and organize work process	C2: Lags and large costs in passing from innovation to effective production	P2: Tentative full integration of research, development and production
F3: Indirect and mediated links with consumers via marketing studies and strategies	C3: Loosing touch with choosy consumers, failures in launching new products	P3: Close and long lasting ties between producers and users, capture learning by using effects
F4: Low cost for standardized products is the first objective, quality the second one	C4: Ex-post quality controls cannot prevent a rising defect rate, consumers more selective about quality	P4: High quality at reasonable costs, via a zero defect objective at each stage of the production process
F5: Mass production for stable and rising demands, batch production for unstable demands	C5: Even mass consumer demands become uncertain: the Fordist production process appears as rigid	P5: Insert the market demand into the production decisions within smaller and less hierarchical units
F6: Centralization of most decisions about production in a special division of a large firm	C6: Sluggish and inadequate reaction of head quarters to global and local shocks	P6: Decentralization as far as possible of production decisions within smaller and less hierarchical units
F7: Vertical integration, mitigated by circles of subcontractors	C7: Given radical innovations, even large firms can no more master the whole techniques needed for their core business	P7: Networking (and joint ventures) as a method for reaping both specialization and coordination gains
F8: Facing cyclical demand, subcontractors are used as stabilizing device, in order to preserve large firms' employment	C8: During the 1970s, bankruptcies and/or loss of competence of subcontractors, now confronted with international competition	P8: Long-run and cooperative subcontracting as far as possible, in order to promote joint technical innovations
F9: Divide and specialize at most productive tasks, main source of productivity increases	C9: Excessive labour division might turn counter-productive: rising control and monitoring costs; built-in rigidity	P9: To recompose production, maintenance, quality control and some management tasks might be more efficient, technically and economically
F10: Minimize the required general education and on the job training of productive tasks according to BABBAGE's and TAYLOR's principles	C10: New technical opportunities (IT), more competition and uncertain demands challenge most of the previous very specialized tasks	P10: A new alliance between a minimal general education and effective on the job training, in order to maximize individual and collective competence
F11: Hierarchical control and purely financial incentives to manufacture and implicit consent to poor job content	C11: Young generations, better educated and with differrent expectations, reject authoritarian management styles. Too much control becomes counter productive	P11: Human resources policies have to spur workers' competence and commitment and work out positive support for firm's strategy
F12: Adversarial industrial relations converge towards wage demands; collective agreement codify a provisional armistice	C12: Firms employment might be hurt by the lack of cooperation and exclusive concern for wage. *A contrario*, concession bargaining does not necessarily provide any advantage for wage-earners	P12: An explicit and long term compromise between managers and wage earners is needed to reap a general support to this model: commitment versus good working conditions and/or job tenures and/ or a fair sharing of modernization dividends

Source: Boyer (1990:8)

(in particular) remains an open question. Alternative arguments abound – the role of culture as a source of work motivation, the macroeconomic management of the economy by government ministries, the abundance of cheap capital, the lower transaction costs that Japanese-style 'relational contracting' brings. But the prima facie case is a strong one. The advantages of Japanese work organisation seem to centre on a more flexible and efficient use of human resources, and the capacity to motivate and channel extraordinary high levels of workforce discipline and commitment (Lincoln 1990:38).

It is this lack of certainty about the necessary features of the model or the viability of different variants that makes assessments of different industry sector and national models so important. Although general interpretations of "national trajectories", or the influence exerted by "national patterns of skill formation" or "national systems of innovation" are necessarily generalized and speculative at this stage, they represent an important enterprise. Most significantly, from the point of view of European work humanization concerns, some of the key issues concern the compatibility or alternative effectiveness of more 'craft-oriented' approaches to work organization, and the types of supportive institutional structures that make these a possible competitive option. As discussed recently by Janice Klein, in a study of the application of Japanese techniques by US firms, traditional socio-technical views emerging from Europe of job enrichment through semi-autonomous groups were premised on a reduction in task interdependence using buffer stocks. Yet, as new Japanese systems of JIT and standardized work techniques are used to improve quality, reduce costs and speed up throughput time, a key aim is the reduction of such buffers and increased task interdependence. On the one hand, this reduces the degree of discretion over work methods and the pace of work in individual *task execution*. On the other hand, it has been accompanied within these factories by greater input of employees into collaborative *task design*, as suggestions for changes are made and discussed from the lower to the higher levels of the organization (Klein 1991). In this way, direct individual autonomy in task execution is sacrificed but replaced by indirect collective autonomy in task design.

Important questions are raised here concerning not only the reality and value of this trade off for the quality of working life, but also the extent to which greater autonomy can and should be delegated to shop floor operators to enhance experimentation and learning in a search for continual and dramatic improvements in process quality and efficiency, and to what degree this is restricted within the Japanese system and may be more viable in the production cultures and product market conditions of a number of European industries and countries. As Jürgens (1990) has argued in a specific discussion of the automobile industry, a meaningful contrast can be drawn up between what could be seen as a 'German model' of skilled worker oriented labour regulation and a 'Japanese model' of group-oriented labour regulation. In comparison to the Taylorist approach, both models increase the degree of self-regulation on the shop-floor, and both the competence and willingness of production workers to accept responsibility. However, they also differ in important ways.

The central characteristic of the German model is the role of the skilled worker and a specific understanding of skilled work as a "profession". In conditions of a 'surplus' of highly trained workers (to a degree imposed on firms by the vocational training system), job design and technology use to address problems of absenteeism (again strongly influenced by legal and contractual regulations), and union/works council demands for lengthening work cycles, the emphasis is on the use of qualified labour, "uncoupled from the production cycle and the rhythm of the machines" (Jürgens 1990:13). The emphasis in production system design is upon the stimulation of work interest, comprehensive responsibility and a large degree of self-regulation. This is, however, to a significant degree still restricted by the direct pressures of production, and such workers are increasingly required to perform such simple production tasks as setting dies, feeding parts, transporting materials, etc.). What emerges is a new type of hybrid German skilled worker.

The Japanese model gives a central role to qualified labour, but not to uncoupled skilled labour. The key feature of the Japanese model is "self-regulation under the pressure of the assembly line and the production pace" (Jürgens 1990:13). Work groups are of central importance in providing integrated job understanding, flexible work and expanded labour deployment within the production system. This occurs, however, under a regime of ongoing coordinated assessment of best possible system performance and permanent improvement through the continuous review of established time-standards (this contrasts strongly to the Western industrial engineering practices that aim for 'normal' performance and are restricted in their time-standards review). Self-regulation is not based on skilled worker competence and professional ethics but the activities of the work group in the more tightly coupled and continuously improved systems outlined by Klein.

These models, as Jürgens (1990) emphasizes, exert strong influence on other countries. It is probably of little surprise to learn that increased vocational training in British factories hints at the adoption of the German way, while American companies are clearly more influenced by Japanese concepts with a search for the formation of production groups in unskilled areas, and a flexibilization and expansion of workers' areas of deployment. There are, however, two clear constraints on the wider adoption of the German model. Firstly, as Jürgens argues, the strong anchoring of the German model in a structure of co-determination and a corporatist system of vocational education means a degree of institutional and legal limitation on company headquarters' scope for action and decision-making that is liable to make it unpopular with management in other countries, even if it is socially and economically possible. Secondly, Japan has strong cost advantages over German manufacturers and the competitive strategy of German car companies to go "up market" is facing increased competition by Japanese competitors now producing higher valued cars and equipment. In the face of such threats, and competition from low-wage and increasingly efficient car producers, a key issue concerns the degree of importance of future technological development in ensuring production efficiency. If the effective implementation of computer integrated manufacturing (CIM) models becomes more significant for productivity, and their combination of flexibility and integration is strongly supported by highly skilled and responsible workers, then the German model may have a strong chance. It is possible, Jürgens concludes, that a mixture of the Japanese and German models will emerge, with group formation, job integration and extreme work efficiency in manufacturing mass production, and skilled-worker-oriented team formation and professionalization in high-technology and service areas.

The challenge provided by the Japanese model does not, however, only exist at the level of the work process. It also extends into broader assessments of national systems of innovation and skill formation. There is, for example, an emphasis in the TPS model on the improvement of inter-firm collaborative arrangements and the establishment of integrated networks and learning processes. However, as observed by a number of commentators (Perrow 1990), the new balance of competition and collaboration in new networks may take on a variety of different forms, ranging from networks of interdependent and innovative small firms (e.g. Third Italy), sub-contracting links of Japanese firms, to forms of corporate decentralization, divisionalization and internal competition as practiced by a number of American and German firms. The range of alternatives and their relative effectiveness in different conditions is far from a closed issue. The existence of present alternatives should not, however, lead to a complacent recognition of diversity: for with the astounding competitive success of the Japanese system and the increased internationalization of production, any examination of the viability of alternative forms must directly face the issue of the extent to which alternative production models will be able to continue to survive.

This whole discussion and debate raises the key issue of how the relatively recent collapse of the American Fordist mass production model of production efficiency and its at least partial replacement by the Toyotist flexible mass production model is to be interpreted. Is this best seen in terms of the replacement of one dominant national exemplary model or "paradigm" of production by another, with other countries forced to adapt to 'best practice'? Or is the present best seen as an era of uncertainty in which a variety of different models are in conflict? Even more importantly, perhaps, there are two further issues to be addressed: firstly, the degree to which there is room for either significant different variations within the Japanese production model or alternatives to that model; and, secondly, the degree to which the success of the Japanese model extends its "flexible mass production" techniques into areas of industry previously far less influenced by the US mass production model that required large standardized markets. The extension of the focus on 'comprehensive innovation processes' to incorporate these issues is a key requirement for the further development and effectiveness of new technology policy concepts. This is an issue that is currently being explored within innovation theory (e.g. Columbia University Project on National Systems Supporting

Technical Innovation), regulation theory (e.g. CEPREMAP Research on National Trajectories), labour market and labour policy (e.g. WZB Research on National Patterns of Skill Formation), and both trade and state theory (e.g. BRIE Research Group, Berkeley). If NTPCs are to be effectively developed and taken seriously, they need to build upon the findings of such groups.

5 National Conditions of NTPCs

Even in their present form, however, NTPCs in Europe represent an important start in the attempt to respond to the Japanese challenge and overcome the previous limitations of work design and work reform programmes. Yet the successful introduction and further development of these concepts in the technology policy arena faces substantial institutional obstacles from established policy processes. They directly challenge the traditional ministerial 'division of labour' between, for example, research and technology, education and training, industrial relations, and industry ministries: comprehensive innovation concepts require the integration of these spheres in the effective promotion of increased productivity. It is not surprising, therefore, that the content and success of the initiatives that are being developed vary depending on the politics of different countries, and take on different and controversial forms. If NTPCs are to be successful, the political and national contexts within which they are developed and deployed need to be explicitly considered *within* more comprehensive innovation promotion strategies.

Yet the nature, source and significance of these NTPCs have so far received little sophisticated analysis. It is not possible, for example, to simply 'read off' these concepts in the policy arena as an 'instrumental' reflection of either changing trade union strengths and strategies or a 'scientific' recognition of the objective efficiency of new socio-technical production systems (den Hertog/Schöder 1989) or a new world of work (Mathews 1990b). Nor is it possible to interpret them solely as one partial attempt to legitimate, as "socially oriented", a global trend observable throughout the 1980s to subordinate government R&D and technology policy to business requirements and industry control (Dickson 1984; Couchman 1989a, 1989b). An examination is needed of the way in which the emergence of NTPCs in different countries both reflects and impacts upon the political, social and economic conditions of specific nations. Only in this way can these concepts be developed and deployed in a manner that maximizes the potential for utilizing national conditions to maximize the overlap between stimulating productivity and the quality of working life.

At one level, a broader view of effective policy measures requires a more considered reflection on the influence of governmental politics on the promotion of NTPCs. 'Instrumentalist' perspectives underemphasize the importance of both party and ministerial politics in defining and promoting NTPCs. One important issue for consideration, for example, concerns the clear overlap brought about by NTPCs between science and technology ministries and education and training ministries, and the manner in which they collaborate, conflict and negotiate over the nature of "technology" promotion in this area. As intervention in this field involves directly addressing the highly politically charged area of working conditions, it is inevitable that "legitimatory" as well as 'accumulation' issues are raised (Offe 1985). How these issues are defined, weighed and played off will depend upon the political conditions within different states; and it cannot be 'read off' from international changes in technology and markets.

At another level, the broader national socio-economic conditions affecting the nature and possible scope of technology and work humanization initiatives is an important area of study for those concerned with effective political intervention in this area. Some general comparisons may, for example, be made between contrasting initiatives in Sweden, West Germany and Australia. In the Swedish case, the existence in the 1970s of full employment and increasing educational levels led to increasing managerial concern with labour turnover, absenteeism and recruitment problems. Faced with these problems, many new initiatives were created in Sweden to enrich production work and improve working conditions. The emphasis upon group work, employee responsibility and worker participation was a direct result of addressing these conditions in the context of a strong trade union movement and a social-democratic government. In addition, both influencing and influenced by these developments has been the dominance within the Swedish economy of large internationally competitive engineering firms characterized by rapid and effective introduction of process innovation, despite comparatively much lower rates of product innovation (Edquist/Lundvall 1989; Edquist/Glimell 1989). This has resulted in what, to paraphrase, can be called a welfare-democratic/Fordist orientation towards job enrichment to secure or increase job satisfaction amongst workers involved in rapid process innovation and in conditions of full employment.

In Germany, work humanization takes a different form influenced by the different national socio-economic context. One of the most prominent features of the German federal initiatives, when compared to Sweden, is the greater orientation towards technology as a key factor in work humanization and a stronger orientation towards the utilization and development of high levels of craft-production skills. This phenomenon can be related to a number of conditions in the German system of innovation and skill formation. These include: the comparative strength of the German 'dual system' of vocation education and training (Sorge et al. 1983); the general orientation of German industry towards the production of more customized and higher quality products in a side range of industries when compared to other European countries such as the UK[3]; the deliberate promotion by the government of a viable domestic electronic and computer industry capable of serving assisting export intensive German industries; the dominant role within the economy of a set of production equipment suppliers (machine tool industry) and high technology engineering "users" (automobile companies); influence that extends to an R&D system for production innovation that promotes production system concepts assisting in the sale of high quality and customized machine tools; and, in the creation of flexible quality oriented production systems utilizing comparative advantages in workforce skills, oriented towards "diversified quality product" markets (Sorge/Streeck 1988; Ruth 1990; Badham/Lennartz 1990). The result of these conditions is, for example, that the promotion by the Research and Technology Ministry (BMFT) of production technology and work humanization is strongly oriented towards larger companies, advanced technologies, and research and investment in production models strongly influenced by the machine building and automobile industry. This is clearly apparent in both the engineering activities of Fraunhofer Institutes and the research on new technology and work organization by a range of industrial sociology research institutes. The key emphasis in contrast to Sweden has, consequently, been more towards an approach of Professional/Diversified Quality Production that emphasizes the promotion of advanced technology and the support and utilization of a high level of craft skills for the creation of new high-quality and customized products, with probably less diffusion of organizational models of group work or extensive participation and learning models.

In strong contrast to both countries, the present commitment in Australia to improvements in working conditions through "award restructuring" and "skill formation" changes is primarily a defensive response to the 1970s collapse of manufacturing industry and employment, in the metals industries in particular, following a dramatic reduction in tariff levels and increased exposure to international competition. This has encouraged union/management collaboration in an attempt to create more production forms of production organization, a task that has focussed on the exchange by production workers of traditional demarcation lines and forms of work control for new career paths and skill and productivity related pay increases. Although strongly influenced by Swedish and German models, in contrast to Sweden there has been less emphasis upon job satisfaction and absenteeism; and in sharp distinction to Germany, there is less stress upon relevant technological design and the adaptation to present high levels of craft skills. The orientation is far more strongly towards a Reactive/Job Integration Skills Push approach creating and rewarding appropriate skills and removing traditional educational and industrial relations barriers to this initiative (Badham 1991a).

These three models are of course merely speculative sketches of the types of national conditions that may promote work humanization initiatives of a particular form and scope. They do provide, however, the kind of orientation required in developing NTPCs that take into account the conditions under which they are to be shaped and implemented. Clearly, this is directly linked to the advances made in innovation, trade and state theory that address the general issue of national trajectories in response to changes in the form of international competition. They also involve, however, a direct focus on the character of, and options facing, programmes to explicitly improve working conditions as part of stimulating world competitive innovation processes. For this reason, the remainder of this paper will focus on some lessons that can be learnt from the German Humanization of Work programme.

6 Background and History of the West German Humanization of Work Programme

The further development of new technology policy concepts that effectively address these issues is clearly a major challenge. The intention of this paper is not to take these up in any comprehensive manner. Our concern here is more limited, that is to contribute in a small way to research on these issues by revealing some of the conditions that have to be taken into account in directly promoting technology and work humanization initiatives. This is carried out through a brief overview and analysis of the complex tensions within the German Humanization of Work (now, significantly, Work and Technology) programme. An improved understanding of these processes would, we argue, enhance the effective use and development of NTPCs. by giving some insight into the way in which the general issues of promoting productivity and the quality of working life through technology policy are reflected and dealt with in the form of detailed and practical decisions that arise in the course of such programmes in different countries.

6.1 International Background

The rise of a humanization of work programme within Germany oriented in a sizeable degree to the creation of 'humane' technologies can be related to three general international conditions. Firstly, following the "socio-technical" initiatives beginning in the 1950s and the "quality of working life" movement established in the 1960s, the early 1970s witnessed the emergence within Sweden, Germany and Britain of labour/social-democratic inspired movements to extend these principles into effective humanization of the workplace. This was reflected in the establishment of the *Arbetsmiljofonden* in Sweden in 1972 and the symbolically significant Lucas Aerospace Plan of the Shop Stewards Committee set up in Britain also in 1972, as well as the 'Humanization of Work (HdA)' programme in Germany in 1974. This phenomenon was directly noticeable in West Germany with the strong influence on the establishment of the HdA programme of the IG Metall Quality of Work Life Conference in 1972, and the coming to power of the Social-Democratic party (SPD).

During the latter part of the 1970s, however, the effectiveness of the various national humanization initiatives became a source of major controversy. Particular attention was paid to the lack of evidence of substantial economic successes or long-term survival of job redesign experiments, the importance of avoiding management dominated "legitimatory" exercises in job design, the severe restrictions on the participation by trade unions and workers representatives early on in the design of new systems, and the significance of creating humane technologies as well as new forms of work organization if restrictive parameters on job redesign are not to be built into the system. In combination with less favourable economic conditions, these created a less widespread and positive attitude to work reforms.

Secondly, from 1950 to the late 1970s a reaction against contemporary technological trends and for what Bob Kling described in 1973 as a "people centred computer technology" emerged within the disciplines of cybernetics and computer science, particularly within the control engineering and artificial intelligence communities. This trend, as detailed by Cooley (1990) is exemplified in Norbert Wiener's book on *The Inhuman Use of Human Beings* (1950), Ida Hoos' paper "When the Computer Takes over the Office" (1960), Hubert Dreyfus' work from the early 1960s culminating in his book on *What Computers Can't Do* (1972), Mike Cooley's investigation of computer aided design (1972) culminating in his book *Architect or Bee?*, Howard Rosenbrock's book on *Computer Aided Control Systems Design* (1974) and a series of papers beginning with classic "The Future of Control" (1976), and Joseph Weizenbaum's paper "On the Impact of the Computer in Society: How Does One Insult a Machine?"(1972) and his book on *Computer Power and Human Reason* (1976).

With particular focus on numerically controlled (NC) machines, computer aided design (CAD) equipment, and artificial intelligence (AI) programmes, this led to a general critique of the technocratic hybris that exaggerated the power and capabilities of the new machines and programmes, and an emphasis on necessity and desirability of preserving and supporting human "tacit skills", "operator experience", and human "purposiveness" as a central element in the new designs. In the early 1980s, this was to have an important influence on the establishment of the UMIST human centred lathe project in England, the workshop-oriented programming (WOP) project in Germany, the UTOPIA project in Sweden, and the ESPRIT Human Centred CIM project.

Thirdly, an important background condition to these two sets of developments was the emergence in the 1960s and development in the 1970s of a growing movement for "alternative", "intermediate", "soft", "convivial", and non-military forms of technology and growth, in contrast to the environmentally polluting, resource wasteful, socially harmful and globally threatening forms of development characteristic of classic industrialism. The classic criticisms are now strongly associated with such names as the Club of Rome, Ernst Schumacher, Ivan Illich, and Amory Lovins. In combination with more radical critiques of science, this "alternative" technology view inspired a greater degree of intellectual interest in the political nature of science and technology and a social concern with the regulation of industrial "progress". After the initial plethora of "technology assessment bodies" emerging in the early 1970s in response to the concerns of environmental and workers movements, this led to a more widespread interest in the creation of more "proactive" forms of intervention in technological direction. In some cases this led to direct links being established between workers movements and "green bans", anti-nuclear protests, "socially useful production" initiatives, etc.

6.2 Key National Influences

In combination all three sets of developments exerted an influence on the establishment and evolution of humanization of work programmes such HdA, particularly the first development. Yet, as observed in comparative studies of work reform programmes, an understanding of different national programmes depends greatly upon much of the nature and success of these activities depend upon national constellations of industrial relations, institutionalized attitudes, state structures and civil servant interests (Auer et al. 1983).

In the German case, some of the dominant influences on the HdA/AuT programme have been

- the key role of the Social-Democratic party (SPD) in establishing the programme in 1974 as a "socially oriented technology policy" tool with strong trade union influence, and the later modification of that role under the influence of the conservative party (CDU) from 1983 onwards;
- the strength of the Research and Technology Ministry (BMFT) in initially orienting activities towards the extensive 100 million DM per annum direct promotion of "human oriented technologies" as well as R&D into health and safety conditions, and its ambivalent attitudes to the programme given its dominant emphasis on increasingly indirect funding of science and technology oriented towards "non-social" international competitiveness yet subordinate commitment to "socio-economically" oriented goals of energy, environment and work reform;
- the complex constellation of engineering consultancy, work psychologists and industrial sociologists in the German research and development system that, on the one hand, provided a scientific-technical basis for the programme and, on the other hand, infused it with the disciplinary orientations and antagonisms from these sections of the R&D system;
- the extensive legal system of co-determination, workplace health and safety and inspectorates that both provided the preconditions for a substantial degree of employee and trade union involvement in technological change and in a number of ways limited and restricted that involvement (in particular through the separate spheres of activity of works councils and trade unions);
- a production culture tradition and set of employee and trade union attitudes that promote a relatively receptive attitude to technological change and a cor-

poratist structure of employer and employee associations that enhances the ability of government, employers and trade unions to negotiate the design and implementation of new technology and work strategies; and
- the post-war social modernization path followed by Germany throughout the boom period of the 1960s and 1970s as well as surviving the economic problems of the early 1980s – a strategy that involves government, employers and unions co-operating in the promotion of an economy strongly oriented towards international competitiveness yet accompanied by a set of institutions and programmes that cushion the social costs and adjustment to these processes.

6.3 Three Phases of the Programme

The history of the HdA programme's orientation and mechanisms can be loosely classified into three phases from 1974 to 1982, 1983 to 1988 and 1989 onwards, with each of these periods marked by governmental changes and/or government reviews of the programme.

The *first phase* established the overall direction and structure of the programme with a basic orientation toward protective action in the workplace through occupational health and safety research and regulation (with the support of the Ministry of Labour) and the development of new "human oriented" forms of technology and work (with the support of the BMFT). Projects are funded within firms for the development and implementation of improved working conditions, usually involving external consultants who are responsible for the documentation of the project in transferable form. An extension upon this concept in 1977 was the establishment of "branch" projects involving employer and worker organizations in an industry working with experts on new technologies and forms of organization for the branch (see Figure 1). A lesser degree of funding is given to fundamental research in scientific institutions, and after 1977 for the exclusive evaluation and diffusion of the programmes results. Since 1979, this has involved the funding of union and employers organizations for diffusion and in 1979 the unions were funded to establish Technology Advice Centres.

As is the case with other BMFT programmes, it is run by a policy group within the ministry supported by an external administrative group or *Projektträger*, in this case the DFVLR. The minister has a special advisory board to advise on the development of the programme staffed by equal numbers of employers, trade unionists and scientists. Each project application from industry must be presented by management with the agreement in writing from the works council *(Betriebsrat)* and, now further details on information and participation procedures. These proposals are evaluated by expert committees in which employers and unions are equally represented. One of the unique features of the programme's administration, compared to others within the BMFT, is the degree of union participation, which also involves participation in conferences to discuss new plans for the programme.

By the late 1970s, therefore, some attention had been paid to diffusion problems and measures to assist employee input into the projects. However, in 1979, in the face of rising unemployment, some employer opposition to union involvement, and trade union scepticism about the programme, the BMFT followed its dominant preoccupation with international competitiveness in removing the robotics and flexible automation parts of the programme out of HdA and placing them in the separate Manufacturing Technologies programme with less trade union involvement. Continuing these emphases, the *second phase* of the programme followed from the parliamentary review and change of government in 1982. A dominant emphasis was the establishment of a tighter link between humanization and innovation, with a stronger focus on the need for projects to be financially profitable. This was followed, not unsurprisingly, by a discussion of the accounting procedures that should be used in such assessments with concern expressed at the overly narrow criteria normally imposed in judging humanization or organizational changes. During this period, other significant changes were made in the commitment to reorient project funding away from the dominance of larger firms, restrictions against public sector proposals for local government, the decision not to fund autonomous trade union and business association projects, and a limitation on the amount of "transfer research".

The *third phase* of the programme began with parliamentary criticism of the programme in 1987 and culminated in a reorientation symbolized by the change of name to Work and Technology (AuT) and the participation by the Federal Ministry of Education and Science (BMBW) in a programme organized "on a broader understanding of innovation and occupational safety and health" (BMFT 1990b:24). The key emphasis was again the importance of uniting humanization with innovation concerns and the need to ensure adequate diffusion of the results of the programme (see Table 2 and 3). In this case, the change involved a reduction in the funding given to a number of occupational health and safety issues, the promotion of priority areas for the programme in the development of innovative technologies and systems integrating people, organization and technology in a comprehensive and prevention oriented innovation concept, and a further emphasis on comprehensive projects as promotional tools (involving cooperative projects, branch-oriented projects, and transfer projects).

The "broad innovation concept", illustrated in the diagram below (see Figure 2), is a major emphasis of the new programme and is well represented in the priorities given to "the design of software in keeping with human needs and workshop-oriented control concepts", "the development of integrated planning processes for a foresighted design of work and technology", the "early inclusion of all those acting at plant level and representing specific interests in the development and testing of design solutions", and a stress upon the rapidly growing interdependence within and between companies (which) makes it all the more necessary to design and plan processes of change and development on a long-term and comprehensive basis. This applies especially to personnel planning and to the design of work and technology. Today many companies lack the methods and instruments, skills and capacities to accomplish this. Especially in areas in which small and medium-sized enterprises have to take the initiative themselves without being able to simply follow the example of big companies (BMFT 1990b:10).

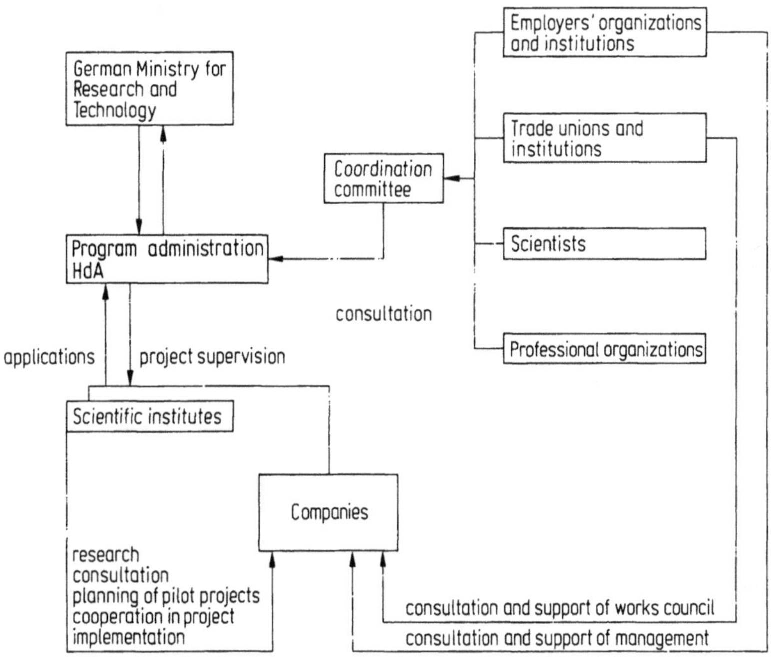

Figure 1: Outline of branch plant programme mechanism
Source: den Hertog/Schöder (1989:35)

7 New Technology Policy Concepts in the West German Humanization of Work Programme

7.1 Overview

A comprehensive examination of NTPCs in the HdA programme would require extensive documentary and interview based research in order to determine the evolution and amalgamation of the views of the different participants in the

Table 2: Diagram of changes in Humanization of Work programme

1984	1985	1986
humane use of new technologies in an office environment	humane design of working conditions in forging industry	humane design of working conditions in coal mining
humane use of new technologies in production	humane design of working conditions in transportation industry	fundamental research and issues of general relevance to the HdA programme
humane design of working conditions in foundries	working conditions and health of workers	
	reduction of noise in working	
	health protection at workplaces with new information and communication technologies	
	risk of cancer at the workplace	

Source: den Hertog/Schöder (1989:37)

programme, and the background of these views within employers organizations, the trade union movement, the BMFT officials, and the research community. Although such a systematic study has not been carried out some general comments can be made here on the basis of experiences in the programme, selective interviews, programme reviews and an assessment of the literature.

The programme began with the fourfold objective of: developing safety norms, minimum standards for machinery and work settings; developing humanized work technologies; developing paradigms and models for work organization and job design; and the distribution and application of scientific knowledge and practical experience. It was generally accepted, however, that most of the funds were to be allocated to ensuring healthier work conditions by minimizing health risks and stressful work situations (Wilpert/Quintanilla 1984).

The interpretation placed upon these objectives by many commentators is somewhat contradictory, reflecting in part the conflicting experiences that led to the programme's establishment. On the one hand, the period appears to be one of optimism. Changes in health and safety and consultation legislation imposed new requirements upon employers. The 1972 Works Constitution Act legally bound

Table 3: Table of the Work and Technology programme priorities (as per July 1989)

Status	Basic principles/cross-selectional issues	Production	Services
Expiring		Reduction/Elimination of stress factors and development of new working structures in the clothing industry Reduction/determination of stress factors & development Improvement of working conditions in the joiner's craft	Reduction/Elimination of stress factors in the hotel
Taken over from the HdA programme and updated	Research on stress and strain factors (e.g. noise, vibrations, combined and mental stress factors) Working conditions and health Carcinogenic risks at the workplace Allergic reactions and diseases Software design in accordance with human needs Holistic planning processes including extended profitability calculations Dissemination and implementation of new scientific findings and operating know-how including accounting Programme-related research	Noise abatement using the trade industry and the sheet metal processing industry as companies Application of new technologies in production in keeping with human needs Improvement of working conditions in forging industry Improvement of working conditions in the foundry industry Improvement of working conditions in the mining industry Application of new assembly line technologies in accordance with human needs	Serving the design and application of new technologies in offices and administration in accordance with human needs
New/Planned	Simple measuring process for noxious work substances Health protection at the workplace in air conditioned rooms Work oriented design, health protection and acquisition of skills when new technologies processes and materials used Work-related design of networked systems (CIM, Logistics, JIT, etc.) Work oriented concepts of organization, acquisition of skills and staff management Concepts of the acquisition of skills by social groups of employees Demographic trend and effects on working conditions and employees structures	Application of new technologies in the printing industry in accordance with human needs Improvement of working conditions in the food industry and in agriculture Transfer of design know-how in certain industries taking the example of the show industry Planning of working conditions in craftsmen's business in accordance with human needs	Design of new technologies in commuter traffic in accordance with human needs Planning of working conditions of office workers in accordance with human needs Planning of working conditions in the trade in accordance with human needs Improvement of working conditions in nursing and providing care for people who are old, sick or who have disabilities

Source: BMFT (1990b:21)

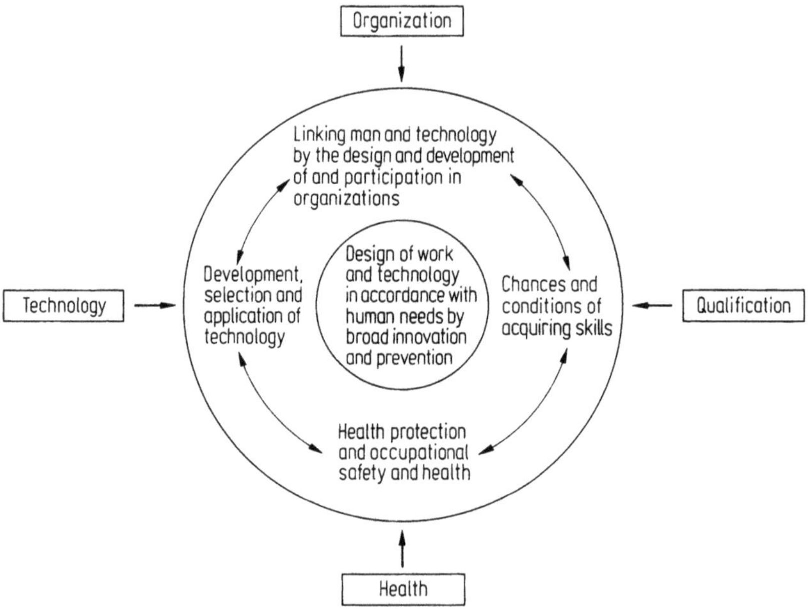

Figure 2: Diagram of broad innovation concepts of the Work and Technology programme
Source: BMFT (1990b:7)

employers (paragraph 90) to inform works councils at the "appropriate point" of planned changes in technology and work organization, and paragraph 91 stated that works councils could modify or stop any changes which went "significantly" against the "accepted findings" of work science. In addition, after giving its approval to quality of working life initiatives, the coordinating German trade union body (DGB) supported the HdA as an important proactive initiative in this area. Moreover, the recently elected Social-Democratic party (SPD) established the programme as part of a "socially oriented technology policy", itself an important feature of a "society oriented modernization" strategy to improve international competitiveness and cushion the social costs while pursuing a "humane growth path" (Naschold 1980). Moreover, in the face of implementing rationalization measures to increase product quality, flexibility and delivery times, employers organizations appeared to recognize the importance of addressing the human and organizational issues that this raised.

On the other hand, however, there had been long debates within the trade union movement before committing itself to action in the area of promoting alternative forms of work organization. The background in the early 1970s led to the spread of "crisis" language in response to the oil shock of 1973, sluggish world markets, the revaluation of the German Mark, etc. Moreover, earlier strike activities peaking in the "September Strikes" of 1969 were part of a multidimensional reaction against rationalization and the worsening of labour conditions, a reaction that included passive resistance, absenteeism, greater labour mobility etc. This phenomenon, particularly noticeable in assembly industries and in many jobs fulfilled by a growing number of "guest workers", led to an important change in the attitude of many workers. Faced by unemployment, stress and threats to incomes and qualifications, new technology was no longer seen as neutral, and the desire for more humane working conditions became a much higher priority (Brugmann 1982). This change could, however, be seen as a defensive response to the "job killing" and "dehumanizing" effects of new technology, the latter associated with reduced qualification requirements and physical problems. It was emphasized in many cases that the initiatives to improve work conditions were employment-*creating* in the face of the unemployment consequences of rapid rates of productivity increase (Markovits/Allen 1981).

In the light of both these sets of ideas and experiences, the NTPCs that emerged should be seen more as a compromise between, on the one hand, a grudging acceptance of the need for change and defensive reactions to the human costs of rationalization, and, on the other, more optimistic interpretations of how these problems could be addressed in the adoption of a more proactive and preventive approach to techno-organizational design and implementation. By the time of the first major formal re-evaluation of the programme in 1982/83, however, the optimism concerning the capabilities of the programme had been considerably weakened. For employers, the economic effectiveness and consequent diffusion of many of the projects had neither been shown nor realized. For trade unions, the hopes for dramatic increases in participation and consultation over change had also not been realized, either in the workplace or in other government technology programmes. For scientists and researchers involved in the programme, the difficulties of real interdisciplinary work and meaningful collaboration between researchers, unions and management had led to less than successful results. For the programme management, the controversial nature of the programme, the problems observed by the participating groups, and the lack of co-operation from other programmes in the BMFT (even the removal of the robotics/flexible automation features from the programme), all represented a far-from-satisfactory outcome.

In the face of these difficulties, the unions adopted a more offensive stance best represented in the IG Metall Action Programme. Sceptical of the significant effect of the HdA programme as it stood, an emphasis was placed on the need for a more radical extension of trade union influence and social factors considerations over the broad range of technology policy. Moreover, greater stress was placed on worker and union control over workplace negotiations, as well as the use of industry level and regional collective agreements to prevent downgrading with new technology, reduce working hours, etc. In terms of direct impacts on the programme, employers demanded that projects be selected with economic viability considerations more clearly in mind. They had been influential in removing the robotics and flexible automation components from the programme and were now partly responsible for the financing of the research made by the scientists as part of the projects. Despite the substantial criticisms made of the programme by both sides, however, there was an overriding agreement that it be allowed to continue.

A similar degree of agreement on the continuation of the programme was also the outcome of an equally critical review of the programme in 1987/88, resulting in the change of name and direction of the programme in 1989. The major reorientation as described above involved a strong focus on NTPCs; however, the form that these take will be strongly influenced by the complex and conflicting pressures that combine to make up and influence the HdA/AuT programme.

7.2 The Complex Nature and Tensions within NTPCs

Over the different phases of the programme, the broad objective have not been substantially changed – the orientation is towards the combination of "humanization" and "efficiency" consideration in *preventive* design to avoid the "costs" of rationalization (due to increased stress, other health and safety considerations, etc.) and the proactive design of new systems to *improve* working conditions as a key factor in 'broader' innovation concepts. It is possible, however, to detect particularly in the more recent changes an increasing emphasis on the latter priority.

Within these broad objectives many different strategies are possible, encouraged by the conflicting character of many of the programme's goals. Concerning the basic assumption of the compatibility between "humanization" and "productivity", it is clearly possible to accept a degree of overlap between attention to human "needs" and reductions in costs due to worker resistance, absenteeism, insurance costs, etc. or meaningful participation of the workforce in new system design and the speed and effectiveness of implementation (through more effectively utilizing 'tacit' skills, incremental innovation and development of the system, consideration of vested interests and consequent support for the new system, etc.). Moreover, the opportunities to exploit these opportunities can be encouraged by shifts in market conditions, technological trends, the establishment of model examples of successful achievements, the creation of participatory techniques, the design of "broader" investment costing and planning tools, etc.

On the other hand, occupational, health and safety measures can be extremely costly; and participatory/responsibility-oriented technology strategies often involve lengthy and expensive processes of training, consultation and organizational change. Market conditions and technological trends are not all 'one way' in a direction that makes such commitments necessary, and the imperatives of competition may make many activities in this area a luxury that many firms either cannot afford or cannot effectively implement given the realities of contemporary organizational structures and cultures.

As a result of this complex combination of overlap and conflict between humanization and efficiency considerations, often in situations of uncertainty about "real" costs and benefits, the extent to which either theme is emphasized will always be a matter of negotiation and interpretation between different interests. It is not surprising, therefore, that despite the presence of different trends and emphases within the HdA programme, each phase of its development is characterized by compromise or balancing between 'optimistic' and 'pessimistic' viewpoints. Of particular significance in this regard has been the influence in shifting the balance of, firstly, the coming to power of the SPD and the establishment of corporatist structures and legal requirements that provide conditions for forcing employers to take actions which they are then pressured to 'make' profitable; and secondly, the presence of economic crisis and unemployment problems that focus attention of both employers and unions on cost reduction strategies and income/job protection; and, thirdly, new rationalization concepts to meet economic challenges that either threaten or enhance workforce skills and autonomy in the workplace.

In addition to this general ambiguity, there is a fundamental conflict between the stated objectives of the HdA/AuT programme to effectively promote new humanized forms of work and the restrictions upon the programme due to continued employer domination of investment and, consequently, innovation in a fundamentally private economy. As frequently observed, HdA funds are relatively small compared to overall private industry R&D; HdA projects often only involve the workforce late on in the innovation process, being used by firms to obtain public money because of economic difficulties rather than humanitarian interests. Projects often collapse, frequently after the pre-phase, and experiments are often restricted or closed down in the face of changing investment strategies. While it is possible to point to numerous important measures taken by the HdA/AuT programme to address these problems, this serves to illustrate rather than undermine the fundamental point: the activities of the HdA programme have to be seen in a continual state of tension, forced to accept the dominance of private enterprise investment and innovation strategy yet using direct promotion measures to often 'indirectly' influence these strategies in directions conducive to the humanization of work.

A third conflict within the programme concerns the disproportionate balance between the optimistic and grandiose aspirations of the programme and the relatively underdeveloped implementation of these aspirations in practice. This feature of the programme may be rooted in its minority and symbolic role within the BMFT, combined with the substantial difficulties that face such a programme. An illustration of such difficulties is, for example, that whereas workers possess "rights" (to information, consultation, etc.), they are often lacking in "ability" to take up these rights (e.g. lack of time, inadequate knowledge and expertise, etc.); and where they are funded by programmes such as HdA to enhance their "abilities", they do not have sufficient "rights" to utilize them effectively (e.g. in meaningful consultation over investment strategies).

	Positive Attitudes	Negative Attitudes
Bureaucrats	Comprehensive job security, protected work capacities (health qualification), participation and control at firm and government level, transparency of democratic social structures. Coordination of workers activities (works councils, trade unions, with researchers, with government officials, etc.), the development of trade union transfer mechanisms	Intensification of work masquerading as rationalization, elimination of jobs through increased productivity, participation as 'ritual', rigidities in payment systems restricting financial gains, insular projects either not taken up elsewhere or with broader negative consequences in polarizing the workforce, increasing subcontracting and undermining of trade union strength, etc.
Employers	Profitable workplace specific and worker specific work structures, flexibility through skill and responsibility amongst the workforce, improved quality through delegation of inspection, etc., increased machine utilization and reduced throughput time, legitimating function, etc.	Unprofitable projects costs, intrusions into managerial prerogatives, excessive control to recalcitrant workers and trade unions, worker and trade union defensiveness, humanization ideology encouraging excessive expectations
Trade Unions	Socio-technical experiments possible with gains to both employers and workers, numerous small improvements achieved, practical examples established of working systems, increased awareness of employers and workers concerning the beneficial and harmful effects of new technology, encouragement of applied scientific research through participation in firm projects key aspect of social modernization strategy	Undeveloped 'torso' character of the programme, lack of diffusion of results, any researcher plans on projects not implementable, lack of national and international publications, limited views of many trade unions, difficulty of adapting to the realities of employer control, unrealistic hopes for change

Figure 3: Attitudes of main participants of the Humanization of Work/Work and Technology programme

* This excludes the opinions of the scientific community and the different advisory boards which should be included in a comprehensive summary of more or less influential participants.

An illustration of the complex tensions characterizing NTPCs within the HdA/AuT programme can be provided by a list of the conflicting perspectives brought to the programme by a number of the leading participants (Figure 3). Although this is somewhat simplistic, it serves as summary important illustrative device in showing the negotiations over the nature and meaning of NTPCs.

It is not surprising, therefore, that as one of us has commented earlier, the programme remains in a 'torso character' with many humanization concepts not fully developed – a phenomenon that emerges not because of the brief history of the programme, the friction points in intra and inter-departmental negotiations, or even uncertainty concerning the future development of the programme, but rather in the continual unstable compromise between "humanization" and "productivity", and the domination of entrepreneurial strategies in resolving these conflicts despite the programme's extensive activities (Naschold 1980).

7.3 The Impact and Significance of the NTPCs

In direct contrast to functionalist approaches that over-emphasize the degree to which politics is adapted to the 'requirements' of techno-economic regimes, the history of the HdA programme reveals the extreme difficulties involved in implementing a human oriented "comprehensive innovation" approach. Some of the key problems have already been mentioned, including domination of economic investment strategies by management with little direct influence from trade unions and state programmes; the difficulty of integrating across programmes both within and beyond the BMFT; and the difficulty of diffusing the results of the projects.

7.3.1 Management Domination of Investment Strategies

At programme level, this factor is clearly apparent in the need for management to apply for programme money for a firm project, despite the need for agreement from the works council. In an attempt to increase the influence of works councils on project, the HdA programme requested from the late 1970s on that a more detailed set of information provision and consultation provisions be provided. In practice, however, these provisions are difficult to implement. One comparative research project on information disclosure in general found in three out of four West German companies studied, that the first information was given only after several years of management planning (Moore 1984). Despite the formal disclosure provisions of the 1972 Works Constitution Act, severe restrictions are imposed on trade union influence by the following factors: information can be withheld on legal grounds; conflicts and communication between central and plant work councils can cause problems, as well as the lack of integration between works councils and trade unions which are legally unable to intervene in plant level negotiations. An additional and crucial factor is observed in a study of the

German machine tool industry: a key problem was not so much the absence of information or consultation opportunities, but rather the inability of the works councils to utilize the information and take advantage of the opportunities due to lack of time, expertise, etc. (Hildebrandt 1987). It is not totally surprising, therefore, that an analysis of 71 collective agreements in West Germany (all recognized by the unions as leading examples) revealed that scientific knowledge of new forms of work organization and humanization of work ideals were hardly ever drawn upon (Pornschlegel/Schütte 1981).

An additional factor revealed in reviews of the HdA programme is the extent to which particular HdA projects can be seen as 'moments' in management investment strategies as many firms used the money as part of prior cost reducing rationalization strategies (Altmann/Düll 1980), or ended the project after the 'pre-phase' as it became out of line with the timing of firm investment strategies (Schallock 1991).

7.3.2 Lack of Programme Integration

As observed in the IG Metall Action Programme, the HdA/AuT programme operates with less than 2% of federal R&D funds. If it is, therefore, to have a significant impact on shaping innovation, it requires integration with other state programmes. The placing of the robotics and flexible automation components of the programme in the separate Manufacturing Technologies programme in the late 1970s indicated a lack of BMFT commitment to this task. Moreover, the increasing direction of BMFT funding towards *indirect* R&D funding lessens the possibility of extending HdA influence through integration with significant direct R&D funding programmes.

The lack of integration with other programmes has long been a major criticism directed against the HdA programme, and attempts have been made to address this issue. In the mid-1980s, the competent HdA panel at the BMFT outlined a 20-point humanization and innovation plan that attempted to outline a "comprehensive notion of innovation" that would provide the theoretical basis for this integration. The BMFT has generally pointed to the communication that occurs at the level of the political community, relevant associations, programme target groups, and working level contracts within the BMFT. In the mid-1980s, links had been formed, for example, in the areas of ergonomics and occupational health and safety (reactor safety, the danger of new energy technologies, laser techniques, agricultural production and forestry, etc.), technology assessment (applied biology and biotechnology, new manufacturing technologies), and within the areas of public transit and the ship of the future (BMFT 1985). During and since that time, further informal links have been developed with the Manufacturing Technologies Programme and the other areas of 'proactive' technology assessment and technology design. However, despite the recent inclusion of the Federal Ministry of Education and Science in the AuT programme, there has not been any development of uniform criteria for identifying relevant practices in other programmes in the BMFT or other ministries; and, despite the development of these and a number of other techniques for improving the diffusion of results, inter-programme collaboration remains weak. This may be a factor influenced by the more traditional technology ministry attitudes of the BMFT at the beginning of the programme, the continuing fundamental commitment of the department to improving international competitiveness, and a more recent emphasis on indirect and indirect specific methods of promotion. As represented in a recent response by the BMFT to a question in the *Bundestag* on R&D policy, public participation, technology assessment, etc., the AuT programme may however come to be accorded a more important role (BMFT 1990a).

7.3.3 Problems of Diffusion

The difficulties of diffusion have been strongly emphasized by both employers and trade unions. Academic critiques of the programme based on project evaluations, interviews and publication assessments have accurately pointed to the lack of published results and the absence of documented real interdisciplinary work or significant influences on the shaping of technology (Dankbaar 1987). They all point to projects that reveal this situation. The additional point has been made that these problems have been exaggerated by the political/disciplinary boundaries in the German scientific community in this area, and the ill-preparedness of this community for the large amount of finance and activity generated by the HdA programme. More importantly for trade unions and employers, however, have been, firstly, the failure of the projects to provide real organizationally successful exemplary models of efficient humanized work solutions and, secondly, the lack of methods for actually diffusing the results that have been achieved. Employers have pointed mainly to the lack of clear economic success of the various projects, and the absence of suitable documentation for diffusion. Trade unions have emphasized the lack of knowledge or expertise of works councils, and the major difficulties involved in informating and training local delegates and works council members in the nature, consequences and available alternatives to new forms of technology and work organization. In order to assist in the diffusion process, the HdA programme has financed both employers organizations and trade unions to diffuse project results. In the case of the trade unions, in 1979, the DGB and IG Metall were funded to set up Technology Advice Centres to assist in this task. In addition many projects have extra money for the writing up of the results in accessible form. More significantly, the present AuT programme is strongly committed to the further development of branch oriented projects involving employers organization and trade unions in an industry sector rather than individual firm projects, co-operative projects combining technology developers, users, public institutions, employers associations, etc., and comprehensive transfer projects involving taking examples of projects and transforming them into specific acceptable form for action in other projects.

Despite the importance of these initiatives, however, severe diffusion problems still remain. If, as the HdA/AuT programme has long assumed, criteria employed by firms are too narrow in judging the economic efficiency of 'humane' work solutions, it is difficult to see how market-based transfer mechanisms can be successful. When this is added to the lack of diffusion of past HdA results, including their techniques for 'broader' concepts of innovation, the diffusion process appears to remain an extremely problematic issue.

7.3.4 Indirect Benefits

It has been accurately observed in a number of assessments of the HdA programme that the direct quantifiable benefits of the programme are probably of less significance than the substantial indirect benefits that have emerged. Apart from concrete gains in the occupational health and safety area, small technological/organizational improvements, and some useful practical examples, the programme has provided an opportunity for a large number of work scientists to obtain industrial design experience. The subsequent creation of a 'humanization of work establishment' both knowledgeable in theory and practice in this area and institutionally and professionally committed to the furtherance of humanization goals may come to appear as a significant achievement as different countries struggle to define, experiment and realize NTPCs. In addition, the problems experienced by the programme and many of the solutions established are the result of the difficult but necessary experiences associated with the implementation of a state-centred social-democratic modernization process. By making the opportunities and difficulties clear and by creating opportunities for new understandings and links to be formed between scientists, trade unionists and ministry officials, the HdA programme may have made an important contribution to the German social modernization strategy. These activities and communities may themselves prove to be an important carrier of new technology policy concepts, or at least a significant influence in the shaping of the future industrial culture.

8 Conclusion

Some relatively clear themes emerge from the history of the Humanization of Work programme.

Firstly, there was a distinctive movement from a phase of 'humanistic' radical perspective to a more business oriented accomodative era, governed by the specific conditions of these time periods. Amongst the most significant influences on the earlier phase are, firstly, the importance of a social-democratic government in an economic boom period in promoting humanization concepts that extended beyond the far more restrained and sceptical acceptance by unions and business; and, secondly, the early implicit shift from a more traditional demand-oriented Keynesian approach towards social modernization to a more supply-oriented approach to socially oriented economic development. The decline of the radical supply oriented approach could be viewed in either classic pessimistic Marxist terms as the inevitable weakening of a radical state programme or in more optimistic Weberian terms as a movement from a "charismatic" founding era to one of "routine" bureaucratic implementation of detailed applications of a number of the early ideas. It is perhaps more appropriate, however, to point to the gradual reaching of a compromise or 'settle point' between business, trade unions and a new conservative political party, influenced by such conditions as a less favourable economic and employment climate, a lack of commitment by trade unions in committing substantial financial and organizational resources, a failure of scientific partners to create an integrated and cumulative set of findings and experiences, and a changed project management[4] – a combination of events that resulted in a decline in project dynamism and innovation and a more accomodative (yet, in the situation, realistic) push for "comprehensive innovation concepts".

Secondly, the development of the programme has to be seen in its particularity as a different type of activity to that of, for example, the Swedish Work Environment Fund. In contrast to the Swedish situation, the programme is more clearly state dominated, directed more strongly by governmental ministries than the labour market partners. In addition, the programme, with substantial spending in excess of one billion DM, has contributed a larger amount directly towards research and development in work humanization. In the process, it has supported and been influenced by a much larger scientific research establishment in the work humanization area. The programme has also been substantially influenced by the political segmentation of West German ministries and their representation on the programme. Firstly, the participation by the Labour Ministry and the BMFT has been important in ensuring the strong health and safety and technology orientations of the programme, and new developments are to be expected in the education and training area as the respective ministry has now been integrated into the programme. Secondly, the direction of research and project proposals has been influenced both by the spending patterns of a technology ministry that has committed substantial funds to the machine building and larger industries and the disciplinary orientations and strategic focuses of the range of powerful Fraunhofer Institutes and psychology, sociology and engineering research centres.

Thirdly, the course of the programme illustrates a number of points about such technology policy initiatives: firstly, the autonomy of the political process (in this case represented in the strong influence upon the programme of the SPD and then the CDU, and the reorientation of the programme after parliamentary criticism in 1987, that itself followed shortly after parliamentary reaffirmation of the programme's activities; secondly, the difference between debates at the ideological level and those at the project level, although we have concentrated very much in this discussion on the more ideological level, a more systematic analysis of the projects themselves is required to determine the real influence and direction of the programme; and, thirdly, the central significance in programme orientation of practical implementation, for example, the ruling that business partners to programmes take part responsibility for paying for scientific research, the decision to not contribute so much funding to large companies with the most powerful trade unions more able and committed to transferring what they see as successful local experiments, the significance of labour market institutions and legislation in enabling or restricting the programme's range of manoeuvre, and the difficulty of ensuring adequate diffusion of promising experiences if immediate productivity gains are not apparent or existing market measures of productivity are being questioned.

What, however, are the implications of these experiences for the understanding and promotion of NTPCs? Firstly, it is necessary to go beyond identifying NTPCs as either a profound threat to work humanization or as the active creator of a 'new world of work'. It is more appropriate to recognize the necessarily complex and conflictual character of such concepts, and the potential that various aspects have for the improvement of working conditions. Those elements that are selected and promoted will be directly related to ministerial and party politics, the strength and strategic orientations of industrial associations, sectoral industrial characteristics and strengths, the research and development system and the orientations of scientific-technological institutions, patterns of industrial relations and skill formation, etc. Secondly, the funding of direct promotion programmes, even as large as the German Humanization of Work programme, are still reliant for their effect on the operations of many such 'indirect' as well as 'direct' influences on paths of workplace innovation, and any design of such programmes that underestimates this fact will inevitably result in disillusionment and ineffective promotion activities. Thirdly, if the traditional limitations of work reform programmes are to be addressed and the Japanese challenge effectively met, new programme forms need to be developed that integrate direct promotion measures more effectively with broader education and training, industrial policy, technology transfer and R&D measures and concerns. As emphasized, these need to be adapted to specific national circumstances rather than imported as a "universal model". In short, the success of NTPCs will not only be based on their 'radical' introduction of social considerations into technology policy, although the recognition and promotion of techno-*organizational* innovation concepts is of major significance. It will, rather, centrally depend on recognizing the socially-located character of such technology policy programmes and their supportive (or non-supportive) context, and building upon the resources available to facilitate the promotion of rationalization paths that conform with cultural aspirations for improved working conditions.

Notes

[1] Also note the central emphasis placed on new visions of comparative advantage and business success. For the BMFT, as outlined in their 1988 annual report, "In the world of work, we now have to an increasing extent the active participation of the users of technology, because employees have to combine expertise in the new technology with experience and knowledge in the tasks that are to be performed with this technology. A lead in expertise with regard to this 'knowledge of integration' is of great importance for our competitiveness, because it cannot so easily be made up elsewhere with the aid of just a few specialists" (BMFT 1988:75). In addition there is den Hertog and Schöder's observation that "One has only to refer to the most recent volumes of the leading business journals in order to recognize the dominant role attributed to the organization and the people working within it. These are not hazy socio-scientific pleas for 'human-centred technologies', but learning experiences based on hard practice" (Hertog/Schöder 1989:2).

[2] This is strongly supported by the preliminary information from the contemporary HdA/AuT review of CIM Gestaltungswissen co-ordinated by the Fraunhofer IPK, Planungstechnik, West Berlin (Schallock 1991).

[3] See the series of articles by the group Hilary Steedman and Karin Wagner, D.W.M.N. Hitchins, S.J. Prais, Valerie Jarvis and A. Daly in the *National Institute Economic Reviews* of February 1985, November 1987, May 1989, and November 1989.

[4] Dr. Lorenzen, the head of the programme management at that time, was strongly instrumental in developing and publishing the ideas and mobililizing the support of the advisory groups for comprehensive innovation concepts.

References

Altmann, N., R. Düll (1980) *Neue Formen der Arbeitsgestaltung und Arbeitsorganisation in der Bundesrepublik Deutschland*, Munich: Institut für Sozialforschung.

Auer, P. (1983) *Arbeitspolitische Reformen in Industriestaaten – Ein internationaler Vergleich*, Frankfurt/Main, New York: Campus.

Badham, R. (1990) The Human Centred Design Constituency, Working Paper, Berlin: Fraunhofer IPK.

Badham, R. (1991a) Introduction, in: *Artificial Intelligence and Society* 4 (5): 1-14.

Badham, R. (1991b) Anthropocentric Systems and the Australian Production Culture, in: F. Rauner, K. Ruth (eds.) Anthropocentric Systems and Production Cultures, unpublished report to EC-Fast, Bremen: EC-FAST, 250-272.

Badham, R., K.-D. Lennartz (1990) Some Notes on the Machine Tool Industry/Automobile Industry Development Block in West Germany, Discussion Paper, Berlin: Fraunhofer IPK.

BMFT (1985) *Reply to a Question in the Bundestag*, Bonn: BMFT.

BMFT (1988) *Annual Report: English Translation*, Bonn: BMFT.

BMFT (1990a) *Answer to the Federal Government*, Drucksache 11/6391, Bonn: BMFT.

BMFT (1990b) *Research and Development Programme 'Work and Technology'*, Bonn: BMFT.

Boyer, R. (1990) New Directions in Technical Change, in: OECD (ed.) New Directions for the 1990ies, Paris: OECD, 1-45.

Brugmann, W. (1982) Einkommenserhöhungen werden erst an vierter Stelle genannt, in: *Frankfurter Rundschau*, 15 January.

Couchman, P. (1989a) Opening Black Boxes: The Social Dimensions of Science and Technology and their Relevance to the Current Policy Debates, in: *NZ Science Review* 1-3 (46): 1-13.

Couchman, P. (1989b) Technology and the Australian State: The Changing Political Discourse on Technology in Australia 1975-1985, Doctoral Thesis, Wollongong: University of Wollongong.

Cooley, M. (1972) *Architect or Bee?*, London: Abacus.

Cooley, M. (1990) Comment, in: H. Rosenbrock (ed.) *Designing Human Centred Technology*, London: Springer, 24-37.

Dalum, B. (1990) National Systems of Innovation and Technology Policy: The Case of Denmark. paper presented to the Workshop on "Socially Oriented Technology Policy", Vienna: Institute for Advanced Studies.

Dankbaar, B. (1987) Social Assessment of Workplace Technology – Some Experiences with the German Program 'Humanization of Work', in: *Research Policy* 10 (16): 45-58.

Dickson, D. (1984) *The New Politics of Science*, New York: Pantheon.

Ebel, K.-H. (1990) *Computer-Integrated Manufacturing: The Social Dimension*, Geneva: ILO.

Edquist, C., H. Glimell (1989) Swedish Frontiers of Change – A Guide to the Impact of New Technologies, Work Designs and Management Practices, Report prepared for the Directorate for Social Affairs, Manpower and Education, mimeo, Paris: OECD.

Edquist, C., B.-A. Lundvall (1989) Comparing Small Nordic Systems of Innovation, paper presented to the seminar on "National Systems Supporting Technical Progress", Maastricht.

Elam, M., M. Börjeson (1991) Strategies of Flexibility or Rewording the World of Work in Sweden?, in: B. Jessop, H. Kastendiek, K. Nielsen, O. Pedersen (eds.) *The Politics of Flexibility*, London: Edgar and Allen, 240-258.

Fruin, W., T. Nishiguchi (1990) The Toyota Production System: Its Definition and Diffusion in Japan, paper presented to "The Organization of Work and Technology: Implications for Country Competitiveness" conference, Brussels: EIASM.

Grootings, P. (ed.) (1990) *New Forms of Work Organisation*, Berlin, New York: de Gruyter.

Gustavsen, B. (1987) Report on the LOM Programme, Ontario: Ontario Centre for the Quality of Working Life.

Hertog, J.F. den, P. Schöder (1989) *Social Research for Technological Change: Lessons from National Programmes in Europe and North America*, Maastricht: MERIT.

Hildebrandt, E. (1987) Computer Aided Process Planning in the German Machine Tool Industry, unpublished WZB Discussion Paper, Berlin: WZB.

Jürgens, U. (1990) National and Company Specific Differences in Organizing Production Work in the Car Industry, paper presented to the Workshop on "The Organization of Work and Technology: Implications for Country Competitiveness" conference, Brussels: EIASM.

Kelly, J.E. (1982) *Scientific Management, Job Redesign and Work Performance,* New York: Academic Press.

Kidd, P. (ed.) (1990) *Organisation, People and Technology in European Manufacturing: Interdisciplinary Research for the 1990s*, CEC/FAST Consultancy Report, Brussels: CEC/FAST.

Klein, J. (1991) A Re-examination of Autonomy in Light of New Manufacturing Practices, in: *Human Relations* X (43): 23-31.

Knights, D., H. Willmott, D. Collinson (eds.) (1985) *Job Redesign: Critical Perspectives on the Labour Process*, London: Gower.

Leydesdorff, L., P. van den Besselaar (1987) Squeezed between Capital and Technology: On the Participation of Labour in the Knowledge Society, in: *Acta Sociologica* 3/4 (30): 80-94.

Lincoln, J. (1990) Work Organization in Japan and the United States, paper presented to "The Organization of Work and Technology: Implications for Country Competitiveness" conference, Brussels: EISAM.

Loveman, G., W. Sengenberger (1991) The Re-emergence of Small-Scale Production: An International Perspective, in: *Small Business* 3: 1-18.

Markovits, A., C. Allen (1981) Trade Union Responses to the Contemporary Economic Problems in Western Europe: The Context of Current Debates and Policies in the Federal Republic of Germany, in: *Economic and Industrial Democracy* 3 (2): 45-57.

Mathews, J. (1990a) Towards a New Model of Industrial Development in Australia, Industrial Relations Working Papers 6, Sidney: School of Industrial Relations and Organisational Behaviour, University of New South Wales, Sydney. (Apply to this Department for copies of J. Mathew's Australian Financial Review article reviewing TQM.)

Mathews, J. (1990b) *Tools of Change*, Melbourne: Pluto Press.

Moore, R. (1984) *The Control of Frontiers: Workers and New Technology, Disclosure and Use of Company Information*, Oxford: Oxford University Press.

Naschold, F. (1980) Problems of a Society-Oriented Research and Development Policy – The Humanization of Work Program at the Crossroads, WZB Discussion Paper IIVG/dp/80-213, Berlin: WZB.

OECD (1988) *New Technologies in the 1990s: A Socio-Economic Strategy*, Paris: OECD.

Offe, C. (1985) *Disorganised Capitalism: Contemporary Transformations of Work and Politics*, Cambridge, GB: Polity Press.

Perez, C. (1983) Structural Change and the Assimilation of New Technologies in the Social and Economic Systems, in: *Futures* 1 (5): 131-154.

Perrow, C. (1990) Small Firm Networks, Draft Discussion Paper, Connecticut: Yale University.

Pollert, A. (1990) Conceptions of British Employment Restructuring in the 1980s, in: I. Varoe, M. McNeil, S. Yearly (eds.) *Decyphering Science and Technology*, London: Macmillan, 74-103.

Pollert, A. (1991) *Farewell to Flexibility*, London: Gower.

Pornschlegel, H., H. Schütte (1981) Arbeitswissenschaflich bedeutsame Vorschriften in Tarifverträgen – Eine Analyse ausgewählter Aspekte, in: *Zeitschrift für Arbeitswissenschaft* 3 (35): 7-12.

Ruth, K. (1990) Industrial Culture and Technology Development in the German Machine Tool Building – Preliminary Findings of a Research Project, prepared for the International CAPIRN conference "Industrial Culture and Competitiveness: The Social Organisation of Production and Innovations", Gelsenkirchen.

Schallock, B. (ed.) (1991) CIM-Gestaltungswissen Projekt, unpublished report, Berlin: IPK.

Sorge, A., G. Hartmann, M. Warner, I. Nicholas (1983) *Microelectronics and Manpower in Manufacturing*, Aldershot: Gower.

Sorge, A., W. Streeck (1988) Industrial Relations and Technical Change: The Case for an Extended Perspective, in: R. Hyman, W. Streeck (eds.) *New Technology and Industrial Relations*, Oxford: Blackwell, 1-22.

Warner, M, W. Wobbe, P. Broedner (eds.) (1990) *New Technology and Manufacturing Management: Strategic Choices for Flexible Production Systems*, Chichester: Wiley.

Wilpert, B., A. Quintanilla (1984) The German Humanization of Work Programme: Review of its First Twenty Publications, in: *Journal of Occupational Psychology* 4 (57): 9-21.

Womack, J.P., D.T. Jones, D. Roos (1990) *The Machine That Changed the World*, New York: Rawson Associates.

Designing Sustainability of Industrial Society

Udo E. Simonis

> There is something fundamentally wrong in treating the Earth as if it were a business in liquidation (Herman E. Daly).

1 Introduction

Garret Hardin, in his essay on "The Tragedy of the Commons", remarks (1968:1245) "In a reverse way, the tragedy of the commons reappears in problems of pollution. Here it is not a question of taking something out of the commons, but of putting something in – sewage, or chemical, radioactive, and heat waste into water; noxious and dangerous fumes into the air; and distracting and unpleasant advertising signs into the line of sight". Hardin states that utility considerations are much the same as before: "The rational man finds that his share of the total costs of wastes he discharges into the commons is less than the cost of purifying his wastes before releasing them. Since this is true for everyone, we are locked into a system of 'fouling our own nest', so long as we behave only as independent, rational, free-enterprisers". He concludes: "The air and waters surrounding us cannot readily be fenced, and so the tragedy of the commons as a cesspool must be prevented by different means, by coercive laws or taxing devices. We have not progressed far with the solution of this problem".

In this paper, I shall focus on three elements of such a solution. First, I will present the potential of ecological structural and technological change of the economy and will provide some empirical evidence on the relationship between structure and scale of the economy and their environmental impacts. Secondly, I shall point to some of the shortcoming of environmental policy and will ask for a shift toward waste reduction or pollution prevention technology. Thirdly, I shall put forward some ideas on how to integrate ecological dimensions into practical economic policy.

"Ecological structural and technological change of the economy", "preventive environmental policy" and "ecological orientation of economic policy" thus seem to me to be major strategic elements for decreasing the pollution load, or the points of departure for escaping from the (reverse side of the) tragedy of the commons and reaching sustainability of industrial society.

2 Ecological Structural and Technological Change of the Economy

In both the East and the West, economists, planners and engineers are seeking to change the traditional patterns of resource use. "*Perestroika*" and "modernization" are two actual catchwords in this process, and new environmental priorities play a part in the envisaged structural and technological conversion of the economy (cf. Enyedi et al. 1987). Reconciling ecology and economy in a specific sense relies on the premise that a reduction in the input of energy and material resource will also lead to an *ex ante* reduction of emissions and wastes that have a negative impact on the natural environment (*ecological structural change*).

In order to clarify the relationship between economic structural and technological change and environmental impacts, one needs suitable information concerning the material side of production, for long-term sustainability cannot appropriately be described in such terms as income, investments and consumption. One possibility is to select and compare some relevant indicators describing the environmentally significant feature of the production process.

The availability of environmental indicators such as emission data relating to 'representative' pollutants – like sulphur dioxide (SO_2), carbon dioxide (CO_2) and nitrogen oxide (NO_x) – has rapidly grown in the last decades (e.g. the annual reports on the Environment by several industrial nations, by UNEP and the OECD). These indicators concern certain negative environmental effects of production. Less is known on the environmental significance of the input factors in industrial production, or on the question of which indicators provide environmentally significant information about the structure of the economy. Given the present state of statistics, only a few such indicators can be tested in a cross-national comparison of Eastern and Western countries.

2.1 De-linking Economic Growth from Environmentally Significant Inputs

Using a set of four indicators (input factors), Jännicke et al. (1989; the following data and arguments rely on this article) have studied 31 countries of both COMECON and OECD with regard to the relationship between structure and scale of the economy and environmental impacts. The four factors whose direct and indirect environmental significance is thought to be self-evident are: *energy, steel, cement,* and *freight transport.* Regarding their patterns of production and consumption these are environmentally "hard" factors, characteristic of a certain structure, technology, and/or scale of the economy.

The main hypothesis of the research therefore was a rather simple one and reads like this: positive environmental effects of structural and technological change of the economy are to be expected by actively de-linking economic growth from the use of environmentally significant inputs (resources). Such active *de-linking* (or *ecological structural policy)* would

- result in a decrease of *resource depletion* and/or *environmental pollution;*
- mean *ex ante* instead of *ex post* environmental protection;
- promote *integrated, less polluting technologies* which touch upon several environmental effects (pollutants) at one and the same time.

Ecological structural and technological change as a shift of input factors can thus be conceived as a process of *successive de-linking:* The contribution of traditional (hard) input factors to the national product decreases, i.e. they change or lose their function in the development process.

2.2 Examples of Successful and Unsuccessful De-linking

Taking the Federal Republic of Germany as an example, Figure 1 illustrates a five-fold de-linking from the growth of the Gross Domestic Product (GDP): The de-linking of energy and cement consumption and weight of freight transport from GDP became apparent during the 1970s; regarding steel, the de-linking process began already in the 1960s. In this way, structural change of the economy generated *environmental gratis effects* of various kinds:

- the stagnating consumption of primary energy led to a reduction of harmful emissions (pollutants);
- the relative decline in the weight of freight transport indicates that the volume of materials employed was reduced rather than increased;
- the fall in the use of *cement* represents a direct gratis effect as far as the emissions from *cement* factories are concerned; this decrease coincided with the trend towards labor-intensive renovation of the housing stock, as compared to new construction;
- the decrease in *steel consumption* accounts for a considerable reduction in harmful emissions as far as processing is concerned; this drop was strongly marked and partly due to increased recycling activities.

Environmental gratis effects occur when the rate of usage of the input factors (resources) having a (strong) negative impact on the environment remains below the growth rate of the GDP. Comparing the rates of usage of the four selected input factors with the GDP growth rate, three distinct development patterns emerge (Jänicke et al. 1989):

- The factors having impacts on the environment decline absolutely; i.e. *absolute structural improvements* are induced, corresponding to absolute environmental gratis effects.

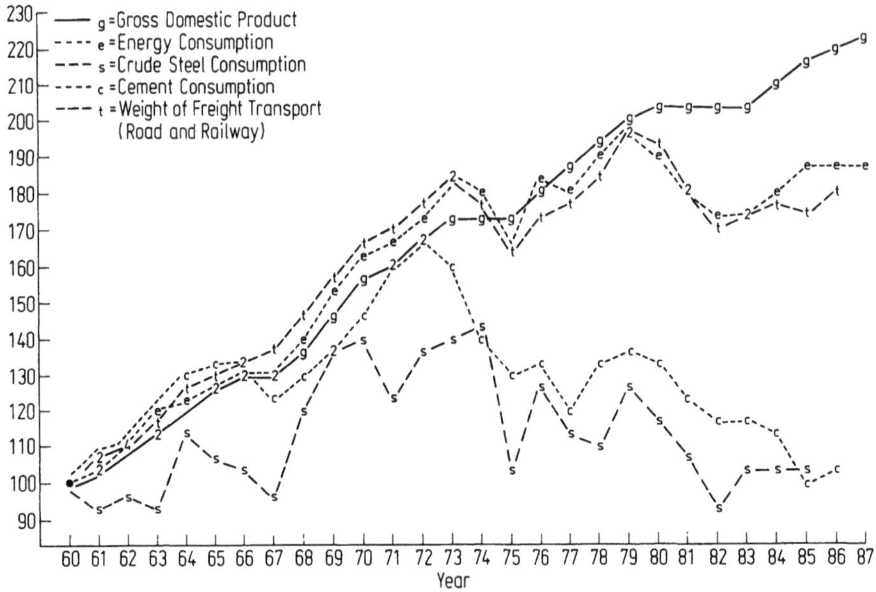

Figure 1: Structural economic change in the Federal Republic of Germany, 1960-1987 (1960=100)

Source: Jänicke et al. (1989)

- The factors having impacts on the environment remain constant, or increase, but with a lower growth rate than the GDP; i.e. *relative structural improvements,* corresponding to relative environmental gratis effects.
- The factors having impacts on the environment increase at a higher growth rate than the GDP; i.e. *structural deterioration*, corresponding to absolute negative environmental effects of economic growth.

In Table 1, 16 out of the 31 countries studied are grouped according to these three development patterns.

Of all the industrial countries investigated, *Sweden* went through the most rapid structural and technological change. The drastic reduction in cement production (-41%), the decreasing use of crude steel (-38%), and the decrease in the weight of freight transport (-21%) add up to notable environmental gratis effects ("absolute structural improvement").

In *Japan*, the de-linking process was partly neutralized by the rapid growth in industrial production and thus only resulted in "relative structural improvement".

In Czechoslovakia, by contrast, no de-linking of economic growth from the four input factors took place (see Figure 2). The development profile of this country,

Table 1: Environmental impacts deriving from structural change – Percentage changes, 1970-1987

Country	Consumption of Primary Energy	Consumption of Crude Steel	Cement Production	Weight of Freight Transport	GDP*
Group 1: Absolute Structural Improvement					
Belgium	7.1	-24.5	-17.6	-2.2	42.7
Denmark	-2.7	-15.6	-33.2	20.1	40.8
France	30.3	-34.8	-23.4	-14.5	51.6
FRG	13.4	-26.3	-32.8	4.4	38.4
Sweden	26.4	-37.9	-41.2	-21.4	32.7
UK	-2.3	-43.5	-28.7	-18.2	32.4
Group 2: Relative Structural Improvement					
Austria	32.1	-33.9	-6.0	21.3	54.3
Finland	39.6	14.8	-11.2	12.2	65.7
Japan	37.3	-2.3	27.4	7.5	90.2
Norway	51.1	-21.6	-40.3	34.7	87.5
Group 3: Structural Deterioration					
Bulgaria	74.9	24.9	42.3	77.5	37.3
Czechoslovakia	31.5	22.5	37.3	62.9	33.9
Greece**	119.3	67.3	162.9	43.1	69.1
Portugal**	89.0	34.2	133.1	27.4	69.0
Soviet Union	76.3	33.4	35.9	70.2	47.7
Turkey	218.8	184.4	173.2	118.6	118.2

* Calculation of the Gross Domestic Product percentage changes on the basis of constant (1980) US$. Bulgaria, Czechoslovakia and Soviet Union data refer to percentage changes, 1970-1983, of the Gross National Product.

** Transport data only take railway transport data into account.

Source: Jänicke et al. (1989)

with sluggish structural and technological change, is also representative of the other economies of Eastern Europe.

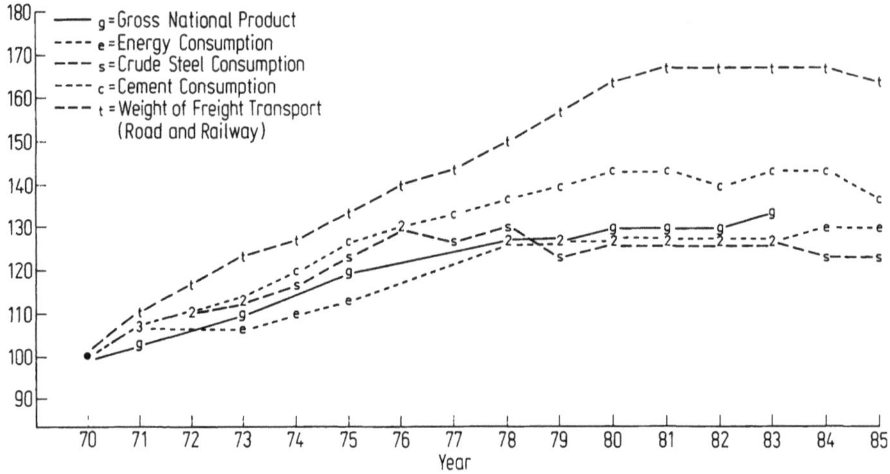

Figure 2: Structural economic change in the CSSR, 1970-1985 (1970=100)
Source: Jänicke et al. (1989)

2.3 Trends Towards Industrial Restructuring

Despite certain analytical limitations of such empirical research (e.g. the selection of only four input factors), several conclusions can be drawn from his international comparison as regards past trends of industrial restructuring (cf. Jänicke et al. 1989).

- Structural and technological change in the form of de-linking economic growth from environmentally significant inputs was evident in most but not all of the countries studies.
- Several countries enjoyed environmental gratis effects as a result of active structural change. In some cases, especially for Sweden, these effects were quite considerable.
- In other countries, the possibly beneficial environmental effects of structural change were levelled off by the rapid industrial growth pursued. This was especially true for Japan.
- The strong correlation between the level of production (GDP) and environmental impacts, still evident in the 1970s, had dissolved in the 1980s. The high-income countries featured fairly rapid structural change.
- In the medium-income countries, a distinct pattern emerged in that there were cases of rapid quantitative growth and cases of qualitative growth, i.e. economic growth with constant or decreasing resource input.

Summing up, it is not yet possible to speak of one dominant trend towards industrial restructuring. However, the environmental gratis effects of active structural and technological change are highly evident and thus provide one strategic element for designing sustainability of industrial society.

3 Preventive Environmental Policy

Theoretically speaking, environmental policy can be defined as "the sum of objectives and measures designed to regulate society's interaction with the environment as a natural system; it comprises aspects of rehabilitation, conservation, and structural adjustment" (IIES 1982:6). Practice, however, does not conform to such a comprehensive definition. Only parts of the interaction between society and environment have become the subject of policy. So far, environmental policy has mostly been designed as react-and-cure strategies concerning air and water pollution, noise and waste, with emphasis on the rehabilitation aspect and on end-of the-pipe technology.

For a variety of reasons, this conventional environmental policy was, and is still, meaningful and very much necessary. It has, however, a number of shortcomings, some of which are addressed in the following along with some suggestions for overcoming them through preventive environmental policy, i.e. anticipate-and-prevent strategies and integrated, low emission technology.

3.1 Environmental Expenditures – Environmental Damages

Since the beginning of the 1970s, when systematic records first began to keep track of the funds allocated for environmental protection, the sum of the respective public and private investments in the industrialized countries has reached large proportions. Industrial society thus appears to be 'paying through the nose' its back payments for the negative environmental impacts of production accumulated in the past.

In the Federal Republic of Germany, for instance, the sum of expenditures aimed at rehabilitating and conserving the natural environment has come to the handsome total of over 275 billion DM (or about 175 billion US$). In a detailed study, Leipert/Simonis (1989) from the International Institute for Environment and Society (IIES) have first established the respective data which in the *Statistical Yearbook of 1989* were published as official data by the Federal Statistical Bureau. (The following data are from them.)

Table 2 shows the total and sectoral *environmental protection investments* for the manufacturing sector of the German economy for the years 1975 to 1987.

Table 3 shows the *total expenditures* for *environmental protection* for both industry and government for the years 1975 to 1987.

Figures like these, however, are ambivalent. On the one hand, they give cause for proud political statements about the successes of environmental protection, according to the motto "the more, the better". On the other hand, they are – presumably – not sufficient to secure industrial society's sustainability. At the same time, they may symbolize a serious inherent deficiency of industrial society: Environmental protection expenditures are spent when damage to the natural environment has occurred and can no longer be denied. They are, in fact, repairs to the process of economic growth, signs of a "post-fact" policy that reacts at a rather late stage to damages (and must react to them) but does not, or cannot, prevent them. Therefore, it is necessary to always confront the 'success stories' of environmental protection *expenditures* with figures on the environmental *damages* themselves.

Again taking the Federal Republic of Germany as an example, an estimation by Lutz Wicke (1986:123) from the Federal Environmental Protection Office says that the *annual damage* to the natural environment in monetary terms may be over 103 billion DM, or in the order of 6% of GNP, and not 3%, as the OECD had estimated for the industrial countries some years ago.

Table 4 is based on differing estimation methods, combining data on actual damage costs and findings from willingness-to-pay studies. Though the results therefore must be taken with some care, the table leads to the conclusion that, despite high annual environmental protection expenditures, enormously high environmental damages still occur annually. Of course, this conclusion may not only be true for Germany but for all other industrial countries as well.

There are more shortcomings to conventional environmental policy (Simonis 1985). To name the more obvious ones: environmental policy often identifies the given problem too late, so that the ecosystems affected cannot survive. As it is pursued as a *media-specific* policy, i.e. separately regulating air or water quality, noise or waste, it also lacks coordination among its specific goals, measures and institutions. This then may result in shifting a problem from one environmental medium to another – e.g. from air to water or soil, or from one place to another, as is the case with long-range, trans-boundary pollution. In addition, environmental policy often becomes entangled in a debate on principles. If measures must instantly be taken, the argument is twisted by the *polluter-pays principle* – which is advocated in general – to the taxpayer-pays principle, thus shifting the burden of environmental protection costs from the individual polluter to the community, the government or society at large.

Thus, innovations in policy planning and implementation are needed. Preventive environmental policy, it seems, can counter the shortcomings of conventional environmental policy. But in order to switch towards preventive policy, several conceptual as well as practical constraints have to be overcome (in the following, I rely on the arguments put forward by G. Scimemi 1988).

Table 2: Environmental protection investments, manufacturing sector, Federal Republic of Germany, 1975-1987, in million DM

	Total Investments		Waste Disposal		Water Pollution Control		Noise Abatement		Air Pollution Control	
Year	Current Prices	1980 Prices	Current Prices	1980 Prices	Current Prices	1980 Prices	Current Prices	1980 Prices	Current Prices	1980 Prices
1975	2,480	3,090	170	210		1,110	200	240	1,210	1,530
1976	2,390	2,830	200	230			220	260	1,150	1,380
1977	2,250	2,560	200	230			210	230	1,100	1,250
1978	2,150	2,370	170	180			200	220	1,100	1,220
1979	2,080	2,190	260	160			200	210		1,020
1980	2,650	2,650	210	210			240	240	1,290	1,290
1981	2,940	2,810	250	240			210	200	1,530	1,460
1982	3,560	3,250	390	360	1,130	1,030	230	210	1,810	1,650
1983	3,690	3,270	290	260	1,100		230	200	2,070	1,820
1984	3,500	3,100	270	240	1,040		230	190	1,960	1,750
1985	5,620	4,940	330	280	1,060		260	220	3,970	3,530
1986	7,300	6,400	420	350	1,150		250	200	5,480	4,900
1987	7,890	6,900	...	...	...	...	...	...	...	...
Annual Change in %										
1975-1979	-4.3	-8.2	-1.5	-6.6	-4.1	-7.9	0	-3.3	-5.6	-9.6
1979-1984							2.8	-2.0		
1975-1984	3.9					-2.1	1.6	-2.6		

Source: IIES research project. Data for 1986 and 1987 from Statistical Yearbook (1989:587)

Table 3: Total costs of environmental protection investments for industry and government, Federal Republic of Germany, 1975-1987, in million DM

Year	Industry Current Expen-ditures	Industry Depreci-ations	Industry Total Costs	Government Current Expen-ditures	Government Depreci-ations	Government Total Costs	Industry and Government Current Expen-ditures	Industry and Government Depreci-ations	Industry and Government Total Costs
At Current Prices									
1975	3,200	1,520	4,720	3,000	1,920	4,920	6,200	3,440	9,640
1980	5,160	2,250	7,410	4,690	3,390	8,080	9,850	5,640	15,490
1985	7,930	3,160	11,090	6,430	4,340	10,770	14,360	7,500	21,860
1987	9,050	3,780	12,830	7,580	4,790	12,370	16,630	8,570	25,200
At 1980 Prices									
1975	4,050	1,870	5,920	3,790	2,570	6,360	7,840	4,440	12,280
1980	5,160	2,250	7,410	4,690	3,390	8,080	9,850	5,640	15,490
1985	6,230	2,640	8,870	5,340	4,030	9,370	11,570	6,670	18,240
1987	7,260	3,080	10,340	6,330	4,270	10,600	13,590	7,350	20,940

Source: IIES research project. Data for 1987 from Statistical Yearbook (1989:587)

Table 4: Environmental damage in the Federal Republic of Germany ("measurable damage" in billion DM per year)

Environmental Sectors	Environmental Damage
Air Pollution	ca. 48.0
Health hazards	between 2.3 and 5.8
Material damages	more than 2.3
Degradation of vegetation	more than 1.0
Forest blight, etc.	between 5.5 and 8.8
Water Pollution	far more than 17.6
Damage to rivers and lakes	more than 14.3
Damage to North Sea and Baltic Sea	far more than 0.3
Contamination of ground water, etc.	more than 3.0
Soil Contamination	far more than 5.2
Costs of Chernobyl accident	far more than 2.4
Rehabilitation of 'yesterday's waste'	more than 1.7
Costs of preserving biotopes and species	more than 1.0
Other soil contamination, etc.	far more than 0.1
Noise	more than 32.7
Degradation of residential amenities	more than 29.3
Productivity losses	more than 3.0
'Noise rents', etc.	more than 0.4
Grand total of damage	more than 103.5

Source: Wicke (1986)

One constraint has to do with the particular history of an environmental impact. In cases of yesterday's wastes, when damage has already occurred, a curative strategy obviously is the only conceivable option. However, in all cases where damage must be expected for the future, the choice between a preventive and a curative strategy is basically open. The anticipatory principle strongly leads to encourage the first option. As practice often is a mixture between the existing and the new, most policies actually will also include a mixture of prevention and cure. Demanding preventive environmental policy will then mean seeking, and at last finding, a *better balance* between the anticipatory and the reactive component within policy action and, particularly, between end-of-the-pipe technology and integrated, low-emission technology.

3.2 Basic Conditions of Preventive Environmental Policy

One can conceive three factors as concomitant policy-relevant processes in time: the accumulation of environmental damage, the acquisition of technical knowledge, and the rise of public awareness. The time sequence of these processes, especially the relative timing of their *critical levels*, may be decisive for the whole issue of preventive environmental policy.

To illustrate the relationship between these three factors, Scimemi (1988:34) has redrawn a diagram suggested by Winsemius, using three separate functions: *Level of Environmental Damage, Level of Technical Knowledge, Level of Public Awareness*. The relative position and the shape of these functions depends, of course, on the specific circumstances (country, environmental sector, historical phase) under consideration; the common case may be the one illustrated in Figure 3.

Line $D_1 - D_2$ indicates the *accumulation of environmental damage* over time. The accumulation of damage starts at a given point in history (in the diagram somewhere between time 0 and time 10). At that point, neither the scientific community nor the general public is yet aware that anything of importance is happening. Line $T_1 - T_2$ indicates the process of acquisition of *technical* knowledge. This process may not start until some time after damage has begun to accumulate (in the diagram somewhere between time 10 and 20), and proceeds gradually. During that phase, the public may still be unaware of the hazard. Somewhere between time 20 and 30, *public awareness* starts to rise, as is indicated by line $P_1 - P_2$.

Within these concomitant processes, a certain stage becomes important *(critical level)*. As illustrated, the technical understanding of the issue reaches a critical level, *t*, thus ensuring the first of two conditions required for effective policy action, i.e. *technical* rationality. Public awareness also reaches a critical level, *p*; at that time, the second condition for effective decision-making, i.e. political viability, is fulfilled. It is at this stage that action will be taken to avoid the occurrence of further damage.

Recalling past developments in environmental policy, at the national or the international level, one may easily discern that these developments evolved very much in conformity with Scimemi's theoretical interpretation. Thus, the diagram may be considered to be a true representation of real events. (There may be cases, however, where public awareness, line $P_1 - P_2$, started earlier than technical knowledge, line $T_1 - T_2$.)

What then are the opportunities to influence these basic conditions of policy action to favor preventive environmental policy? Figure 3 helps to formulate three general and two more specific options (cf. Scimemi 1988:38-39).

General options:

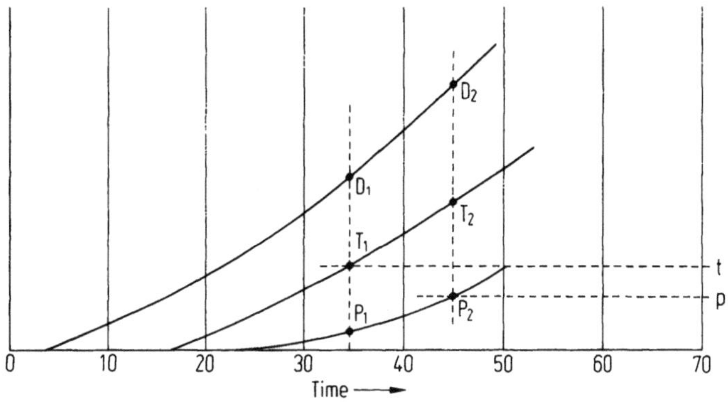

Figure 3: Factors of the environmental policy life cycle: Damage, Technical Knowledge, Public Awareness
Source: Scimemi (1988:38-39)

- retarding damage accumulation (i.e. sliding the $D_1 - D_2$ curve towards the right);
- accelerating technical knowledge (i.e. sliding the $T_1 - T_2$ curve towards the left and/or raising its slope);
- increasing public awareness (i.e. sliding the $P_1 - P_2$ curve towards the left and/or raising its slope).

Specific options:

- dynamic environmental standard setting (i.e. lowering threshold *t*);
- active public participation (i.e. lowering threshold *p*).

All these various options make policy decisions possible at a stage when the level of environmental damage is still relatively low, or even non-existent (for a detailed discussion, see Scimemi 1988:39-45). Two of these options are illustrated in Figure 4.

3.3 Environmental Impact Assessment as Part of Preventive Policy

Accelerating technical knowledge and public awareness can, of course, be promoted through a variety of approaches and methods, but it depends a great deal on the specific environmental issue at hand. *Environmental Impact Assessments* (EIA) are increasingly being applied, not only for public but also for private investment projects. They entail efforts to learn more about possible environmental impacts, and are intended to allow appropriate action to be taken before damage has occurred. In that sense, environmental impact assessments can be said to be part and parcel of preventive environmental policy (Wathern 1988).

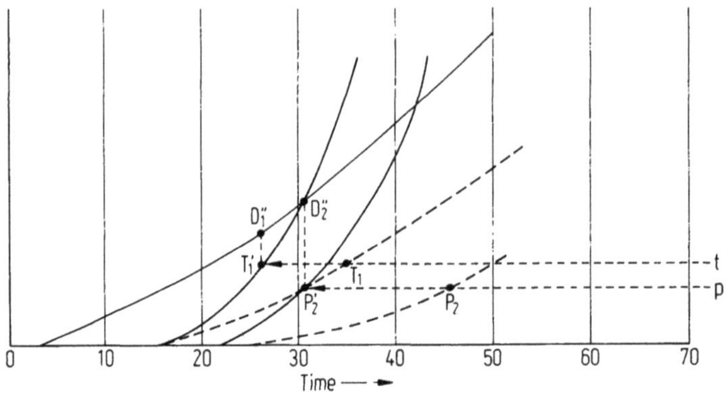

Figure 4: Accelerating the generation of Technical Knowledge and/or the development of Public Awareness
Source: Scimemi (1988)

In the last decade, some headway has been made to institutionalize and standardize EIA procedures, both nationally and to a minor extent also internationally. As the EIA procedure is particularly used for specific investment projects, it allows for the 'accelerating effort' to be targeted, and generally it also permits the burden of such efforts to be imposed upon the project initiator himself, thus conforming to a precondition of preventive environmental policy, i.e. the *polluter-pays principle.* Difficult questions, however, remain in implementing EIA as a preventive procedure in cases of global environmental change, like climatic warming or ozone depletion.

The required levels (thresholds) regarding technical knowledge and/or public participation in environmental decision-making differ widely from one environmental medium and one country to the other. The question of *how much knowledge/awareness is enough* normally falls upon the political decision-maker (the government, the environmental agency, the institution in charge of the problem), even if the scientific community (or parts of it) is ready to say "we know enough" and the public (or parts of it) is demanding "something must be done". Therefore, stalemates in decision-making on environmental issues are quite frequent (cf. Scimemi 1988:44).

What is "enough knowledge/awareness" for one country (government, agency, institution) may not be enough for the other. The normal outcome of such a situation is a compromise over the *emission standards* to be implemented. They will be weaker than technically/politically feasible because knowledge/awareness on cause-and-effect relationships and social priorities are said to be insufficient.

Eminent cases in point are the emission standards for SO_2, CO_2 and NO_X in the air-pollution field, and the nitrate standard in the water-pollution field. Thus, the dilemma of setting strict emission standards is a serious one. Meanwhile, the forests may continue to die back, the ozone layer may continue to be affected, and water may continue to become contaminated ... or, as Prince Charles said at the Second International North Sea Conference in 1988, "While we are still waiting for the doctor's diagnosis, the patient may die".

The conclusion therefore is that for designing sustainability of industrial society, environmental standard setting must be conceived as a continuous technological process. With growing knowledge/awareness on actual and probable environmental damages, the thresholds for action must be consecutively lowered, i.e. standard setting must be dynamized to make industrial restructuring possible on time.

This need to come to terms with the future is not unique to environmental policy, of course. Implementing the prevention principle is requested in all other domains of policy where collective interests are at stake and/or by which the commons are threatened. One such major domain we have to address when discussing the possibilities and impediments of the *sustainability of industrial society* is economic policy.

4 Ecological Orientation of Economic Policy

"Ecology in essence means the necessary and feasible harmony between man and nature, society and environment" (C.F. von Weizsäcker). Economy in general means disharmony with nature or even war against the environment. Use is made of nature both directly and indirectly when raw materials are processed into products, and the environment is polluted by the emissions and wastes generated by industrial production. These are, then, the two processes in which nature remains the loser; she exchanges natural raw materials for produced waste materials. Besides labor and capital, nature is the truly quiescent and exploited third production factor. How can nature's position in the 'economy game' be strengthened (cf. Simonis 1988)?

4.1 Conflicts between Economy and Ecology

The use of raw materials and the generation of emissions and wastes are of course old, not new issues. Scientific and technological developments, however, have made it possible to increasingly exploit depletable resources and have lead to an ever increasing accumulation of harmful emissions and non-biodegradable wastes. Nature is no longer able to absorb all these substances, many of which are not only toxic for flora and fauna but for human beings as well.

Efforts to hide harmful emissions and toxic wastes – in landfill sites, in transfer stations or permanent depositions, to spread them through high smokestacks and incinerators, or to dump them into the water bodies and abroad – have at best been temporarily successful because many emissions and wastes are "mobile poisons" or reappear in different form. These activities lead to a *linearization of ecological cycles* (Johan Galtung) – i.e. the natural diversity is reduced, the robustness of ecosystems declines, ecological symbioses and equilibria break down. As a consequence, environmental degradation increases and the absorption capacity of the natural environment decreases.

Accordingly, the conflict between ecology and economy can be attributed to two actually or possibly (?) incompatible basic principles: the ecological principle of *stability*, as a precondition for the sustainability of ecological systems, and the economic principle of *growth*, as the inherent logic of economic systems – more precisely, the principles of business profitability, national economic growth, and world market expansion.

Given the actual and the pending ecological crisis, the question on whether these economic principles can be changed, reshaped and finally brought into harmony with ecological principles, on which level, in what way, and at what time, is, of course, a controversial question in both theory and practice; and it represents a specific challenge to the natural, the technical and the social sciences.

The answer to that question not only depends upon the respective individual and the societal constellation of interests. The answer particularly depends upon the ability of and the willingness to plead for social innovations, i.e. on (i) whether and how the potential of an ecological self-regulation of the economy is used, and on (ii) whether and how the option of an ecological economic policy is implemented.

4.2 Ecological Self-Regulation of the Economy

Generally speaking, only a small fraction of the present environmental problems would exist if the economic contexts would have remained so comprehensible that producers and consumers would personally be able and liable to recognize and perceive the consequences of their own decisions towards depleting resources and polluting the environment. Or, if business profitability, national economic growth, and world market expansion could not be increased by externalizing parts of the ensuing costs. This is the old but still relevant – because unresolved – problem of the *external effects of production*. Scientific and technological developments have been, and still are, coupled with negative external effects – i.e. the shifting of costs to society, future generations, and nature. With respect to the environmental problem, all these cost components are relevant. Take, as an example, acid rain and the ensuing damage to our forests:

- First, this example shows the shifting of a part of the costs of industrial production, i.e. insufficiently-reduced air pollutants, onto nature, which is resistant only up to a certain level: the forests are dying.

- Second, this example shows the shifting of costs onto the succeeding generations, i.e. a future with less and/or regionally even no forests.
- Third, this example shows the shifting of costs onto third parties (i.e. partial expropriation of the forest owners) and onto society, in the sense that economic and technical decisions of individual polluters (especially emissions from power plants, cars, transboundary pollution) impair the well-being and physical health of the population.

Thus, the economic system evidently makes incorrect calculations with respect to the 'ecosystem forest'. Both business accounting and national accounting do not provide adequate signals which may prevent pollution levels that are not tolerable for the ecological systems. Conventional accounting shows favorable balances for the production of energy, for the automobile producers, and for the exporters of pollutants (just to stay with the three sources of pollution mentioned above), although the 'ecosystem forest' is definitely being damaged by the emissions from these economic actors. Losses here – profits there; compensation does not take place nor is liability provided for.

One of the pending tasks both for theory and practice can thus easily be prescribed: "Internalize the external effects of production; shift the costs back to the economic units that cause the environmental problem; include the ecological perspective into all technological decision-making!".

To re-organize the economy towards a materially *integrated cycle* would, first of all, mean reducing systematically the use of depletable resources and the generation of polluting emissions and wastes – but this is in contradiction to the prevailing "throughput economy". In practice, recycling and clean technology are still at an incipient stage and the step from simply disposing or handling wastes towards systematically avoiding wastes ("low waste economy") has not been made.

Certainly, this is in part because many waste products cannot be recycled or only at high cost. But it is also true because the right price signals have not been set ("Prices should tell the ecological truth") (von Weizsäcker 1990:141). So far, no adequate incentives exist for preventing waste generation and conserving depletable resources. This state of affairs has also to do with the above mentioned shortcomings of the economic accounting procedures which do not adequately measure the diminishing stocks. Therefore, two contradictory trends prevail: increasing monetary income – *decreasing natural stock.*

Promising proposals for *ecological accounting* on the factory level and on the national level have been made. With ecological accounting the volume of energy, materials, wastes, land use, etc. are being computed and, by simulating the given shortage, accounting units are determined which then enter *the* business accounts. Thus a measure can be developed which may guide private investment decision-making and at the same time provides a public information instrument for promoting qualitative economic processes. At the national level, *satellite systems* are being developed which complement the traditional national accounts.

In industrial society, another ecological principle that is no longer adhered to is the *sustainability or resource* use. Traditionally, for instance, forest owners have followed the rule: "Do not cut down more wood than you regrow". This rule is substantially undermined: in industrial countries, externally produced acid rain collides with internal resource conservation; and in developing countries, accumulated external debt lead to an overexploitation of the given national resources. Sustaining the yield of the forest capital stock thus is being replaced by indirect expropriation and by resource depletion.

Another basic principle to be re-established in industrial society is that of *liability*. With respect to environmental problems, the legal system, and also economic behavior, is marked in most countries by strict proof of causality. Only when the injured (damaged party) can prove who caused the damage (polluting party), is the polluter held liable for compensation. Instead, in some countries (for some cases) statistical probability is sufficient for obligating industry to compensate for damages *(collective liability)*. In Japan, once this principle was established by the courts and through legislation, it quickly helped to improve environmental quality through ecological self-regulation of business activities.

In general, the liability principle would strengthen the anticipate-and-prevent strategy in environmental policy and shift the technical solutions for environmental problems from *ex-post* to *ex-ante* approaches, i.e. from *controlling or end-of-the-pipe technology* towards low-*emission* or *integrated technology*. To implement this principle in practice, small steps or big leaps could be taken: from continuous reporting on wastes or automatic monitoring of harmful emissions, to collective funds or strict individual environmental liability.

4.3 Ecological Economic Policy

Confronted with serious environmental damage, conventional economic policy is increasingly being challenged. Its guiding principles, goals, instruments, and institutions are questioned, and a new concept is emerging: *ecological economic policy*.

Conventional economic policy is based on the *guiding principle* of maximizing flows: volume of production, income, profits, turnover. (Kenneth Boulding, nearly twenty years ago, called this the "throughput economy".) Instead, the "ecological economy" would have to be based on a different guiding principle, i.e. "increasing efficiency and maintaining substance". Aspects such as environmental compatibility and resource conservation would become important, and structural adjustment of products and technologies according to ecological considerations would become the task (cf. WCED 1987).

Regarding goals, it seems necessary to redefine the growth targets and to include "environmental stability" into *the* catalogue of economic policy goals. The conventional policy goal indicators were *developed* at a time when environmental pollution was already a problem but not yet a public issue; and since then they have not really been readjusted. Economic growth is still being measured in terms of goods and income categories only (GNP – Gross National Product), while their effects on the stock and the quality of the resources (natural capital) are not adequately being considered. In the conventional concept of economic growth, all monetary transactions are summed up independently of their specific function; also, increasingly more expenditures are included which are solely being spent for the (necessary) compensation for damage originally caused by the given technology (*compensatory* or *defensive expenditures*).

Qualified goal indicators for economic policy could be defined in various ways: Computations of compensatory expenditures, i.e. assessment of an environmentally adjusted net product (ENP – Eco National Product); combined growth and distribution indices (RWG – Redistribution With Growth); integrated system of economic and environmental indicators, or attached "Satellite Systems", etc.

Regarding *instruments,* conventional economic policy relies mainly on two instruments, i.e. variations of interest rates and of tax rates. From an ecological point of view, new taxes and charges are required which, to some extent, should replace traditional taxes. In a situation of structural unemployment and environmental pollution, the introduction of resource *tares* (as e.g. an energy tax) and *emission charges* (as e.g. charges on SO_2, NO_X, and CO_2 emissions), and a decrease of wage taxes, are all called for. Such a structural tax reform would change the existing incentives in the economy towards accelerating resource efficiency *and* increasing employment opportunities at one and the same time.

Economic policy manifests itself in and works through particular *institutions.* An ecological orientation of economic policy therefore suggests creating new economic institutions and abolishing or redefining old ones (cf. Ausubel et al. 1990). (The debate on the negative environmental effects of decisions by the World Bank and the IMF is just a case in point.) The actual and the pending environmental crisis requires structural reforms by which economic and technological institutions would have to incorporate the ecological perspective and environmental institutions to improve their conceptual and practical competence, and by which environmental impact assessments (EIA) would have to be integrated into all major economic and technological decision-making (cf. Wathern 1988:5-25).

5 Conclusions

According to the above deliberations, designing sustainability of industrial society is obviously a demanding concept, both theoretically and practically. It requires a far-reaching conversion of economy and technology and a reorientation of environmental policy. The natural, the technical and the social sciences have to improve *the* methodological foundations and the institutional arrangements for a successful practical implementation of such a concept.

References

Ausubel, J.H., H.E. Sladovich (eds.) (1990) *Technology and Environment*, Washington, DC: National Academy of Sciences.

Enyedi, G., A.J. Gijswijt, B. Rhode (1987) *Environmental Policies in East and West*, London: Taylor Graham.

Hardin, G. (1968) The Tragedy of the Commons, in: *Science* 1: 1243-1248.

IIES (International Institute for Environment and Society) (1982) *Research Program 1983-1987*, Berlin: IIES.

Jänicke, M., H. Mönch, T. Ranneberg, U.E. Simonis (1989) Structural Change and Environmental Impact. Empirical Evidence on Thirty-One Countries in East and West, in: *Environmental Monitoring and Assessment* 2 (12): 99-114.

Leipert, C., U.E. Simonis (1989) Environmental Protection Expenditures, in: *Rivista Internazionale die Scienze Economiche e Commerciali*, 3 (36): 255-270.

Scimemi, G. (1988) Environmental Policies and Anticipatory Strategies, in: U.E. Simonis (ed.) *Präventive Umweltpolitik*, Frankfurt/Main, New York: Campus, 27-48.

Simonis, U.E. (1985) Preventive Environmental Policy Prerequisites, Trends and Prospects, in: *Ekistics* 313: 369-372.

Simonis, U.E. (1988) Ecology and Economic Policy. in: A. Vlavianos-Avanitis (ed.) *Biopolitics*, Athens: Biopolitics International Organisation, 163-167.

Wathern, P. (ed.) (1988) *Environmental Impact Assessment Theory and Practice*, London: Unwin Hyman.

WCED (World Commission on Environment and Development) (1987) *Our Common Future*, Oxford: Oxford University Press.

Weizsäcker, E.U. (1990) *Erdpolitik*, Darmstadt: Wissenschaftliche Buchgesellschaft.

Wicke, L. (1986) *Die ökologischen Milliarden*, Munich: Kösel.

Constructive Technology Assessment: A New Approach for Technology Assessment Developed in the Netherlands and its Significance for Technology Policy

Joey van Boxsel

1 Introduction

Technology Assessment (TA) is a class of policy studies originally intended to counteract the adverse effects of technology on society and the environment. The concept was developed on both sides of the Atlantic Ocean in the late 1960s. Whereas institutionalization of TA had already taken place in the United States in the early 1970s, Europe had to wait for the 1980s for the first small TA-institutions to appear. In the Netherlands NOTA, the Netherlands Organization for Technology Assessment, was set up in 1986.

In the government white paper (1984) preparing the foundation of NOTA, the idea was presented that it should be possible, taking into consideration all the lessons from previous unwanted side effects of technology, to already take some of these effects into account during the development and introduction of technology. The possibilities and limitations of such an approach were investigated by analyzing and comparing theories of technology development and case studies. From the theoretical part of the work, a new discipline was seen to emerge: technology dynamics. But also a practical approach resulted from this work. It was called "Constructive TA". NOTA has applied Constructive TA on a number of topics with interesting results. Whereas in the 'classical' form of TA, research of the impact of already-existing technologies, TA can get in a somewhat antagonistic position towards technology policy, Constructive TA offers numerous opportunities for fruitful interactions with technology policy and technology management of all the relevant actors in society. In this paper, it will be explored what already has been accomplished and what contribution to these technology policies reasonably can be expected from Constructive TA.

2 Social and Environmental Consequences of Technology: The Emergence of Technology Assessment

Science and technology, twin brothers for a few hundred years now, have grown in an accelerated way since the middle of this century. So have the impacts on society. A lot of these consequences were welcomed by most people in the industrialized countries; but undeniably we have also seen a lot of unwelcome, unexpected effects: the arms race, detrimental effects on the environment on a worldwide scale, consequences for employment and working conditions, the risks from nuclear energy and big chemical plants. New technologies like recombinant DNA and biotechnology raised high hopes but also scared people. Even the progress in medicine appeared to have its dark side. To counteract these adverse effects and to choose the best technologies for society, numerous types of policy studies were developed. Technology Assessment – the acronym TA is used worldwide – is an important category of these.

The idea originated in the late 1960s in scientific circles both in the United States and Europe (Hetman 1973). The researchers who were involved mostly belonged to research institutes, advisory bodies or professional organizations that had been asked for advice by government organizations or regarded it as their responsibility to advise government and the public at large about these important matters.

However, institutionalization of TA did not spread the same way on both sides of the Atlantic. In the USA, the Office of Technology Assessment (OTA) of the US Congress was founded in 1973. In about five years, OTA developed an effective form of TA that fitted with its position in the American political system. Of course several organizations belonging to the executive branch of government were also active in supporting their policies with research and outside advice: to mention but a few, the Environmental Protection Agency and the National Institutes of Health.

In their methodology, the TA (or TA-like) activities from government agencies resembled the usual ways they generate and use research results and advice. The OTA methodology, however, has several interesting characteristics. Topics for TA studies are requested by Congress by means of an official "letter of request". This is mostly the final stage of a careful selection process of the OTA ("issue management") and talks between OTA staff and representatives of relevant Congress Committees. In executing a TA, the OTA keeps full responsibility for the whole process and the way the results are presented to Congress. An important part of the actual research is within this framework carried out by external research institutes in the form of mostly rather small research contracts. All relevant parties, the "stakeholders", play an important role in the various stages of the TA: from the generation of the problem definition, the evaluation of the intermediate results to the review of the draft reports. Also the way OTA presents its results is worth mentioning: not in the usual form of the best single recommendations with the scientific information backing these recommendations. The interaction with the broad variety of stakeholders is used to include in the final report to Congress only the relevant "issues" and to present for each issue the most relevant "options". For each option, not only the 'hard facts' are given about the pros and cons, but also differences of opinion amongst stakeholders are included. During the last twelve years, about a hundred of these reports were presented to Congress. The scientific background documents are also publicly available to the interested reader. Furthermore, OTA is active in the dissemination of the reports to a broader audience by publishing executive summaries and actively seeking publicity.

The OTA methodology has proved to produce reports of sound scientific quality that are useful for decision-making by a multipartisan political body like Congress. It has set an example that has been studied and followed by researchers and policy-makers from all over the world.

The development of TA in Europe followed a different path. The example of the American OTA was seriously debated in a number of countries (e.g. Germany, France and the Netherlands), but only in Sweden was some kind of independent organization with a mission resembling TA actually founded: the Secretariat (now Institute) for Futures Studies.

In the same period, a lot of activities were started by ministries, advisory councils and research institutes linked to government in the already-mentioned 'problem areas': (nuclear) energy, the environment, health, nutrition, employment, working conditions, etc. With the exception of the TA department in the German Ministry for Science and Technology (BMFT), the more comprehensive forms of TA mostly remained an academic interest. The first sign of change in Europe was marked by the initiation by the European Communities of the FAST programme (Forecasting and Assessment in Science and Technology). Also the OECD report *Technology on trial* (1979) was important in this respect. In the early 1980s, debate started again in a number of European countries about the necessity of TA organizations, especially for parliaments, now with more success. TA organizations related to parliaments now exist in France, the United Kingdom, Germany, Denmark and the Netherlands. The European Parliament has set up its STOA programme (Science and Technology Options Assessment). These developments have been documented very well on the occasion of the two European Congresses on TA. Country reports (Ministry of Education and Science 1987) and proceedings (de Hoo et al. 1987) were the outcome of the Amsterdam congress in 1987. For the congress in Milan in 1990, documents were prepared with a description and analysis of the development of TA in Europe (Smits/Weijers 1990) and the state of the art of TA organizations working for parliaments (EPTA 1990).

A lot has been thought about these differences in development of TA in Europe and the United States. Originally, this might have to do with the general lack of interest in science and technology policy in general in most European countries. In the US there, were numerous debates on costly programmes for defence, space, nuclear energy, supersonic transport, etc. But also after the first and certainly after the second oil crisis, when the climate became much more in favour of technology, it had no immediate influence on TA. In those days, Europe compared itself mostly with Japan and looked at technology and Japanese technology policy as a way to get out of the crisis. TA was seen as hampering this policy (TA as "Technology Arrestment") or not taken into consideration at all. A third reason might have to do with the differences between the political systems in Europe and the USA. In the United States, a clear distinction exists between the "legislative" and the "executive" branches of government. In all European countries, much closer ties exist between governments and government parties in Parliament. An independent organization working for Parliament might interfere with this delicate relation. Indeed, in some instances it was seen as strengthening the parties in opposition, whereas in other instances the opposition parties feared strengthening of an already-too-powerful government. In several recent publications about the future of Technology Assessment, a closer co-operation between TA and technology policy is seen as a crucial prerequisite for further improvement for both of them (see e.g. Smits/Leyten 1991).

3 Science and Technology Policy in the Netherlands and their Broader Concerns

The funding of universities and the founding of specific research institutes are classical instruments for science and technology policy. But in today's context, these will not usually be regarded as science or technology policy because the policy is not aimed at influencing the content of the research. Sometimes the term "conditions providing policy" was used to set it apart from the more explicit forms of science and technology policy. In the case of universities and fundamental research institutes, it was regarded to be against the principle of academic freedom to interfere with the authority of professors and the mechanisms of peer review. Also, other longer-existing forms of science policy, such as the Dutch foundation for Pure Scientific Research (ZWO), operated conforming to these principles. In the case of applied technological research institutes, their mission was given at the time of their foundation or their programme is for the greater part determined by the specific government department responsible for the institute.

Science and technology policy in their more explicit meaning are deliberately set up to give government an instrument to direct science and technology to the goals of society. In this form, science and technology policy in the Netherlands dates back from the 1960s: a good milestone is the foundation of the Advisory Council for Science Policy (RAWB) in 1966. In this period, no distinction was made between science and technology policy: both were seen as directed to government-sponsored pure and applied research. The connections with e.g. industry policy or public health policy were indirect. In the early 1970s, this trend continued. In 1973, the Dutch cabinet of ministers welcomed a minister for Science Policy – be it a co-ordinating minister "without portfolio". (This meant that he had no full-fledged ministry at his disposal but only a small staff; for management services, he had to rely on the ministry of Education and Science.) After an extensive round of talks with almost every party with an interest in science policy, he presented a white paper that formulated the goals and set the stage for a number of institutional innovations. Not only were the quality of the research and the scientific infrastructure seen as important goals, but also "adaptation to the needs of society" and "democratization of research". The most important instrument for government policy-making and involvement of Parliament was the annual Science Budget. In this Science Budget, a description was given of the content, organization and financing of the research programmes of all the ministries involved; special committees were set up to discuss the science budget with ministries, within the Cabinet and in Parliament. A number of additional instruments was developed to complement this policy-making at the level of policy preparation and policy implementation. The following are but a few of these instruments:

- sector councils, councils consisting of representatives from science, government and society at large with the task to formulate a research policy for specific sectors;
- stimulator programmes, multidisciplinary and/or application or policy oriented research programmes;
- fact-finding committees, to evaluate the state of affairs, analyze trends and formulate recommendations in specific areas; and
- public-awareness activities.

In the 1970s, the Netherlands witnessed intensive debates on a number of issues, nuclear energy, the environment, micro-electronics and recombinant DNA. In universities and other research institutes, special programmes or departments were founded to study the relations of science, technology and society, e.g. "environmental studies", "science and society courses" and "science shops (to answer practical questions of the general public). In these debates, a large amount of social concern about responsible technology became visible. In general, the climate was rather critical of technology. In several cases, government installed special broadly composed committees to advise the government about these issues.

But at the same time, a swing to the other direction became visible. As in most other western countries, the second oil crisis demonstrated that the economic recession had a structural character and that, for that reason, structural measures became necessary. Japan, seemingly not suffering at all from the oil crises, was seen as the great example. Especially the Japanese way of production and the close collaboration of industry with government and the research and development community, co-ordinated by the Ministry of Trade and Industry (MITI), were seen as remedies for our economic problems. This point was also strongly put forward by the Scientific Council for Government Policy (WRR), which was founded a few years earlier not as a 'policy for science', but as a 'science for policy' organization for the government (WRR 1980). In the Netherlands, the ministers of Science Policy, Economic Affairs and Education and Science joined efforts and produced in 1979 a white paper on Innovation Policy (Minister for Science Policy 1979). It systematically analyzed the feasibility of a number of policy instruments. The most successful of them would become the "innovation oriented research programmes" and a number of subsidy schemes for industry. It also stressed the necessity of complementary social innovations.

In the early 1980s, the importance of technology policy grew rapidly. During the Cabinet formation of 1982, the decision was taken to transfer the responsibilities for technology policy to the Ministry for Economic Affairs and to make the Minister of Education and Science also responsible for science policy. This separate technology policy was shaped by a number of instruments. Initially, these focused on the generation of new basic technologies; later on, the emphasis shifted to the transfer of knowledge and to the application of basic and other technologies. Also the advisory and implementation system was adjusted to the new policy. (See e.g. the report of the Advisory Committee on Technology Policy Reinforcement, the Dekker Committee 1987.) The perspective on the integration of technology in society gradually changed from promoting "acceptance" to promoting the societal "embedding" of technology.

Meanwhile, the policies towards the universities also changed. A number of instruments was developed to evaluate the quality of university research. On the basis of this information, large-scale transformations, i.e. mostly fusions and reductions, were carried out. Important tasks new in this respect were given to the old and respectable Royal Dutch Academy of Arts and Sciences (KNAW). The Foundation for Pure Scientific Research was transformed into an organization for all university research, pure and applied, for every discipline and with a number of special programmes for interdisciplinary research areas of social significance (NWO). Finally, the relationship of government with mostly technological research institutes was adapted to the 'new times'. The strong influence on their research programmes was diminished, their 'missions' were formulated and sometimes renegotiated, and they became to a larger extent responsible for the execution of their programmes and the acquisition of contracts. The 25-year-old Advisory Council for Science Policy (RAWB) was restructured into the Advisory Council for Science and Technology (AWT), serving both ministers. This overall change of Dutch science policy was concluded with a white paper with the telling name "Towards a New Research Landscape. Science Policy for the Nineties" (Ministry of Education and Science 1989). In the final chapter of this white paper, the importance was stressed of the proper integration of science and technology in society. A number of measures was proposed to give substance to the responsibility of the scientific community and to promote technology assessment, public information and awareness, and the debate on ethical questions relating science and technology.

4 Technology Assessment in the Netherlands

The debates in the Netherlands on nuclear energy and on environmental problems, which from a government point of view focused on the Minister for Economic Affairs and Environmental Affairs, respectively, were important for the general climate on science and technology. However, it were the debates on recombinant DNA and micro-electronics which really broke the ground for Technology Assessment in the Netherlands. These debates were held on the platforms that were created for political discussions on science policy in the Netherlands. They offered a good opportunity for also discussing the wider implications for society of science and technology in general.

Recombinant DNA was disputed very intensively. As a matter of fact, when the worldwide moratorium on recombinant DNA had already ceased and there was almost general agreement on guidelines for safe experiments, the moratorium in the Netherlands continued for the more risky categories of experiments. There really was an impasse. While the Dutch government wanted to start a biotechnology programme including recombinant DNA applications, the Dutch Parliament had decided that this moratorium could only be lifted after a discussion in Parliament. This discussion was prepared by a committee with a very broad constituency. Of course, it counted members that were experts in recombinant DNA research; but only a few. But in addition to them, the committee had numerous other members: scientists from all kind of disciplines that could give information on the risks and application areas, policy-makers and representatives from the churches, trade unions, employers organizations and a representative of a union of concerned scientists. The committee also organized a hearing for all kinds of local organizations in the neighbourhood of recombinant DNA facilities. The committee was assisted by a professional staff and invited experts and policy advisers from abroad. The committee worked for almost two years and then came with a very authorative report (Broad DNA Committee 1984). The fact that the report contained minority views on several topics added to its strength. Both in the preparatory stage of the committee and during the debate on the final report Parliament not only discussed recombinant DNA, but also the wider issues of social and ethical aspects of new technologies. Propositions were debated for topics like an "enabling legislation" for science, "research impact statements" and an "ombudsman for science".

A similar situation occurred with the issue of "microelectronics", as the chip technology was called these days. Government wanted to stimulate the research and introduction of the new information technologies, but there was strong opposition in politics and society. Especially the trade unions feared big job losses. Also in this case, a committee was formed with a very broad composition and task. Their analysis (Rathenau Committee 1980) proved very convincing. The committee did not deny that the large-scale introduction of the new information technologies might have serious consequences, but it showed that not joining these new developments would certainly have very bad consequences, whereas the introduction could be accomplished in such a way that the unwanted effects could be kept to an acceptable level. Therefore, they recommended not only a number of measures to stimulate the introduction of chip technologies, but also stressed the importance of technology assessment. The committee stressed the point that the institutionalization of technology assessment would also benefit other new technologies.

As more often is the case in real life, these broadly applauded recommendations were in themselves not enough to spark off a new development. An unexpected event triggered finally the foundation of TA in the Netherlands. During the Cabinet formation of 1982, it was decided to split technology policy from science policy and to transfer it to the Ministry of Economic Affairs. Parliament agreed after the provision was made that the Minister of Education and Science, remaining responsible for science policy, would set up an organization for Technology Assessment. In 1984, the minister presented his plans to Parliament in a white paper, *Integration of Science and Technology in Society* (Ministry of Education and Science 1984). The overall philosophy was that, for a satisfactory integration of science and technology in society, three societal processes, very well geared to each other, were necessary: decision-making, information-gathering and enabling the public to form its opinion. To this end, the recently-adjusted mechanism for decision-making was seen as sufficient; but organizations were proposed for the other two processes. To raise the awareness of the public, an independent foundation for public information and awareness was proposed. This proposal was completely supported by Parliament; and in 1986, the Foundation for Public Information on Science and Technology (called PWT) was formed. For the process of information-gathering, a small semi-independent unit for Technology Assessment close to the Ministry of Education and Science was proposed. Although Parliament very much applauded the idea of a unit for TA, they thought the proposal as not independent enough: that is, independent from government. Parliament discussed several possibilities for a more independent construction, among others having a Dutch analogue of the American OTA. They finally agreed on a proposal in which they had an important say on the programme, but where the independence from government was guaranteed by anchoring the unit in the scientific world. So in 1986 NOTA was founded, the Netherlands Organization for Technology Assessment. At the start NOTA existed of a steering committee of nine persons and a secretariat. The steering committee members have various backgrounds, the chairman and half of the members appointed by the minister after nomination by the Royal Dutch Academy of Arts and Sciences, the other half after nomination by the Scientific Council for Government Policy. The Royal Academy is responsible for the secretariat (9 permanent staff, several temporary) and the organization budget (3 million guilders a year). Recently, the secretariat was changed into a normal institute, the steering committee becoming a governing Board.

In the first five years of its existence, NOTA has been active on a number of issues: telecommunication and information technology in the public sector, environmental, ethical and patent aspects of plant and animal biotechnology, pollution prevention and a number of smaller studies on other technologies; defence, medical, energy, materials. Furthermore, NOTA has been very active in promoting the methodology for TA, new ways to disseminate the results of TA and international co-operation in TA. NOTA is responsible for Dutch contribution in the FAST programme of the EC and organized in 1987 with FAST the first European TA congress in Amsterdam. In 1990, NOTA summarized its experiences of the first three years in a report for Parliament, "Technology Assessment: To Adjust or to Channel?" (NOTA 1990). In the report, NOTA stressed the importance of looking at TA as a process directed towards supporting decision-making by Parliament. This means that scientific research is not the only input into this process, but that discussions with all kinds of stakeholders are necessary. The output of this process consists of more than scientific reports, but also includes policy summaries, conferences, video movies, etc. Even the formation of networks of parties involved and the frequent reformulation of the problem definition and the setting of the agenda for public debate should be seen as important outcome of this process. It was shown what kind of demand this poses to the scientific infrastructure for TA. The report presented "three goals for TA in the nineties": improved international co-operation in TA, more explicit attention for the cultural and ethical presuppositions in technology development and assessment, and a stronger effort to narrow the gap between innovation and societal evaluation of new technology. The approach NOTA developed to this last goal is called "Constructive Technology Assessment" (CTA). The remaining part of this chapter is devoted to describe the idea of CTA, the first experiences, its possible future and its significance for technology policy.

5 The Idea of "Constructive Technology Assessment"

Constructive Technology Assessment: what is meant by this new word combination and especially the word "constructive"? (With definitions of "Technology Assessment" a bookshelf can be filled; but to develop my argument, I take this definition for granted.) "Constructive" means, directed to the construction, to the design and introduction of technology. The intention was to find a descriptive term, but we stuck to the term although it also has positive connotations.

How did this idea emerge? There are several coinciding reasons, but to explain this it is necessary to describe the 'context of invention'. The idea was first launched in the government white paper, *Integration of Science and Technology in Society*. To set up TA in the Netherlands according to the international state of the art, a number of country studies and methodological analyzes was carried out. We stumbled across an impressive literature discussing the possibilities for "early warning" and "anticipation". But we also discovered that most of this literature was implicitly written from the viewpoint of a society that is concerned and a government that wants to regulate. In other words, the "early listeners" that are implied in "early warning" are groups in society, government agencies and Parliaments. Of course, there were also experiences with references in the literature of attempts to develop clean technologies, to improve working conditions, safer technologies or "user-friendly" information systems; but we were struck by the virtual absence of connections between TA and the design of new or improved technologies. We were surprised by the simplicity of the idea that emerged: by now, after more than twenty years of experience with TA in the 'classical' sense, it should be possible to use this experience to take environmental and social aspects into account already during the design of new technologies. The Dutch Parliament reacted enthusiastically to this new idea, perhaps just because of this uncomplicated simplicity. They encouraged us to develop this idea further because they thought it was indeed very important, in the words of one member of Parliament, "to narrow the gap between innovation and evaluation of new technologies".

Once the idea of CTA had occurred – it helped to have a mobilizing name for it – all kinds of information got a new perspective. CTA could be a framework to analyze all kind of practical experiences, compare them and draw general conclusions, that could be useful even in other technology areas. There was a rapidly growing body of literature in the economics, history and sociology of (science and) technology, that could be analyzed for mechanisms to influence the development of technology from a societal point of view.

But right from the beginning, our enthusiasm was moderated because we realized that such an approach would also meet tremendous limitations. One category of limitations had to do with the enormous complexity and concomitant uncertainty – to the point of ignorance – about the development of technology. Already in the white paper *Integration of Science and Technology in Society*, we made reference to the analysis of technology development of David Collingridge in his book *The Social Control of Technology* (1980). Key concepts in his analysis are the "control dilemma" and "entrenchment". The control dilemma stems from the fact, that during the early stages of development, steering the development of technology is possible; but because of the lack of reliable predictions, you don't know in what direction to steer – whereas in later stages, it becomes clear what the real effects of a technology are, but then it has become very difficult to change technology. The culprit for this is "entrenchment": all the technical, financial, institutional and cultural investments and inflexibilities that make it very difficult to change direction without huge costs in some way or another. The other category of limitations had to do with the difficulties to influence science and technology in its early stages: the development of science is difficult to influence altogether because of the unpredictability of its usefulness, because it is largely an international affair and because of academic freedom. When a new factory is built or technological products are sold on the market and there are laws prohibiting certain specified hazardous or unwanted effects, then there are opportunities to regulate. But when the technology is still in its development stage, only mechanisms working on a voluntary basis are effective; and secrecy kept for competitive reasons is a big obstacle.

Encouraged by moral support from Parliament, NOTA set up a working party together with a number of experts from various technological backgrounds to analyze the possibilities and limitations of CTA along these lines. The original idea proved to be very productive. The current boom in the economics, history and sociology of technology proved to be a gold mine for people looking for 'levers' to exert influence on technology. They definitively did away with the classical linear "invention – innovation – diffusion" scheme. Technological innovation is not a one-off affair, not a straightforward process in which limited numbers of actors and criteria are involved. Instead, they showed at how many points in time during the complete lifecycle of a technology decisions are taken that guide its further path, and how many different criteria and actors can take part in that process. This emerging 'language' to analyze technology development from a societal point of view gave, as we hoped for, an added value to the analysis of several cases. Four of them were 'positive' cases, analyzing the experiences and possibilities in the areas of factory automation, design of production systems, recombinant DNA and the manufacturing of detergents. Two other cases from the energy field demonstrated that a CTA-like approach could have been possible and could have had interesting results. But, not unexpectedly, the severe limitations were also amply illustrated. The working party, however, managed to steer away from an all-too-pessimistic view; and in their report to NOTA (Daey Ouwens et al. 1987), they presented a realistic course for action:

- to make a real start with CTA experiments, using the idea of a socio-technical map to connect the network of researchers and innovators to other parties in society and improving the mechanisms such as 'discussion platforms';
- to devote 1% of technology programmes to CTA; and
- to introduce this approach in engineering courses and to involve professional unions and employers organizations.

However, the working party also emphasized the experimental character of CTA and the fact that CTA is aiming at stimulating a social learning process. For this reason, they also recommended

- stimulating further research on the development of technology and especially the combined effort of the various disciplines. For this, the term "technology dynamics" was coined;
- analyzing more cases where attempts had been made to design new technologies or adjust existing technologies to environmental and social criteria.

The report was presented to Parliament and again it reacted enthusiastically. It was discussed in the parliamentary debate in 1988 on the new set up of technology policy. Whereas NOTA had decided not yet to support the "1% recommendation" because it thought the CTA methodology had to prove itself somewhat more, Parliament asked the Minister of Economic Affairs to take this recommendation into serious consideration. Also the Science Policy committee, the 'home base' of NOTA, very much liked the report and asked NOTA to incorporate the CTA approach into its projects wherever possible.

6 CTA: Experiences of NOTA

Since 1988 NOTA, together with the researchers participating in the NOTA programme, has made significant progress in all of the directions mentioned above.

6.1 CTA Case Studies

Because NOTA gives priority to TA studies on concrete issues, it was only possible to start a programme of relatively small studies and workshops. The first step was the organization of a workshop for researchers from various disciplines to discuss a paper about the possible topics for technology dynamics, especially from the point of view of social and environmental aspects. After the workshop, a call for tenders was announced and small projects were granted on the following topics:

- Economic Incentives for Environmental Policy (Klink et al. 1991)
- The Manufacturing Automation Protocol (Dankbaar/van Tulder 1989)
- Expectations regarding New Materials (Vergragt et al. 1989)
- The Role of Information Technology Consultancy (Moulaert et al. 1990).

Furthermore, a grant was given for the organization of a workshop:

- Small Countries and Technological Development (van Tulder 1989).

In addition to this, NOTA sponsored studies on the practical opportunities for CTA in industry (Poelman/Kauer 1990) and the Dutch experiences with the development of technical aids for persons with disabilities (Swarte 1992).

6.2 Technology Dynamics

The suggestion of the CTA working party to continue to bring together researchers in the various disciplines studying technology development was taken seriously. NOTA sponsored a study on the relevance of history for TA (Kranakis 1988a and 1988b) and co-sponsored a research and awareness project called 'The Technological Culture' (Schwarz/Jansma 1989). As an offshoot of the same workshop that started off the programme of CTA case studies, a grant was given for a workshop on technology dynamics. The position paper was written by Schot (1989).

Especially in the workshop on technology dynamics, a very thorough attempt was made to bring historians, sociologists, economists and a few philosophers and policy-makers together to discuss the possibilities for joining efforts in making suggestions for practical use of CTA. A number of the Dutch researchers were attracted by the subject and continued studying it. In 1991, they were granted a subsidy to organize an international workshop on the same topic at the University of Twente. On the basis of an earlier article of Rip/van den Belt (1988) and an adapted version of the 1989 paper of Schot (1991), participants were invited. They prepared 5 papers of a theoretical nature and 13 papers about specific topics that could be regarded as CTA or on the basis of which possibilities for CTA could be discussed.

6.3 Constructive TA Projects by NOTA

Up to now, still no manual has been written "how to do" Constructive TA; but at the time Parliament asked NOTA "to put more CTA in its programme", the first CTA report of 1987 was the only basis on which to proceed. As a result, the first series of CTA projects carried out by NOTA departed from a rather intuitive approach to CTA:

- making a 'socio-technical' map of the various 'actors': the technology developers, government departments and groups and organizations in society;
- making a state of the art and a technological forecast;
- organizing a debate between the various actors, beginning with informing the various participants separately and at the end bringing together a representative cross-section of the socio-technical map in a strategic conference;
- writing a synthesis report on the basis of the studies and the strategic conference and disseminating the results.

Now that a number of CTA projects has recently been finished, the time is ripe for an evaluation of the outcomes, to confront these with the recent insights from technology dynamics and case studies of CTA projects carried out elsewhere, and to draw conclusions about practical guidelines for doing CTA.

NOTA has gained CTA experience with the following topics:

Telecommunication: The development under scrutiny was ISDN (Integrated Services Digital Network), a new type of transmission technology with a capacity intermediate between normal telephone lines and such high-capacity modes of transmission as local or wide area networks. This offers new opportunities for services, but at high-investment costs. Furthermore, the international dimension is very important in this technology (Slaa and Rijkers 1988). The important questions for society are, who is benefiting from ISDN, who is paying for it, and are there risks involved (privacy, vulnerability). A number of constructive discussions took place during the strategic conference between representatives of various organizations which until then had never talked with each other or only in a polarized way via the media. The results were presented to Parliament.

"Smart Housing" ("Domotics" or Home Automation): This project was also prepared by a study (van Diepen/Fokkema 1990) and by a survey on who was, and more importantly, who would become involved in this still futuristic technology. A novelty of this project was that the parties involved were carefully clustered in 12 categories of actors and that each participant (two from each cluster plus independent experts) was asked before the conference about his/her own role and the preferred role of each of the other participants. This proved excellent material to spark off a discussion, especially the reciprocal 'after-you' situations that were discovered this way. A summary report was presented to Parliament (NOTA 1991).

Biotechnology for the Third World: This project followed the grassroots approach in every sense of the word. Starting point were contacts between a university group for "Biology and Society" and a few organizations of concerned farmers. These were concerned not only about their own future, but also on the future of farmers in the third world if biotechnology would strengthen the position of the rich countries as earlier technologies had already done. This network of concerned organizations grew and a big conference (400 participants from all over the country) was organized to discuss their concerns with representatives from the Ministries of Agriculture and of Development Aid and with agricultural researchers (Sarink et al. 1988). The conference resulted in the preparation of a new biotechnological programme as part of the Dutch development programme.

Introduction of Genetically-Modified Organisms into the Environment: The possible environmental risks are of great concern, as much to the research community and the biotechnological industry as to environmental organizations and the ministries responsible for regulating the safety of this recent application of recombinant DNA. Crucial for the further development of this promising technology is knowledge how to prevent risks and how to have a reliable procedure for licensing the use of new organisms. In a well-prepared conference, it appeared to be very well possible to have a good discussion about a risk research programme between experts and concerned non-experts with an effective follow-up.

Prevention of Waste and Emissions (the "PRISMA" Project): This was a big project and different from other CTA projects in an number of respects: the core of the project consisted of demonstration projects in ten different companies, there was an extensive involvement of two ministries (for the Environment and for Economic Affairs), the overall project included several background studies on economic incentives and environmental management of companies, numerous dissemination activities took place during the project and international activities were an integral part of the project. Also in the view of the participating companies, this was a successful project: in every demonstration project, practical and costeffective measures for pollution prevention could be found and partly even be implemented during the project. The overall project is described in a report that also contains the full references to all the other reports that resulted from this project (de Hoo et al. 1991).

Beside these projects, a number of smaller studies was carried out to explore the possibilities for CTA in other areas (artificial intelligence, information technology for government agencies, new materials, energy, technology for persons with diabilities, home-care technology and neonatal care).

6.4 Dissemination Activities

Various channels were used to discuss and promote the CTA approach and the preliminary results of the CTA projects, case studies and insights from technology dynamics.

In the three above categories, many more researchers were involved than those who actually carried out the research. Several universities started CTA projects themselves and included the CTA approach in their teaching programmes. NOTA supported several of these activities.

The ideas and results were presented to audiences, varying in size from small workshops or minisymposia annex press presentations to full-scale conferences with several hundred participants. Special attention was given to presenting and discussing CTA with members of Parliament, government officials and representatives of companies, employers organizations and professional unions.

Both NOTA and the researchers and organizations that collaborated in the CTA activities published in scientific magazines, professional magazines, general magazines and newspapers. Interviews or documents were given to journalists to help them write articles.

In two cases, video movies were made by professional film-makers, sponsored and supported with the writing of the scenario by NOTA and other organizations: the ISDN project (telecommunication) and the PRISMA project (pollution prevention). In the PRISMA project part of the material was published in the form of computer programmes.

Finally, both in the course of almost every CTA project, at international conferences and in the international relations of NOTA as an organization due attention was given to the CTA approach. As a result, several references were made to this approach in documents resulting from international co-operation (e.g. Ministry of Education and Science 1987; OECD 1988; Commission of the European Communities 1991).

7 The Future of Constructive Technology Assessment

For NOTA, CTA is already a success. It certainly helped to do more interesting and more useful TA projects than would have been done otherwise. The CTA approach both gave practical suggestions that could be immediately included in projects, but there also is a general feeling that it helped on a conceptual level. Working with networks and (re)defining the problem definition in an iterative process was right from the beginning the cornerstone of the NOTA methodology. But doing this 'from the outside' directed to Parliament, government agencies and groups in society is quite different from forming productive networks with the designers and developers of technology themselves. The insights gained from technology dynamics helped to find a way through this new territory. So NOTA will certainly proceed with CTA.

This statement also makes clear that NOTA always has had very realistic expectations about CTA: an improvement in TA practice worth the additional effort. We never looked at it as an panacea preventing all unexpected and unwanted effects of technology, as a cookbook guaranteeing 100% success. In our discussions with others, however, it often appeared that we had made the impression that we did and we had to convince them that we did not. Sometimes another organization jumped to the conclusion that CTA was the final solution and that one could do away with the oldfashioned, 'technology harassment' type of TA. In these cases, we had to disappoint them and had to rob them of an illusion. 'Management of expectations' was certainly an important part of disseminating CTA.

But of course, we still have longer-term ambitions and we see many opportunities for improvements. As a matter of fact, as I explained, NOTA is right now completing its first 'learning cycle' and is in a process of evaluating and improving its own methodology.

This methodology will be improved along the following lines:

- In CTA practice until now, a 'socio-technical map' and a technological state of the art and forecast were made; and these were combined to organize a dialogue in the network of developers, users and other parties involved. The purpose of this dialogue was to take as much societal aspects into account and to do this as early as practically feasible. Experience has been gained how to organize such a process in practice. This can be translated into improved methodology.

- Upon closer scrutiny of a large number of concrete cases of technology development, two things become clear. Firstly, in technology development, many more stages can be distinguished than invention, innovation and diffusion: innovation occurs in several stages, but especially the diffusion phase can have very many different trajectories. Secondly, the introduction of a completely new technology is an exception; most efforts are devoted to make more or less drastic changes in existing technologies. A lot of these later changes concern social and environmental aspects. The lesson for CTA is that there are many more moments to influence technology effectively than during its first introduction, and it is not necessary nor even wise to direct all available resources to this point.
- Technology dynamics and case studies have shown the importance of "entrenchment" – both in a technological and a socio-institutional sense – to understand the development of technologies. On the one hand, this explains the relative success of technologies; on the other hand, it explains the inflexibility of existing technologies and the difficulties that new technologies can meet when they are introduced into an area where they have to compete with existing technologies. The lesson that can be drawn from this is that it is not only difficult to forecast the intrinsic qualities of a technology, but it can also be worthwhile to plan the 'embedding' both of existing technologies and of new technologies.
- Technology dynamics and case studies have indicated that there is enough room for interesting improvements in the taking into account of societal aspects during the design and redesign of technologies, but that nevertheless – despite all efforts – new and unexpected effects can show up, both of a positive and a negative kind. It might seem that by definition it is impossible to take these effects into account before they are detected, but this is not the case. Technologies can be designed in such a way that unexpected effects can to a certain extent be detected earlier and/or better. This property that makes monitoring easier could be called "transparency" and can be corrected earlier and better; the usual term for this is "flexibility". These very general properties of a design may take totally different concrete forms. They can be incorporated not only in the technology themselves, but also in the technological or the socio-institutional surroundings of the technology.
- The CTA experience thus far has shown that each CTA project is a learning process on (at least) two levels. Around the developing technology a network is evolving that learns to take more societal effects into account earlier than it would have taken otherwise. But looking at the project as a CTA project also helps to use more generalized heuristics and to learn from experiences with similar technology in other places or even with other technologies. This 'learning to learn' is sometimes called "double-loop learning". These learning processes can be planned as part of a CTA project.

The experiences so far have also elucidated the position of the various forms (strategies) of CTA that Rip and van den Belt distinguish in their 1988 paper:

- supporting the development of desirable technologies;
- technology forcing (by regulation); and
- influencing ongoing transformations of technology.

Promoting the development of technologies that are desirable from a social or environmental point of view runs the risk of covering too wide an area, of 'spreading too thin'. Should (government) support extend to all technologies that are developed because of social or environmental reasons, to all technology that is developed by government or to all technology – and if so, what kind of instruments should be used in these various instances? This is a very worthwhile discussion, but it should not be mixed with the discussion about the prevention of unwanted side effects of technology, no matter what its origin is. To the contrary, it is very important to be aware that all these socially-desirable technologies, e.g. the so-called clean technologies, should benefit from a CTA approach to ensure that preventing one undesired effect does not cause another! So CTA can be very effective as an integral part of the development of socially or environmentally desired technologies, but it should not be equalled to the whole process of actually developing and introducing these technologies.

Regulating technology was, by definition, seen as "classical" TA. Therefore, technology-forcing is more like a hybrid of 'classical' TA and CTA than CTA in a strict sense. It will probably only be effective in those cases where good predictions can be made about the feasibility of improvements in existing technology. Otherwise, the arrangements will end up in 'strategic games' when it becomes clear that the original targets cannot be reached in time or cannot be reached at all. Technology-forcing is CTA in the sense that it is directed to the (re)design of technology, but it is not CTA because the voluntary basis for persuading companies to do CTA is abandoned.

Attempts to influence technology transformations that occur for any good reason with the purpose to create the best balance between social and environmental costs and benefits, therefore come closest to what was intended with the idea of CTA. In combination with the fact that redesign of existing technologies is the rule, not the exception, and that studies during the diffusion phase can be done both to evaluate the previous phase and to plan for the next, the conclusions about desirable technologies and technology-forcing lead to an important conclusion about the relation of 'classical TA' and CTA. They are not approaches that exclude each other or an approach that should be replaced by the other. CTA is the addition of a new (set of) instrument(s) to the 'toolbox' of TA, no more but also no less.

The further improvement of CTA along these lines will proceed along the same four paths as pointed out earlier: technology dynamics, case studies, CTA projects and dissemination activities. In each path, the lessons from the last years can, as was described above for CTA studies, be included and these paths can be connected more tightly to each other.

A crucial role in producing this synergy can be played by technology dynamics. The efforts of students of (science and) technology in the various disciplines have offered us a number of models and concepts to analyze technology development and the possibilities to influence technology from a societal point of view. What we need now is a sorting out of the various models and concepts and look for a common denominator. As a matter of speech, we are looking for a kind of 'esperanto' and a platform so that people from the various disciplines can more easily communicate with each other and with TA practitioners and policy-makers. To use another metaphor, we are not suggesting that a new church, the church of technology dynamics or the church of Constructive TA, be founded. We are asking for "oecumene" coming out of the various churches on the streets to discuss the problems – and opportunities – of technology for society.

The workshop at the University of Twente in September 1991 already provided an agenda for further research. As a result of research until now, it has become clear that two more topics have to be studied and other disciplines have to be called in for help in order to understand the development and diffusion of technology more fully, and possibly to influence technology development from a societal point of view.

The first topic has to do with the actual – also informal – decision-making during the implementation of technology. There is already a fair amount of literature (and experience) about the implementation within companies. This material and the relevant disciplines (management of change, innovation science, socio-technical methodology) have hardly been explored by technology dynamics to find clues for CTA. But in a number of other cases, the unit of analysis is not a single company but a network of companies, financial and consulting services, various types of government agencies and a wide variety of other organizations like trade unions, consumer organizations, environmental groups, etc. Often the media, public opinion and politics play a role. In modern societies, these networks are not directed by one actor in one place. There is great need for clarification of these processes. Perhaps some disciplines which, until now, have not thought about these processes themselves, can offer valuable insights.

The second topic is about norms, values and criteria. A lot of the analyses so far take for granted the norms and values on the basis of which decisions are taken during development and diffusion of technology. The recent attention for ethical aspects of technology proves that this is not justified. In several cases, the question has been asked whether the discussions about the goals and values should not have been carried on before or during the development. This is a crucial but very difficult problem, because it combines several other tricky problems. It questions the role of politics and democratic processes on other levels of society, the use of ethics, the relation between ethics and legislation and even the cultural presuppositions on which our whole society is built. Extremely difficult questions indeed, but ignoring them while concentrating on the practice of CTA would be 'penny wise and pound foolish'. Philosophers – not only ethicists – political scientists and other social scientists have to be asked to join the discussions.

Further analysis of case studies is also worthwhile. Good case studies offer insights that more theoretical analyses cannot provide. Firstly, theoretical analysis cannot do without case studies without running the risk of missing the 'tacit knowledge', the importance of context and even important social issues. Secondly, speaking the language of technology dynamics, case studies of completely different technology areas can be compared, and practitioners of CTA – but also of TA and specific areas of interest – can communicate with each other and learn from each other. Thirdly, in the practice of doing CTA, especially in the stage of convincing others that CTA can be useful to them, 'good stories' (of course well documented) are far more convincing than elaborate theoretical analyses.

8 The Significance of Constructive TA for Technology Policy

Technology Assessment and other types of policy studies to study the impact of technology, such as 'environmental impact assessments', have grown impressively the last twenty years in virtually all Western countries. Not only the volume of these studies has greatly increased, but also numerous methodological improvements were made, enhancing the quality and utility of these studies. Many valuable results have been obtained.

Nevertheless, if the question is asked whether all these improvements really have diminished the occurrence of unwanted and unexpected side effects, it is difficult to give a positive answer. We have only recently recognized the world-wide scale and interdependence of the environmental problems, and no one believes that all the necessary adjustments have already been made to prevent further deterioration. Information and telecommunication technologies are spreading in an increasing pace: there is still growing concern about the worldwide consequences on the amount and quality of work and on everyday life and culture. Also the concern for applications of biotechnology and other technologies in medicine, agriculture and industry is widespread.

A lot of thinking has been devoted to this problem of 'doing better and feeling worse'. In its first summary report to Parliament (NOTA 1990), NOTA paid much attention to this question. During the Second European TA conference in Milan, this was one of key issues (Smits/Weijers 1990).

Four reasons have been put forward to explain these somewhat disappointing outcome of the efforts in TA and comparable approaches.

1. The amount of TA and comparable studies has increased significantly but not sufficiently: the size of the infrastructure for TA is still subcritical in relation to the size of the budgets spent on technology development. This is, of course,

not a surprising conclusion from a professional community; but the argument is supported by a lot of policy-makers and groups in society. The remedy is obvious: more money for TA. A lot of proposals have been made and were in some countries put into practice to spend a certain percentage of money for technology development to TA or TA-like activities. Nevertheless, there is consensus that this is not the only and not even the most important reason. A lot of experts in the field agree that other improvements are necessary.

2. There is still a gap between innovation and evaluation of technology. These two activities are still largely carried out in separate communities. Even in governments and Parliaments, it are often not the same ministries or spokesmen who deal with these issues. The conclusion is that every possible effort has to be made to narrow this gap. The initiative for this has to come from both directions. From the part of the TA community approaches like CTA could bridge the gap, whereas authorities responsible for science and technology policy from their part could help to create the conditions and the climate to introduce these approaches in the R&D community and in industry. In the Netherlands, both trends are emerging; but they are clearly still in their infancy.

3. The normative and cultural basis for evaluating but also for developing technology has to be made visible and should be made a topic for public and political debate. Normally, the criteria for developing technology are taken for granted. Only when the technology is introduced does the discussion start about the question whether a specific technology is really the one society wants or needs. Sometimes the discussion only starts after the technology has spread to a certain extent and unwanted effects have already occurred. In this way, controversies can arise that have damaging effects on the whole of society.

4. As technology development is becoming more and more an international affair, TA has – in order to remain effective – also to be done more in an international framework. Science and technology development are international by nature and have been for a long time. But the effects of technology and the way to deal with them have also become an international affair. In Europe, authorities like the European Commission and the European Parliament can and should play an important part, because their responsibilities both for technology programmes and supranational regulation are growing.

These four reasons are not independent of each other: to the contrary. From the perspective of this chapter, the actions resulting from the second reason – the gap between innovation and evaluation – are the most urgent. But all the actions resulting from the other three conclusions have to be incorporated. So the CTA approach has to integrated into technology policy, but this approach should allow for opportunities to discuss the underlying goals and values and the necessary strengthening of the infrastructure needs a strong international component. The climate for such a line of action seems to develop in the right direction. The Dutch TA community is willing to co-operate with other countries.

A last important remark should be made about the kind of technology policy that can benefit from these improvements. I have used "technology policy" with lower-case letters. With this I mean that not only the explicit "Technology Policy", e.g. as carried out in the Netherlands by the Ministry of Economic Affairs, but also that every policy of every government agency, company, research institute or any other organization involved in developing or applying research could benefit from this approach in their handling of technology in a more socially and environmentally concerned way.

References

Broad DNA Committee (1984) *Recombinant DNA*, Den Haag: Staatsuitgeverij.

Collingridge, D. (1980) *The Social Control of Technology*, London: Pinter.

Commission of the European Communities, FAST (1991) The Social and Economic Implications of New Technologies. First Biennial Report. Society and Technology towards the Future, Brussels: MONITOR/FAST report DIO.

Daey Ouwens, C., P. van Hoogstraten, J. Jelsma, F. Prakke, A. Rip (1987) Constructief Technologisch Aspectenonderzoek. Een verkenning, NOTA study V4, Den Haag: NOTA.

Dankbaar, B., R. van Tulder (1989) The Construction of an Open Standard. Process and Implications of Specifying the Manufacturing Automation Protocol (MAP), NOTA Working Document W8, Den Haag: NOTA.

Dekker Committee (1987) *Wissel tussen kennis en markt*, Den Haag: Ministerie van Economische Zaken.

Diepen, P. van, J. Fokkema (1990) De Intelligente Woning, NOTA study V12, Den Haag: NOTA.

European Parliamentary TA Network (EPTA) (1990) *European Parliamentary Technology Assessment – EPTA* -, Brussels: CEC.

Hetman, F. (1973) *Society and the Assessment of Technology*, Paris: Organisation for Economic Co-operation and Development.

Hoo, S.C. de, R. van Berkel, H. Brezet, S. Cramer, F. Reijenga, H. Dielmann, J. Schott (1991) Kiezen voor preventie is winnen, NOTA study V16, Den Haag: NOTA.

Hoo, S.C. de, R.E.M.H. Smits, R. Petrella (eds.) (1987) *Technology Assessment: An Opportunity for Europe*, proceedings Vol. 1-2, Den Haag: Staatsuitgeverij.

Klink, J., Y. Krozer, A. Nentjes (1991) Technologische ontwikkeling en economische instrumenten in het milieubeleid, NOTA study V13, Den Haag: NOTA.

Kranakis, E. (1988a) The Relevance of History for Technology Assessment, NOTA Working Document W6, Den Haag: NOTA.

Kranakis, E. (1988b) Technology Assessment and the Study of History, in: *Science, Technology and Human Values* 3/4 (13): 290-307.

Minister for Science Policy (1979) *Innovatie. Het overheidsbeleid inzake technologische vernieuwing in de Nederlandse samenleving*. Den Haag: Staatsuitgeverij.

Minister of Education and Science (1984) *Integration of Science and Technology in Society*, Den Haag: Staatsuitgeverij.

Minister of Education and Science (1987) *Technology Assessment: An Opportunity for Europe*, Den Haag: Staatsuitgeverij.

Minister of Education and Science (1989) Zicht op een nieuw onderzoeklandschap. Wetenschapsbeleid voor de jaren negentig, Tweede Kamer (1989-1990, 21319, nr. 2), Den Haag: Staatsuitgeverij.

Moulaert, F., F. Martinelli, F. Djellal (1990) The Role of Information Technology Consultancy in the Transfer of Information to Production and Service Organizations, NOTA Working Document W10, Den Haag: NOTA.

NOTA (Netherlands Organisation for Technology Assessment) (1990) *Technology Assessment: To Adjust or to Channel?*, Den Haag: NOTA. (This publication contains a full bibliography of NOTA.)

NOTA (1991) Op de drempel van de intelligente woning, NOTA study R6, Den Haag: NOTA.

OECD (1979) *Technology on Trial*, Paris: OECD.

OECD (1988) *New Technologies in the 1990s: A Socioeconomic Perspective*, Paris: OECD.

Poelman, W.A., F.A. Kauer (1990) Ontwerpers in de industrie. Van locals tot cosmopolitans, NOTA Working Document W15, Den Haag: NOTA.

Rathenau Committee (1980) Rapport van de Adviesgroep Microelectronica, Den Haag: Staatsuitgeverij.

Rip, A., H. van den Belt (1988) Constructive Technology Assessment: Toward a Theory, paper, Enschede: University of Twente.

Sarink, H., J. Bunders, H. Mabelis (1988) Biotechnologie en Wereldvoedselproductie, NOTA study V6, Den Haag: NOTA.

Schot, J.W. (1989) Technologiedynamica: een verkenning van de potenties voor technology assessment, NOTA Working Document W11, The Hague: NOTA.

Schot, J.W. (1991) Technology Dynamics: An Inventory of Policy Implications for Constructive Technology Assessment, NOTA Working Document W35, Den Haag: NOTA.

Schwarz, M., R. Jansma (1989) *De Technologische Cultuur*, Amsterdam: De Balie.

Slaa, P., D. Rijkers (1988) ISDN as a Design Problem, NOTA study V5, Den Haag: NOTA.

Smits, R., J. Leyten (1991) *Technology Assessment. waakhond of speurhond?*, Zeist: Kerckebosch. (With English summary: Technology Assessment: Watchdog or Tracker? Towards a Comprehensive Technology Policy.)

Smits, R., T. Weijers (1990) *State of the Art of Technology Assessment in Europe*, Brussels: CEC.

Swarte, V.P.P. (1992) Ontwikkelingen in de markt van hulpmiddelen voor gehandicapten, NOTA Working Document W39, Den Haag: NOTA.

Tulder, R. van (1989) Small Industrial Countries and Economic and Technological Development, NOTA Working Document W9, Den Haag: NOTA.

Vergragt, Ph., K. Mulder, A. Rip, H. van Lente (1989) De matrijs van verwachtingen ingevuld voor de polymeren tenax en twaron, NOTA Working Document W12, Den Haag: NOTA.

WRR (Scientific Council for Government Policy) (1980) *Plaats en taak van de Nederlandse industrie*, Den Haag: Staatsuitgeverij.

Integrating Social and Environmental Costs into High-tech Industrial Development Planning: Experiences and Expectations in the United States

Gregory A. Daneke

1 Introduction

In the United States, the era of Reaganomics brought a double-whammy to states and localities. On the one hand, federal programs (particularly environmental management) were given back to the state governments to fund as well as administer. Meanwhile, industrial development initiatives (except for defense industries) were radically curtailed. Amid this abandonment and in the wake of increasing economic uncertainties of the 1980s, many regions – especially the arid regions of the Southwest – sought to develop their own alternatives to their declining single industries (e.g. copper, cattle, agriculture in Arizona and New Mexico, oil in Texas, autos in the upper midwest, and financial institutions in New England). In many states and localities, economic revitalization has come to mean high-tech industrial development. High-tech industries, especially those associated with advanced electronics, are hardly a panacea. They are neither environmentally sound, nor recession proof. Furthermore, they place extreme demands on public infrastructure and service systems. Mitigating these adverse impacts without foregoing economic benefits has required new approaches to development planning. Such approaches strive to establish more integrated socio-ecological understandings en route to institutional designs which facilitate meaningful trade-offs between diverse life-quality (including economic well-being) elements.

Drawing upon extensive interviews with government officials, public-interest groups, and high-tech industrialists in Arizona, California, Michigan, Massachusetts, and Texas, this study reviews the various social costs associated with high-tech industries – particularly advanced electronics. These observations are combined with insights from various disciplines to substantially recast the conventional wisdom of economic development. The elements of a new approach are outlined, one which relies upon adaptive-systems concepts and attempts to generate a more integrative picture of the social ecology of state and local industrial policy.

2 Regional Economics and the High-tech Solution?

In recent years, state and local government efforts in the area of economic development have shifted significantly from traditional 'smoke-stack chasing' to an emphasis on high-tech innovation and entrepreneurship. Given a general lack of understanding of the dynamics of the innovation process and the intense competition between regions, much of this microlevel industrial policy has been largely frenetic, ill-conceived, and ineffectual (note Daneke 1988; Feller 1988; Luger 1985; OTA 1984b). Furthermore, elements of these new strategies seem to be predicated upon the ill-founded, yet continuously popular, belief that economic development in itself will reduce many accelerating social ills such as youth and minority underemployment, deteriorating infrastructure and public services, disintegrating communities, crime, and so forth. More importantly, they tend to ignore that high-tech industries contribute to some of these traditional problems, and generate a number of negative social and environmental consequences of their own.

3 The Problematic Panacea

The inordinate amount of emphasis placed upon high-tech industrial development by state and local planners of late becomes increasingly curious when one begins to consider the difficulties inherent in becoming a high-tech 'hot bed'. Since the primary 'raison d'etat' of economic development is job generation, it is understandable that many state governments fell prey to near-terminal cases of 'Silicon Valley Fever'. That is, they wanted to duplicate the elements found in California's high-tech region located near Stanford University and the San Francisco Bay. In the recession years of the early 1980s, national and regional studies continued to illustrate how the 'Fortune 500' were losing jobs, and how various small high-tech start-up companies were the only sources of employment growth (e.g. Birch 1981). These studies, however, tended to overstate the importance of spin-off firms, and failed to deal with the highly rarified environments in which they are likely to flourish.

To begin with, the type of firm generation experienced in the Boston Route 128 and Santa Clara County (Silicon Valley), California regions may have been highly dependent upon a particular time frame within the development of a particular technology. For example, many small chip operations spun-off during a special period of material advance in which massive government (mostly defense and space programs) orders could not be filled by existing firms (Markusen/Bloch 1985; Saxenian 1985). Likewise, many 'reluctant entrepreneurs' started their own companies in order to explore their own research during an era in which no single firm could pursue all options and where government research support was readily available. Furthermore, the 'critical mass' of resources necessary to foster this chain reaction of auxiliary electronics industries may have taken a half-century to build.

Interestingly enough, other less direct factors are usually isolated when explaining the success of the various high-tech regions. These usually include such items as first-rate research universities and the existence of associated industrial parks. More central, perhaps, is the availability of a skilled labor force and the access to venture capital resources (see Glasmeier 1985). Given a focus on these indirect causal elements, state and local initiatives have generally been limited to the following:

- creating "incubators" (office and lab space and support systems) for research-oriented firms;
- developing networks of venture capitalists, innovators, and community leaders;
- providing public research funding and/or insuring small business access to existing R&D dollars;
- setting up seed and venture capital pools, using vehicles such as pension funds;
- establishing additional tax and/or other incentive systems; and
- setting aside research park or similar facilities adjacent to universities.

While the jury is still out on many of these efforts, preliminary assessments suggest that they may be far less successful than hoped for (see Daneke 1988; Feller 1988; Luger 1984). Consider that the oldest of the second generation centers, the Research Triangle in North Carolina (which began in the 1950s along with the war on cigarettes) has yet to generate much in the way of high-tech entrepreneurship (Luger 1984). Even Texas, which has also been the recipient of large-scale government-funded research consortia (MCC and SIMATECH) has a long way to go to replicate the leading regions (see Farley/Glickman 1986).

Admittedly, numerous regions have been relatively successful in attracting existing high-tech industries. This "chip chasing" phenomena, however, is not at all the same as being able to engender "home-grown" high-tech start-ups. While a few ancillary firms will arise in manufacturing regions, success in acquiring assembly and/or fabrication plant, is not highly correlated with spin-offs. The research operations and/or corporate headquarters of these high-tech companies tend to remain in existing centers. Manufacturing, with its low skill/low wage jobs, has gravitated to "right-to-work" (nonunion) regions, as well as overseas. Some engineering work force accompanies these essentially nonresearch operations, yet they are rarely a source of start-up pressure. Neither are branch operations a stable element of the economic base, since they are easy enough to shut down or relocate in response to highly competitive global markets.

It is also interesting to note that many high-tech industries are now experiencing the same kind of global competition which caused the decline of US smokestack industries. The balance of trade in microelectronics continues to augur in favor of Japan, and the burgeoning biotechnology industry is gradually shifting to Europe. High-tech in general – and microelectronics in particular – may no longer be the United States' 'competitive advantage'. While the USA continue to have a scientific edge, actually technological development is becoming increasingly globalized. Multinational joint ventures in R&D, as well as manufacturing and marketing, have made geographic boundaries irrelevant. Thus, the scientific intensiveness of a region's industrial base is no longer a guarantee of continued exclusive prosperity.

4 A Mixed Blessing

While economic uncertainty has not necessarily been ameliorated by successful chip chasing, the mere fact of diversification has benefitted several regions. However, this success may prove to be a partial victory, indeed. Contrary to popular belief, high-tech is not environmentally 'clean'. Recent studies of industries such as microprocessing describe a number of significant environmental hazards (Sherry 1985; Siegel/Markoff 1985). These include:

- groundwater contamination involving highly toxic trichloroethane (TCA) and trichloroethylene (TCE) in arid regions highly dependent upon groundwater for potable supply;
- overtaxing of municipal sewage and land-fill systems with toxic liquids and solids;
- worker exposure to hazardous materials (e.g. toxic solvents, acids and dopants, and arsenite, arsine, phosphine, and diborane), and radioactive substances; and
- air pollution via "reactive organic gases", and other toxic substances.

Less apparent, but perhaps equally perplexing, are the various social costs of high-tech. These costs are often highly intangible, yet eventually these adverse social undercurrents will manifest themselves in a variety of severe social service delivery and infrastructure demands (Miller 1988; Rogers/Larsen 1984; Larsen/Rogers 1988). High-tech research areas are especially prone to a number of conflicting employment patterns. These include:

- in general, high-tech industries provide mostly low-paying, low-skilled job opportunities with little upward mobility;
- the few high-skilled technical jobs are also very high stress, and have been associated with a number of family problems;
- land and housing costs around high-tech research centers are extremely high, adding to individual, organizational, and community stress; and
- infrastructure (roads, sewers, and energy systems) and other capital investment requirements are high, coupled with rising life-quality and social amenities costs (parks, recreation) for industries which are highly volatile in terms of "boom/bust" cycles.

Several regions have developed viable responses to some of the above social and environmental costs of technology. Their experiences provide vital lessons and ingredients for an integrated approach to economic development. As will be illustrated below, an integrated assessment cannot be divorced from the basic chores of institutional design, particularly when designs arise fairly spontaneously through the process of social problem solving.

5 The Curious Case of Groundwater Management

By far the most challenging of these social and environmental costs is "groundwater contamination". It is exacerbated by the lack of adequate institutional safeguards upon these vital, yet hidden, resources, and by the fact high-tech industries have increasingly located in regions like the Southwest, which are highly dependent upon groundwater. Groundwater constitutes over 50% of all the drinking water used in the United States and, in some arid regions (e.g. Tucson, Arizona), it is virtually the sole source of culinary supply. Moreover, for many US communities, it supplies nearly 80% of domestic and livestock needs, about 40% of irrigation needs, many commercial activities, and almost 25% of self-supplied industrial needs (OTA 1984a). In the past half-century, groundwater uses of all types have increased dramatically and these trends are likely to continue.

Despite the importance of groundwater, its management has been rather lax and generally ad hoc in nature. Serious efforts usually await a supply crisis due to "mining" of an aquifer at levels far in excess of its recharge rate. Quantity crises have only recently given way to quality crises, and the major threat to quality has turned out to be the high-tech firms, especially computer, aerospace and microelectronics companies.

A key case in point is the Santa Clara Basin or California's famed Silicon Valley. While active water management in the region began in the 1930s when agricultural interests banded together to combat salt water intrusion, by the mid-1970s, monitoring had become fairly haphazard. To the extent that crops weren't dying, the water was assumed to be of adequate quality. During this period, of course, agriculture and ranching were being replaced by the high-tech industries and suburban development which grew up near military facilities and Stanford University.

By the early 1980s, a new and more profound threat to the groundwater in the region was discovered. Underground storage tanks containing highly toxic trichloroethane (TCA) and trichloroethylene (TCE) were found to be leaking into the drinking water systems of the Los Paseos neighborhood near San José, California, from the Fairchild Semiconductor plant. Estimates suggested that 14,000 gallons had escaped over the previous decade (Olivieri et al. 1985).

In a later "class action" suit, residents of Los Paseos blamed Fairchild for the unusually high rates of cancer, birth defects, and miscarriages but direct causal linkages were difficult to substantiate. Since the original discovery, Fairchild spent over $15 million in attempts to clean up the aquifer and the Los Paseos factory was voluntarily closed very early on. However, much of the initial corporate cleanup proved ineffective and government response was practically nonexistent.

Not until a few years after discovery of the Fairchild leaks did the Bay Area Regional Water Quality Control Board order a comprehensive testing program. As it eventually became clear, Fairchild was not alone in its difficulties. Leaks, of some sort, were found at virtually all of Silicon Valley's large production sites by the mid-1980s. Fortunately, the natural phenomena of a 200-foot clay band prevented many of the leaked solvents and other toxins from immediately entering deeper wells. However, the total plume presented a ticking toxic time bomb which might eventually enter supplies as the aquifer was drawn down. The initial government response was to stop the leakage and short-term repair of the storage facilities.

Regional agencies delayed cleanup, thinking that the federal government would play the major role. After all, Santa Clara sites were among the 'worst' (most hazardous) on the US Environmental Protection Agency's (EPA) "superfund" list. Superfund was a huge scheme to pool "polluter payments" and focus resources on the 'worst' regions. However, partisan politics, at the national level, delayed targeting and actual cleanup for decades. Merely being high on the bureaucratic priority list was no guarantee of funding, as the program evolved into an instrument of political patronage.

Local governments had to eventually take matters back into their own hands and quickly encountered new political and bureaucratic difficulties at the regional level. They initially struggled with monitoring and enforcement. In order to get new storage requirements, a task force of local fire marshals allowed agreements with several companies to keep old tanks in place, as long as they were not currently leaking. The manpower for follow-up monitoring simply was unavailable. Public dissatisfaction with these early negotiations caused local labor and environmental groups to form their own 'toxins coalition' to police future agreements and push for more rigorous regulatory activities.

Basically it was this 'grass-roots' organization which finally began to gain control of the crisis. By pulling together community activists and executives from high-tech firms and focusing on immediate problems (e.g. leaking tanks), this civic group was able to build a basis for continued action. Moving from problem identification and amelioration to prevention, the coalition gathered momentum. Gradually they pushed for the adoption of new and vastly improved storage ordinances. Moreover, this citizen/industry coalition was able to encourage corporations to place nearly 100 million into site cleanups and storage upgrades. They also set up a budget to fund research into methods for clearing the aquifer itself. With the help of students in the Water Quality Engineering Program at Stanford, the coalition tested "air stripping" techniques and the use of biological agents. In sum, this ad hoc agency began to lead the way in both active management and long-term planning.

At first blush, lessons from the Santa Clara case tend to coincide with the "public choice" models developed by Elinor Ostrom (1989) in her own studies of groundwater management. While her aim is to explore theoretical issues related to the "prisoner's dilemma" character of "common pool resource" problems, the policy implications of her work are fairly clear. Stated simply, Ostrom believes that these spontaneous civic associations are a form of 'microconstitutional' policy-making worth emulating on a larger scale.

While this study basically agrees with the position, it is not immediately clear how these reinstatements of market motives will emerge without some type of nonmarket intervention. The paradox of public choice is that the market itself rarely recognizes noneconomic social responsibilities until after the failure is initially being readdressed by some type of government agency. The mammoth scope of many of these problems often dwarf individual action. Also, the issue of economic development pressures complicates decision-making considerably. However, as will be addressed below, strategies which ignore the potential energy of community self-organization are obviously myopic. For now, suffice it to say that building "sustainable" economies when resources and technologies collide is not a matter of either conventional economic or environmental planning; it requires understanding the complex political economy of resource management, including certain natural institutional processes.

6 Additional Environmental and Social Costs

Another point revolves around the fact that "special districts" and other forms of "single issue", quasi-governmental bodies (e.g. the water districts studied by Ostrom), rarely take a comprehensive look at ecosystem interactions. Controlling pollution of an aquifer often means that residuals are merely transferred to another medium. Groundwater contamination is only a small part of the toxins picture. Microprocessing manufacturers pose a serious burden to municipal facilities as well. In 1982, over 25% of the firms in Silicon Valley were cited for failure to pretreat their sewage. Being a haven for small start-up companies has unforeseen drawbacks. Numerous 'small generators' can legally avoid pretreatment, and the aggregate chemical buildup can easily overtax municipal systems. San José had to spend nearly US$ 200 million to upgrade its treatment plants. Meanwhile, solid wastes readily overwhelm existing transport and landfill systems. Highly toxic wastes, of course, demand specialized remote dump sites. In the year 1980 alone, California's electronic firms dumped nearly 65,000 tons of toxic materials into special hazardous waste landfills. By the 1990s, these toxic waste sites are in very short supply, and their management has become a national problem.

In addition to liquid and solid wastes, high-tech firms produce a variety of air pollutants. In the San Francisco Bay Area Air Quality Management District, high-tech firms contribute nearly 10 tons of "reactive organic gases" per day to the airshed. These include various "ozone precursors", such as hydrocarbon solvent vapors, etc. While few high-tech firms actually dispose of wastes through incineration, there are routine emissions through work space venting, spills, and leaks. Thus, once again the absence of large, belching smokestacks creates the illusion that high-tech firms do not contribute to air pollution.

Direct releases, of course, are only a small part of the air quality equation, and, in turn, air quality is merely a portion of the general life-quality picture. Traffic congestion, a major contributor to air pollution, is perhaps no worse for high-tech industries than any other. However, given their propensity to locate in relatively rural areas, they can quickly overwhelm existing transit networks. In fact, the 'catch-22' of high-tech development is that it requires an extremely rich natural environment to entice high-tech firms; and yet, they threaten these relatively fragile environments with their presence. Research personnel, in particular, demand high levels of environmental amenities, as well as cultural opportunities. The movement of a couple of high-tech research shops to Oregon was explained by the favorable natural setting (Simmons 1985). Maintaining these settings amid significant in-migration provides a formidable task for state and local governments.

Another major problem is land and housing speculation. The mere rumor of a research intensive enterprise coming to town can cause huge housing and office space cost escalations. Austin, Texas has experienced wildly fluctuating real estate prices in the wake of its rather modest inroads into high-tech development, and these unrealistic swings have caused associated economic dislocations. Places like Boston and Santa Clara County have the highest real estate prices in the country, and individuals relocating from these regions can easily outbid indigenous populations. Moreover, cheap lab space needed to spawn homegrown spin-offs may be unavailable unless it is subsidized by governments.

Beyond these wealthy outsiders with their high life quality demands, most of the jobs are low-wage, assembly line and clean room activities which promise very little upward mobility. This gulf, combined with the high stress levels – especially for technical personnel – create a number of "quality-of-work-life" problems. Technical staff, especially those who must attempt to match the process as well as product innovations of the Japanese, often work around the clock for days on end. Meanwhile, assembly line personnel are under the constant pressure of working in and around deadly chemicals and extremely stringent quality control standards.

Labor unions usually serve to address these types of working conditions; however, unions are virtually nonexistent in the high-tech field. High-tech firms routinely argue that maintenance of "flexible work rules" necessitates a nonunion shop; thus, firms must cope with numerous work life issues on their own or through already overtaxed community resources. In increasingly competitive markets, quality-of-work life tends to be deferred. If job stress and/or lack of job satisfaction are deferred indefinitely, they eventually surface in ways that place increased burden on social service systems. As Lenny Siegel and John Markoff (1985:114-115) summarize:

> At a time when many Americans are espousing a new reverence of human life, our regard for human labor is falling. For the average worker today, work is no longer a fulfilling enterprise, and for an increasing number of people, labor is an inadequate source of income. Our economy may flourish as technology downgrades human activity into a mere factor of production, like silicon, electricity, or money; but our society will certainly suffer.

In sum, high-tech firms are no bargain. While they offer a valuable supplement to ailing commercial inventories, especially in single industry communities, they bring just as many environmental difficulties as traditional manufacturing. Yet, the myth of cleanliness coupled with less apparent patterns of pollution and personnel problems tend to cause public promoters to overlook these difficulties until the costs mount to astronomical levels. Of course, many of these costs could be drastically reduced if mitigation measures were introduced at the onset and/or burden factored in alongside benefits early on in the economic development planning process.

7 Towards a Sustainable Perspective

As Robert Ayres (1989) argues, technology and environment can be viewed as an integrated system which he calls an industrial metabolism. Viewed in such a fashion, trade-offs between economic benefits and environmental costs, as well as the balance between production and consumption, become more apparent. It is Ayres' hope that such perspectives will eventually lead to "materials balance" and/or "sustainable processes" designed to reduce overall global environmental loadings. Historically speaking, many industrial processes were designed to maximize the use of virtually free natural resources (e.g. 100 gallons of water per gallon of beer in the Pacific Northwest region of the US). High-tech industries do not necessarily exhibit these extreme levels of "through-put", however, toxic build-ups suggest that air, land, and water requirements were not adequately accounted for. Engineering substitutions might abound, if the true ecological (including energy) flows were well known at the onset, and vastly improved environmental accounting should guide industrial and institutional design. At this stage, however, it is perhaps more realistic to explore how such perspectives can be used to extend the concept of development to include mitigation of social and environmental impacts. This means, essentially, "load reduction" en route to "sustainability".

As suggested above, mitigation is usually an afterthought, at best. To the extent that problems are identified and dealt with, it is generally after the fact, where true mitigation is either impossible or extremely expensive. In the midst of declining federal support, state and local governments are accepting increased responsibility for such activities as environmental protection. Thus, it is incumbent upon regional planners to not only develop institutional arrangements which replace and/or augment federal processes, but also to devise strategies which integrate mitigation measures into economic development policies. Elements of these strategies must effectively inspire increased responsibility on the part of private sector interests, as well as utilizing remaining federal systems where appropriate. These types of integrative strategies require both new analytical mechanisms and an overall reconceptualization of the economic development process from a socioecological perspective.

The theory and practice of economic development planning has barely kept pace with the shift to technology-based industries let alone been able to account for their adverse impacts. As suggested above, economic development has had a rather myopic focus on short-term job generation. While many economic development policy makers are aware of long-term elements, it is difficult to operationalize them in terms of programs, especially given the emphasis on measurable performance (see Feller 1988). Most approaches are completely inadequate to the task of designing programs which promote long-range evolutionary processes of sustainability. Even when economic development theories are more or less satisfactory for purposes of understanding long-term changes in a regional economic base, and so forth, they are still inadequate to explain the role of advanced technology and in mitigating adverse impacts. As Papunek (1983) points out, theories tend to center on single factors such as capital formation and saving rates, human resource potential, and, occasionally, sociopolitical factors. But integrative theories are few and far between. He describes these partial elements as "the earnest search for a mirage" (Papunek 1983:17). The more complete approaches are linked to concepts of "industrial location" and/or "spatial production relations" (see Storper 1985). However, as the "neoclassical hegemony in location theory" has broken down conceptual integration between geography, economics, and urban planning has also become strained as well (see Walker 1985). Economic development theories can barely comprehend the ongoing "urban restructuring" (Bergman 1986).

Meanwhile, conceptual efforts in the area of technological development are usually too narrowly drawn to be very useful. There are a couple of rare exceptions, however. For example, Smilor et al.'s (1988) characterization of Texas as a "technopolis" is instructive. Building on Japan's concept of high-tech city/states, they develop a 'wheel' of critical sectors and factors, including roles for federal, state, and local governments, support groups, universities, large corporations, and emerging companies (see Smilor et al. 1988:146). Their normative elements include:

- the achievement of scientific prominence;
- the development and maintenance of new technologies for emerging industries; and
- the attraction of major technology companies as well as the creation of home-grown companies (1988:147).

A similar but perhaps more useful perspective on critical dynamics is provided by Hart and Denison (1987). Drawing upon a wide spectrum of empirical and theoretical discussions they reconceptualize the highly interactive and iterative process of "creating new technology-based organizations" through the use of a "systems dynamics" approach (Hart/Denison 1987:512). However, while an extremely useful picture, it does not effectively distinguish short-term from long-term or, in some cases, even causes and effects. However, they provide a basic ecological model and identify clues to the primordial ooze of ingredients necessary for successful technology-based economic development such as "access to technical expertise, availability of support services, and quality of life" (Hart/Denison 1987:525). Yet for these types of systems dynamics models to be more than static descriptions, they need to be tied to concepts which unlock the processes of evolutionary change.

From a policy perspective, the most instrumental approach to understanding the evolution of human systems are "adaptive-learning" concepts and one of the most useful conceptualizations is that of Edgar Dunn (1971). Dunn, a resource economist attempts to expand the neoclassical as well as allometric growth models, while providing an open systems exchange model. His key insight into institutional dynamics is that systems must adapt and evolve through social learning, and that meaningful policy interventions require enhancing natural learning processes. He explains that the equilibrating optimizing concepts that dominate the methodological base of the growth model are, in effect, machine system concepts. If we are to move beyond the uncomfortable restrictions they impose we must come to understand how social systems behave as learning systems (Dunn 1971:21)

Adaptive-learning concepts not only facilitate a more fluid and interactive approach to policy design, they mesh fairly well with reemerging evolutionary visions of political economy of technological entrepreneurship (see Schumpeter 1961; Nelson/Winter 1982; Rosenberg 1982).

In tandem, "adaptive-learning" and "systems dynamics" or adaptive systems provide an active planning model capable of viewing economic development in broader socioecological terms. In other words, the long-term viability of the system might – and usually does – require "trading off" long-range environmental and social well-being against short-term economic gain. Viewing unemployment and so forth as elements of the overall regional life-quality and mitigation as a natural part of institutional/societal learning, opens upon new opportunities for building sustainability into the industrial policy.

8 Re-orienting Economic Institutions

A perspective of sustainability through adaptive-learning provides vital clues to the design of alternative institutional arrangements. In areas such as groundwater management, alternative institutions are especially important in light of declining federal roles. In their survey of decentralization patterns, Davis and Lester (1987) illustrate how state and local governments have tended to merely reduce their commitment to environmental concerns in the wake of the Reagan administration's "new federalism." With specific reference to high-tech generated pollution, viewing environmental costs as the isolated consequences of doing business makes them far less likely to be effectively mitigated. Moreover, waiting for "superfund" cleanups becomes like "waiting for Godot". Conversely, viewing potential pollution as an element of an overall economic development strategy might encourage the establishment of up-front mitigation measures and institutions for monitoring (learning) and control mechanisms.

Returning to the issue of groundwater management, a number of institutional opportunities become apparent. As suggested above, groundwater is a paradigm case of "market/institutional failure"; yet it is by no means apparent that traditional regulatory (e.g. command/control) strategies are warranted in every instance. EPA economist, Robert Raucher (1986) suggests that market simulations which compare the relative costs and benefits of prevention versus remedial responses do not necessarily always support prevention. Furthermore, since most states do not have the funds to support full-blown regulatory measures, a multi-objective, multi-institutional approach may be both more effective and less costly. In establishing groundwater management, a state must reconcile its competing objectives. As noted, a state's goal for industrial development may be opposed to its goal of preventing groundwater contamination, and in order to reconcile these competing goals a state needs to assess its present and future groundwater needs and the relative importance of competing goals. According to the authors of *Groundwater Strategies for State Action* (Gallagher et al. 1984), existing and proposed state groundwater protection policies can be designed along a continuum ranging from maintaining groundwater quality at its natural, pristine levels to setting levels of protection based on current and anticipated uses of aquifers or groundwater regions. Along this continuum, three basic types of policies can be considered: "nondegradation, limited degradation, and differential protection" (Gallagher et al. 1984:60).

The type of policy depends upon the level and type of management elements desired; these include (Gallagher et al. 1984:71):

- aquifer classification based on existing quality, use, or vulnerability to pollution;
- uniform application of standards to all useable groundwater resources;

- containment or source classification where regulatory priorities are based on pollutant toxicity or on discharge volume; and
- recharge zone protection (typically a local government management option).

Specific protection techniques might include the establishment of groundwater quality standards, groundwater use restrictions, land use controls, public education efforts, and source-oriented controls, such as permits and facility standards. It can also include approaches that will reduce the capital costs of compliance through low-interest loans, tax credits, or exemptions. Furthermore, governments can provide programs which encourage smaller firms to pool their resources in implementing a joint waste reduction strategy or constructing joint resource recovery facilities and/or subsidies to firms that actively work on new methods of reducing hazardous wastes. Other inducements can be provided which eliminate problem chemicals and/or change chemicals to other forms and reduce the amounts requiring disposal. Financial incentives may be an effective addition to regulation when encouraging the development of innovative control technologies. However, it must be determined the firm is under financial hardship and the community wishes to retain the economic benefits offered by the firm. In other words, the market mechanisms can be utilized to weed out unsustainable industries.

Another type of financial incentive that could be used by local government is the Industrial Development Bond. These can be applied to augment capital when a firm cannot afford to borrow at the market interest rates. The local government can strike an agreement by which local banks agree to lend the money at a reduced interest rate if the firm joins in the quest for sustainable processes. Along these lines, Susan Sherry's (1985:244) citizens' guidebook to *High Tech and Toxins* recommends the use of a variety of innovative nonregulatory mechanisms, such as community-supported technical assistance to aid firms (especially small and/or start-ups) with their chemical handling.

These various community-based strategies do not necessarily let either the federal government or the corporations off the hook. While agencies, like the EPA, have tended to fall back to more of a facilitator role, they can be used to underwrite regional administrative coordination as well as to provide technical assessments. In shifting from an emissions to a 'risk' focus, EPA has itself begun to experiment with multi-objective planning. In fact, Santa Clara County was selected as a site for this new "Integrated Environment Management" strategy. Via this approach, EPA sets up the following: an Intergovernmental Coordinating Committee, consisting of county supervisors, mayors, city council members, and members of the elected boards of regional regulatory agencies (Ayres 1984:17).

With specific reference to groundwater management, EPA describes its role as "building and enhancing institutions at the state level" (Eikenberry 1984:14). In short, federal agencies continue to provide a valuable management resource despite deregulation and decentralization. If federal agencies begin to see themselves as a catalyst for institutional experimentation and learning at the regional level, integrated planning and design could proceed in earnest.

Likewise, increased corporate involvement promises significant mutual benefits. Given the types of conciliatory programs alluded to above, it is certainly in a firm's best interest to take part in planning and implementation. However, communities which can inspire an ethos of corporate social responsibility may also find that its own resources can be extended dramatically. Beyond the mitigation of environmental problems, good corporate citizenship has been identified as a critical element in high-tech industrial development in general. In fact, the Congressional Office of Technological Assessment described this community culture element as the key to a successful regional economy. This included:

- an organizational culture that promotes a common civic perspective and a positive attitude about the region's attributes and prospects;
- an environment that nurtures leaders, both public and private, who combine an established track record for innovation with a broad view of their community's resources and promise; and
- a network of business/civic advocacy organizations that attracts the membership of top officers of major companies and receives from them the commitment of time and effort to work on issues of mutual concern, including cooperation with the public sector (OTA 1984b:8).

In other words, in successful high-tech communities, economic well-being and life-quality enhancement have already been more fully integrated. Norris (1984) and Ouchi (1984:chapter 9) describe how such a culture of cooperative and creative capitalism evolved over time in the city of Minneapolis, and sustained high-tech entrepreneurship in times of a general economic decline.

9 Conclusions

Building these cooperative communities is, of course, easier said than done. Beyond the mere fact that many state and local governments still fail to appreciate the social and environmental costs which accompany the overstated economic benefits of high-tech development, there are a number of political and institutional barriers to effective public and private cooperation in pursuit of viable mitigation strategies. These barriers include:

- declining resources amid competing demands for infrastructure investments in systems which support high-tech development (e.g. university research facilities);
- inadequate state regulatory apparatus, and outdated, combative approaches to regulation;

- political pressures to limit regulatory activities which might discourage innovations and entrepreneurship; and
- a lack of effective networks for facilitating corporate social responsibility.

Nevertheless, visualizing economic development in an integrated social ecology at least initiates the possibility for social learning. In turn, institutional mechanisms designed to further learning and adaptation enhance the probability of efficient and effective mitigation. If the major industrialized nations, particularly the USA, are going to set a better example of sustainable development, then those communities who exhibit integrative planning must become the model for regional economic development.

The experience of Santa Clara, California, is being repeated all over the USA. Scottsdale, Arizona's recent (1992) problems with TCE from the Motorola plants and the general institutional malaise suggest that history indeed repeats itself. Yet, communities just beginning to build-up high-tech industrial complexes appear to be learning some of the lessons alluded to above. That is, they tend to see social and environmental consequences as the real costs of attracting high-tech investments. Communities from Portland, Oregon to Austin, Texas have developed more nearly integrated development strategies. Unlike their counterparts in the federal government (e.g. Vice President Quayle's Competitiveness Council), they do not see environmentalism and economics as mutually exclusive or necessarily combative. Hopefully, this type of 'adaptive-learning' will eventually percolate up through the system as improved procedures occasionally do.

References

Ayres, J.A. (1984) Controlling the Dangers from High-tech Pollution, in: *EPA Journal* 10: 14-17.

Ayres, R.V. (1989) Industrial Metabolism, in: J.H. Ausubel, R. Frosch, S.R. Herman (eds.) *Technology and Environment*, Washington, DC: National Academy Press, 23-49.

Bergman, E.M. (1986) Introduction: Policy Realities and Development Potential in Local Economics, in: E.M. Bergman (ed.) *Local Economics in Transition: Policy Realities and Development Potential*, Durham, NC: Duke University Press, 1-14.

Birch, D. (1981) Who Creates Jobs?, in: *The Public Interest* 65: 3-14.

Daneke, G.A. (1988) Short-stack Chasing: The Prospects and Perils of State and Local High-Tech Development, in: R. Rist (ed.) *Annual Review of Public Policy*, Beverly Hills, CA: Sage, 281-300.

Davis, C.E., J.P. Lester (1987) Decentralizing Federal Environmental Policy: A Research Note, in: *Western Political Quarterly* 40 (3): 555-565.

Dunn, E.S. (1971) *Economic and Social Development: A Process of Social Learning*, Baltimore: The Johns Hopkins Press.

Eikenberry, K. (1984) Groundwater Quality: The Next Regulatory Issue, in: *EPA Journal* 10: 12-14.

Farley, J., N.J. Glickman (1986) R&D as an Economic Development Strategy, in: *APA Journal* 52 (1): 407-418
Feller, I. (1988) Positive and Negative Knowledge: An Interim Critique of Assessments of State Advanced Technology Programs, paper presented to the Association for Public Policy Analysis and Management, Seattle, WA.
Gallagher, T., T.R. Henderson, J. Trauberman (1984) *Groundwater Strategies for State Action*, Washington, DC: Environmental Law Institute.
Glasmeier, A.K. (1985) Innovative Manufacturing Industries: Spatial Incidence in the United States, in: M. Castells (ed.) *High Technology, Space, And Society*, Beverly Hills, CA: Sage, 55-80.
Hart, S., D. Denison (1987) Creating New Technology-based Organizations: A System Dynamics Model, in: *Policy Studies Review* 6(3): 512-528.
Larsen, J.K., E.M. Rogers (1988) Silicon Valley: The Rise and Fall of Entrepreneurial Fever, in: R. Smilor, G. Kozmetsky, D.V. Gibson (eds.) *Creating the Technopolis: Linking Technology Commercialization and Economic Development*, Cambridge, MA: Ballinger, 99-116.
Luger, M. (1984) Does North Carolina's High Tech Development Program Work?, in: *APA Journal* 50 (3): 280-290.
Luger, M. (1985) Explaining Differences in the Use and Effectiveness of State Industrial Policies, working paper, Durham, NC: Institute of Policy Sciences and Public Affairs, Duke University.
Markusen, A.R., R. Bloch (1985) Defensive Cities: Military Spending, High Technology and Human Settlements, in: M. Castells (ed.) *High Technology, Space and Society*, Beverly Hills, CA: Sage, 106-120.
Miller, M. (1988) The Low Down on High Tech, in: *Southern Exposure* 14 (5/6): 35-39.
Nelson, R.R., S.G. Winter (1982) *Evolutionary Theory and Economic Change*, Cambridge, MA: Harvard University Press.
Norris, W. (1984) Only Big Business has Resources to Address Unmet Social Needs, in: *Woodlands Forum* 1 (2): 10-13.
Olivieri, A., D. Eisenberg, M. Kurtovich, L. Pettegrew (1985) Ground-water Contamination in Silicon Valley, in: *Journal of Water Resources Planning and Management* 11 (3): 346-358.
Ostrom, E. (1989) Microconstitutional Change in Multiconstitutional Political Systems, in: *Rationality and Society* 1 (1): 11-50.
OTA (1984a) *Protecting the Nation's Groundwater*, Washington, DC: US Congress, Office of Technology Assessment.
OTA (1984b) *Technology, Innovation and Regional Economic Development*, Washington, DC: US Congress, Office of Technology Assessment.
Ouchi, W. (1984) *The M-form Society: How American Teamwork Can Recapture the Competitive Edge*, Reading, MA: Addison-Wesley.
Papunek, G. (1983) Economic Development Theory: The Earnest Search for a Mirage, in M.P. Todaro (ed.) *The Struggle for Economic Development*, New York: Longman, 17-25.
Raucher, R.L. (1986) The Benefits and Costs of Policies Related to Groundwater Contamination, in: *Land Economics*, 62 (1): 33-45.
Rogers, E., J.K. Larsen (1984) *Silicon Valley Fever: Growth of the High-technology Culture*, New York: Basic Books.

Rosenberg, N. (1982) *Inside the Black Box: Technology and Economics*, New York: Cambridge University Press.
Saxenian, A.L. (1985) Silicon Valley and Route 128: Regional Prototypes or Historic Exceptions, in: M. Castells (ed.) *High Technology, Space and Society*, Beverly Hills, CA: Sage, 81-105.
Schumpeter, J. (1961) *The Theory of Economic Development*, New York: Oxford University Press.
Sherry, S. (1985) *High Tech and Toxins: A Guide for Local Communities*, Sacramento, CA: Golden Empire Planning Center.
Siegel, L., J. Markoff (1985) *The High Cost of High Tech*, New York: Harper & Row.
Simmons, B. (1985) Environmental Protection's Help to Attract Business to Oregon, in: *California Journal* 5, 37-40.
Smilor, R., G. Kozmetsky, D.V. Gibson (1988) The Austin/San Antonio Corridor: The Dynamics of a Developing Technopolis, in: R. Smilor, G. Kozmetsky, D.V. Gibson (eds.) *Creating the Technopolis: Linking Technology Commercialization and Economic Development*, Cambridge, MA: Ballinger.
Storper, M. (1985) Technology and Spatial Production Relations: Disequilibrium, Interindustry Relationships and Industrial Development, in: M. Castells (ed.) *High Technology, Space and Society*, Beverly Hills, CA: Sage, 265-283.
Walker, R.A. (1985) Technological Determination and Determinism: Industrial Growth and Location, in: M. Castells (ed.) *High Technology, Space and Society*, Beverly Hills, CA: Sage, 226-264.

Socially Oriented Technology Policy in Germany: Experiences of a North Rhine-Westphalian Programme

Erich Latniak and Georg Simonis

1 Introduction

The following article will describe some of the experiences from the North Rhine-Westphalian programme *'Mensch und Technik – Sozialverträgliche Technikgestaltung/SoTech'* (Man and Technology – Socially Acceptable Shaping of Technology) initiated in 1985. 'SoTech' was the first attempt of a German *Bundesland* (federal state) to manage an integrated technology policy programme, one emphasizing the social aspects of development, implementation and the societal outcome of information and communication technologies (ICTs) by assessing their impact and trying to "shape"[1] these technologies. This type of programme seems to be a new and different approach in technology policy in Germany. It is worth discussing and analyzing with a focus upon its development and political importance. We will therefore primarily discuss the problems of integrating social and economic aims in technology policy by analyzing this programme as an example.

Taking into account the societal changes in Germany during the last three years, the circumstances for every kind of policy changed fundamentally. The economic and financial situation in North Rhine-Westphalia (NRW), Lower Saxony and Bremen (the three federal states with technology programmes of this type) is still increasingly tense; therefore, nobody can presently foresee the policy restrictions in Germany induced by the integration of the former German Democratic Republic. Hence we will avoid forecasts of future policy developments aimed at integrating environmental, social, and economic goals – we can only talk about possible political trajectories. Consequently, we will concentrate on a review of the formulation and implementation period of the SoTech programme and the *Initiative Zukunftstechnologien* (Future Technology Initiative), choosing the time between 1985 and 1989 as programme context. The experiences of this period lead us to the thesis that it is impossible to reach an uncontradictory integration of the different political, social and economic aims – or even an uncontradictory formulation and implementation of this kind of policy. The outcome of these approaches is at least as contradictory as were both the reasons for the programme and the prevailing interests during the period of programme formulation and implementation.

The rationale for selective intervention in and regulation of societal processes still dominates the administrative view of policy-making – even though other types of policies are necessary to reach an integration of social, environmental and economic aims. In our view, an integration of these three groups of aims can only be sequentially achieved by both a broader discussion process with societal actors, and an implementation strategy where these cooperatively-formulated prospects are achieved in cooperation with the actors involved. Hence, the bottleneck of an integrated technology policy is not a lack of support for technology or hardware development, but rather a lack of social innovations organized by social actors. This type of policy would mean a step forward from interventionist regulation to self-regulation of societal processes in technology policy on the basis of public discussion of technological options and practices (Simonis et al. 1990). It might be misleading to talk of deregulation in this case, because the idea is not to abolish regulations but to socially develop a new kind of innovative regulations for economic modernization. We will illustrate the emergence of this kind of policy efforts by discussing the success or failure of integrating social and economic aims on three dimensions: political programme formulation, administrative organization and the projects of the SoTech programme carried out between 1985 and 1989.

First of all, we will give a short characterization of the prospects and obstacles of technology policies in Germany at the beginning of the 1980s, especially of those policies aiming at an integration of divergent social interests and goals. We will do this by describing some of the major changes in German technology policy during the 1970s to illustrate the new conditions for these policies emerging at that time. Following this, we will look at the specific economic and political situation in NRW and mention some of the reasons for the Initiative and the SoTech programme which started in 1984. We will then give several examples of the changing prospects of the programme. From our perspective, it is necessary to take into account that political programmes of this kind are complex societal learning processes, and that the resulting policies are continuously influenced by their political environment. Finally, we will draw some conclusions about additional perspectives of integrational policies with respect to the situation in Germany at present.

2 Integration of Differing Goals in Technology Policy?

Seen from a theoretical point of view, a modernization-oriented technology policy can be conceptualized from at least four different points of analysis, all leading to different types of strategies and instruments to support technological change (Böckler 1990:27-80). Technology policy can be e.g.

- a reaction to a crisis in the economy that would lead to crisis regulation policies, e.g. social and economic restructuring policies supported by a labour-market-oriented strategy;
- a support of the autonomous dynamics of economic development that would lead to supportive strategies – e.g. subsidizing specific branches or bottleneck-oriented strategies to improve the speed of modernization and the diffusion of innovations;
- a social reform objective of the government that would lead to strategies integrating societal, environmental and economic goals based on democratization and participation, with the purpose of producing collective goods and jobs for problem groups of the labour market;
- a kind of symbolic policy to avoid legitimation problems that would lead to an 'easing-off': the aim here would be to induce a social acceptance of change processes without trying to change or influence their direction and impacts.

Seen from an empirical point of view, it is quite difficult to find policies that exactly meet these four ideal types. Normally, programmes depend on a specific mixture of motives and reasons. This mixture is a result of various factors: the administration's problem perception, the aims of government and political parties, the interests and influences of political pressure groups and the imponderabilities of political and administrative bargaining processes. Programmes are modified by changes in the economic and political situation and influenced by the implementation structures. Taking these points seriously, it is evident that the ideal-types of programmes are aspects that can be found in the analysis of technology policies.

Let us first review the research and technology policy in West Germany as developed and executed both by the German federal government and states since the 1950s. It is evident that these policies were first of all efforts to modernize industry and economy by a 'technology-push' strategy in selected industrial sectors. In discussing German research and technology policy, there are some points one can take for granted. After a period of building up a research infrastructure in the 1950s, the main objective for "reducing the technological gap" in the 1960s aimed at an improvement of basic research. In contrast, the 1970s were regarded as a period of "optimizing the economic efficiency" (Bruder/Dose 1986:14-16) by improving the diffusion of "key technologies". For a long time, the main financial subsidies of technology policy were invested in nuclear energy, transportation and aerospace technologies – and, with less emphasis, computers.[2] Until the mid-1970s, the environmental and societal impacts of these administratively-pushed technologies were ignored, or at least not investigated or discussed in society. Nor have the working conditions induced by technological applications or the security of working conditions been an integral part of technology policy. The regulation of these problems were developed in political arenas both apart from and differently structured from technology policy.

At the end of the 1970s, the political situation changed for at least five reasons.

A *first* and internal reason for a change in technology policy was that inefficiency became evident with the failure of several industry-oriented programme efforts. The first example was the failure of subsidy strategies for building up a computer industry in West Germany, one to successfully compete with American and Japanese counterparts. The main points of the critics were threefold:

- the administration's formulated goals reached too far into the sphere of entrepreneural decision without being able to compensate industrial strategy deficiencies (Beckerhoff/Sommerlatte 1982:21, 245): the projects were fully subsidised and industry responsibility for the success was low;
- a subsidy strategy to aid firms by providing significant financial resources was ineffective (Beckerhoff/Sommerlatte 1982:16, 265), especially when most of the resources were given to large enterprises like Siemens or AEG-Telefunken;
- the development of software was neglected when compared to hardware development (as one example of a strategic mistake) (Beckerhoff/Sommerlatte 1982:250).

A similar situation occurred in the aerospace industry. Despite the technological success of the European Airbus project, the Multi Role Combat Aircraft (MRCA) Tornado, or the Ariane rockets, the economic outcome led to continuing subsidy of the firms involved and became a burden on federal finances. Together with the increasing problems of the labour market at the beginning of the 1980s, both the concept of technology-push strategies and the instruments used became, at the very least, controversial: neither the industrial objective (supporting industry to achieve success) nor the strategic aim of modernizing production by diffusion of technology were reached in a satisfactory manner. Due to this, the programmes supporting ICTs changed completely in the years following (BMFT 1984).

The *second* reason was the increasing political role of the themes and results of the federal *'Humanisierung des Arbeitslebens (HdA)'* programme, one primarily stressing ergonomic and work safety problems in different technologies. Based on the results of socio-technical approaches in Scandinavia, Great Britain and the USA, the programme started in 1974 and was financed by the *Bundesministerium für Forschung und Technologie* (BMFT; Federal Research and Technology Ministry), as well as by the *Bundesministerium für Arbeit und Soziales* (BMA; Federal Ministry for Labour and Social Affairs). The financial resources of this programme increased from 9 million DM in 1974 to 95 million DM in 1979. This programme was the Social Democratic government's first effort to integrate the aims and positions of the German unions into a technology programme. Although it was perceived as a union-oriented programme – and therefore opposed by parts of the employers' organizations as a means to implement administrative control on economy – many of its results provided the basis for regulating the security of working life and the ergonomy of work. But there was no integration of quality-of-working-life aspects in other federal technology programmes. Besides this, it is evident that the unions' positions represented in this programme did not spill over to other technology or industry policies. By focusing their internal discussions on the HdA programme, the unions did not develop for a long time further initiatives in technology policies on more than a general level. The main aim of the unions was to avoid job losses when ICTs were used for rationalization purposes; but until 1984, the orientation towards the shaping of ICTs remained in a minority position – even within the unions.

The *third* and politically most virulent reason was the massive resistance against nuclear power plants and the emergence of environmental issues in politics since 1970. This led to a blockade of the ambitious nuclear energy plans of the administration and to the energy industry's using new nuclear technologies such as the "Breeder" and "Thorium high-temperature" types. Along with the conflicts triggered by building nuclear power plants, an anti-nuclear movement grew; and an infrastructure of environmentally-oriented 'alternative' scientists and researchers accrued (Katzenstein 1989 a, b; Junne 1989). Additionally, the inflexibility of administrative and economic strategies seemed to create a societal impact which generated broad scepticism towards new technologies. At the same time, the inflexibility of the political parties led to the foundation of the Green party as a parliamentary-oriented effort concerned with environmental questions (Kitschelt 1980). The result of this situation was that, for the administration, it was nearly impossible to support technological change without reflecting on environmental impact in advance. Evidently, there has never been a political or social conflict in a research and technology policy in Germany. The conflicts emerged when implementation became organized.

One consequence of this problem was to strengthen assessment procedures in technology policies in order to avoid conflicts during the implementation period. But even this strategy became problematic and only led to discussions on how to organize TA. A second effect resulting from the conflicts on nuclear energy, was a blockade of scientific expertise as a means to legitimize technology policy. Instead of legitimizing policies by scientific knowledge, the conflict was imported into the expert councils by integrating environmentally-oriented 'counter-experts'.

Combining these observations on technology policy, it seems as if the federal government's intervention-oriented technology-push strategy in research and technology policy reached its financial and societal limits at the beginning of the 1980s. But that did not immediately lead to a complete reorganization; and these facts did not change the strategies of the BMFT *as a whole*. HdA and ecologically-oriented research in these years never exceeded more than 5% of ministry expenses.

Beyond these reasons, there were structural economic factors for such a change. Due to the fact that the overall economic and labour situation in the Federal Republic of Germany deteriorated at the end of the 1970s, the rationale for a technology policy had to be changed, too. Decreasing growth rates, high and still-increasing unemployment and scarceness in public financing all characterized the frame for political decisions and policies at that time. The impact of crisis on the economic development of the federal states and regions was quite different, according to their specific industrial structure. Therefore, the federal states and the communities began to develop and implement their own technology or technologically-oriented industry policies. This is a fourth major change in technology policy since 1980: the increasing efforts of the communities and federal states to support economic change and to avoid further job losses by a more regionally-tailored policy (Hucke/Wollmann 1989). This can be seen as a result of the lack of appropriateness of the centralized federal programmes (Ellwein/Bruder 1982) on the one hand; and on the other, as a result of the reshaping of all types of modernizing policies by both the federal government and the federal states. We will give some clues about the NRW situation in Section 3.

The last aspect to be mentioned is the impulse given by new ICTs and their diffusion in production. Be it for consumer goods, for the machine industry or for communications and services, computer-, microelectronic- and semiconductor-based technology increasingly became a key factor for industrial development. The success of personal computers and the increasing capacity of the new technologies for production purposes led to an new situation: high-tech was of no interest to scientists any more but it did become an aspect of working life for many people. Resulting impacts on employment, qualification, and work organization were extensively discussed among social scientists, unions and administration. The discussions on automatization and on the use of robots in production, indicated that the adequate use of ICTs might become one of the main challenges for economic success, productivity and work satisfaction on the micro level – and for social security and economic stability on the macro level.

At the beginning of the 1980s, there existed only few alternatives for government administrations with regard to developing another type of technology policy as a response to these factors. The situation can be described by three aspects.

- The experience of the anti-nuclear movement was successful in using demonstrations, civil disobedience and legal means to block new nuclear power plants. There was also an increasing issue-oriented focus on environmental aspects in public debates. Subsequently, it became necessary for nearly every kind of technology policy to take these points into account. This is one of the reasons why governmental investigations on environmental questions were intensified, and why Parliament stressed the importance of technology assessment (Dierkes et al. 1986). This can be seen as developing 'filter systems' for societal dialogue. In such an atmosphere, different scientific opinions and positions could be consolidated to portray a comprehensive view analysing the topics discussed and could therefore broaden the basis of decision-making.

  The counter-tendency is a kind of deregulation, according to the standards of the EEC – e.g. in the environmental sphere. Here, as are the standards for work safety, integration into the EEC and the still-ongoing adoption of European regulations was feared by many to be leading to a lowering of German standards or even to an abolishment of rules stricter than their all-European counterparts.

What is indicated by these two tendencies, is that the definition of goals for technology policy was (and still is) becoming more and more complex – to say the least. For the administration, there is – on the one hand – the complexity of problems and the manifold aspects to be taken into account in order to avoid unintended social consequences. On the other hand, there is a growing number of relevant and powerful actors with their own ressources, opinions, and positions to be integrated into the political process.

- Therefore, the failure of an administrative setting of technological goals could only be avoided by reducing state intervention in entrepreneural decisions and by mobilizing societal actors to define policy objectives. This was one of the consequences implemented by the federal technology policies since the beginning of the 1980s. Through subsidies and *"indirekt-spezifische Förderung"* (Menneken 1986:88, 98; Beckerhoff/Sommerlatte 1982:273), responsibility for both projects and their subjects were granted to industry and research institutions (Väth 1986:85-91). As a consequence of failures in supporting computer industry in the 1970s and of the adaptation of both Japanese and American concepts of R&D subsidizing (Väth 1986:93; Beckerhoff/Sommerlatte 1982:267-268), efforts tended to focus on financial aid for investment strategies – either in industry, as before, and additionally to support technological change within industry through qualification, information and documentation. Evidently, the policy goals shifted to supporting innovative small and medium-sized enterprises instead of subsidizing large firms.
- The organizational options for administration were few. Interventionist strategies were expensive and caused unintended effects. Hence, research and technology policy could be implemented in a different way; and the administration could be restructured according to the changing needs of those policies aiming at an adequate moderation of societal processes. Integrating both the competences and resources of social actors could serve to broaden the legitimization and impact of the policies. Without changing the entire policy, this additional method could indicate a 'self-regulation effort' of societal actors induced by the administration. The concept adopted by the federal administration and by several federal states led to an increasing effort to support a public-private partnership type of implementation, e.g. in order to develop decentralized and innovative policy networks (MWMT 1990).

This perspective caused an increasing mobilization of societal actors in technology policy to define goals and policy implementation. Implemented in Baden-Württemberg and, since the change of government to the conservative and liberal party, in federal technology policy as well, the conservative strategy aims at 'technological-lead positions' and supports industries' world-market-oriented strategies for specific branches or sectors (Staatsministerium 1983; Späth 1987). Entrepreneural efforts and research-oriented high-tech products were partly subsidized within a concept of application oriented research (BMFT 1984:67). Close cooperation between research, industry and administration continued with the foundation of research centres and university institutes. In order to reach an integration of TA efforts and regionally oriented high-tech development structures, Baden-Württemberg's government tried to build a research centre and a TA academy at the University of Ulm; but the integration failed, and only the research centre is in operation (Naschold 1989:146).[3]

In addition, the Social-Democratic strategy, e.g. in NRW, similarly stressed "high-tech" for economic modernization; but it differed in the means and degree of integration of the union's interests. And, most of all, it emphasized the social aspects of modernization and cooperation of different actors.

The specific problem with this kind of socially oriented technology programmes, at the present time is that they are not at all based on explicit objectives, perspectives, or concepts. Rather, they are based on a more or less contradictory mixture of reasons. Claims for political action, intertwined with administrative and political sensibilities towards the social aims of the programmes, lead to a situation where the aims and means of the programme are sometimes disintegrated and restabilized on a different level. In order to derive indicators of success, failure, impact or outcome of a socially-oriented technology policy, it becomes necessary to analyze the conceptualization of the programme – taking into account influences by social interests and bargaining processes. Lacking this kind of specific analysis, we are only able to give some hints and impressions from a nevertheless restricted inside view. They might be helpful for a further investigation of these programmes in finding aspects which must be taken into consideration when seeking an explanation for the development of the Initiative and SoTech programme as an integrational effort. In the following, we will try to illustrate the background and the motivation of NRW technology policy by giving some information on the economic and political situation of this federal state.

3 Reasons for a Socially Oriented Technology Policy Initiative in 1984 in North Rhine-Westphalia

Traditionally, NRW has been one of Germany's industrial centres, with chemical industries along the Rhine valley ("*Rheinschiene*") as well as the Ruhr's industrial district where coal-mining and steel production has been concentrated since the middle of the nineteenth century. The Ruhr district was one of the most important power plant areas in Germany. Since the mid-1950s, there have emerged problems of economic restructuring – especially in the Ruhr region, where the decline of coal-mining was followed by the steel crisis of the 1970s. Supported by the federal government, a coalition of interests – the coal-mining and steel industries, the local communities, the labour market administration, the unions and the Social Democratic government of NRW (since 1971) – all cooperatively managed to lessen the social impact of the decline over a period of 20 years.

It should be emphasized that in NRW the unions and the government worked hand in hand for social moderation of the steel and coal-mining industries' decline. This was fostered by the *Montanmitbestimmung,* a far reaching co-determination law for the steel industry which opened up a great deal of influence for unions and works councils from the steel firms. The close links between the Social-Democratic party and the dominating unions suggests greater input in policy formulation – in contrast to Baden-Württemberg, where unions could not play a comparable role. Unions in NRW are a powerful factor – compared to other federal states – with remarkable influence even on local and regional politics.

The regulation and social moderation of the industrial decline has been supported by several programmes since 1968. The first programme was the *Entwicklungsprogramm Ruhrgebiet* (1968 – 1970) for the ailing Ruhr district; it planned expenditures of 3.7 billion DM for infrastructure, investment subsidies, clearing of land, environmental aims, and image improvement. In 1970, it was expanded to the *Nordrhein-Westfalen* Programme; but the inability to reach employment aims was soon apparent. Due to the steel crisis at the end of the 1970s, the North Rhine-Westphalian government made a new and corporatistic approach with the *Ruhrgebietskonferenz* (May 9/10th 1979 at Castrop-Rauxel), bringing together the chambers of industry and commerce, unions, local and regional administrative authorities to discuss the *Aktionsprogramm Ruhr* (1980 – 1984). This programme aimed at direct job creation and technology promotion; but during the programme, the recession hit and unemployment rates reached more than 10% in the Ruhr district. The interesting point here consists of its being an integrated approach to regional development. A quarter of the 6.9-billion DM programme budget was spent for housing; another quarter, for steel and coal-mining (Petzina 1990:545). Additionally, the government supported the ailing industrial sectors with subsidies.[4] In this way, the large steel companies of the Ruhr district were supported in their restructuring and modernization efforts.

But despite this cooperative tradition in economic restructuring, the rates of unemployment in this region are higher than average in NRW. One aspect is that the steel companies diversified into different sectors and invested outside the Ruhr district where steel production capacities are still concentrated. A second aspect is the ongoing decline of coal-mining. The total amount of job losses in the coal and steel sector ('*Montansektor*') has been estimated about 500,000 since the mid-1950s, mostly in the Ruhr district. But only one successful case of a new production plant with a high-employment effect exists in the Ruhr area (General Motors/Opel at Bochum) (Simonis 1989). Though the social conflicts of economic restructuring are more or less cooperatively regulated, the restructuring of industry in the Ruhr region – and the shift from a coal-and-steel-dominated industrial district to other industries providing increasing employment – is still an unsolved problem (GEWOS/GfAH/WSI 1988). As a side effect of the continuing crisis in the Ruhr region, productivity rates in other industrial sectors (machinery industry, electronics) did not reach the average growth rate for these industries in Germany in the 1980s. There was a continuing necessity for support in the restructuring of the Ruhr district.

At the beginning of the 1980s, the North Rhine-Westphalian government perceived two main impulses for a change in industry and technology policy. Microelectronics was seen as the main strategic sector and as a means to produce economic growth (MWF 1986; Späth 1987; Staatsministerium 1983). In perceiving the success of Silicon Valley, the furtherance of a combination of research institutions, production and service organizations seemed to be an ideal way for a technologically-based restructuring – especially in declining regions like the Ruhr district, even if this effort could not solve the labour market problems.

Therefore, efforts emerged in research, development, transfer and marketing of new technologies, as well as the idea of "science parks" as a nucleus of future economic development. It has been one of the main strategies for regional restructuring since the early 1980s (Dose/Drexler 1988; Hucke/Wollmann 1989).[5] The fascinating model of clean and high-value production, all of it based on the work of highly-qualified scientists and workers, seemed to be the 'bright and clean' model – opposite the dark, dirty and coal-based sweat-shop history of the Ruhr region. According to Grabow et al. (1990:71), there were in NRW in 1988 alltogether 174 local technological and innovation-oriented initiatives – such as technology parks, transfer institutions (53!), vocational training centers or employment initiatives.

The improvement of the research and transfer situation to push new technologies has quite a long tradition in NRW. Since the end of the 1950s, several universities and centres of higher education were founded to support the restructuring processes by establishing scientific research centres (Petzina 1989 1990). For example, the University at Bochum was founded in 1961; Dortmund in 1962; Düsseldorf in 1965; Bielefeld in 1969; Duisburg, Paderborn, Siegen, Wuppertal, and Essen in 1972; and Hagen in 1974. But the impact of these efforts for the industrial restructuring was assessed controversially in the beginning (Weißbach/Niebur 1986:340). Meanwhile – and due to the fact that a closer cooperation between universities, centers of higher education and industry could indeed be reached – the research and transfer infrastructure of NRW has come to be seen as one of the main potentials for modernization (MWMT 1990).

But as mentioned before, the point where technology based policies produce societal impacts has been critical in other industrial sectors. The administrative view at the beginning of the 1980s was dominated by the conjecture that there might be increasing scepticism towards every type of new technology. In 1983, conflict over the decision of the *Bundesverfassungsgericht* (Federal Constitutional Supreme Court) on census, privacy and the legitimacy of collecting data on citizens, indicated an increasing amount of social awareness about the impact of computers.[6] If computers and microelectronics would become an area where a new techno-sceptic movement could emerge, it would then be necessary to avoid a blockade of economic restructuring. The politically-initiated mobilization could be an impulse for a socially-compatible shaping of technology only if integration of the ideas of this movement were possible. The integration of social and economic aims could even become a chance for industrial development.

A second reason for the development of a socially-oriented technology programme in NRW was the political competition after the change of federal government in 1982 between Social-Democratic and conservative governments on the "correct" type of technology policy. At that time, NRW was the largest state governed by a Social Democratic government; while the federal government and most of the federal states governments were formed of conservative and liberal parties. Persuaded by increasing efforts in technology policy, especially in Baden-Württemberg (Staatsministerium 1983; Wildenmann 1985), the Social Democrats tried to develop their own unique distinguishable policy approach: one stressing their social and – at the same time – economic competence, while the conservatives' strategies were criticized for only focussing on economic aims.

This point can be illustrated by both the decline and the political problems of the federal government's HdA programme. There, an evaluation of the programme, coupled with increasing criticism, induced both restructuring and reorganization (HdA-Projektträger 1981; HdA 1982). The programme did not conform to the political aims of the conservative and liberal government. For several years, it was not clear whether the programme would last any longer. Hence, for the Social-Democratic government of NRW, an opportunity emerged to demonstrate that a Social Democratic technology policy covers relevant scientific aspects, political objectives of the unions and social policy aspects by financing a programme similar to HdA. The experiences acquired by HdA could be transformed into guidelines for new efforts in socially-oriented technology policy. Without them, the SoTech programme would have had more problems in finding its field of research. The HdA's advance work provided the research findings from which SoTech could start. A second point is that, in NRW, an infrastructure of work-oriented research had been established by HdA project funding. Consequently, it is not surprising that, beginning in 1985, some of the research groups that worked in the HdA programme continued their work in a SoTech context.

4 Administrative Output: The "Initiative Zukunftstechnologien" and the Programme "Mensch und Technik – Sozialverträgliche Technikgestaltung"

In the last section, we demonstrated some aspects of the political and economic situation in NRW which, in our view, must be taken into account to explain the specific political concepts developed there in 1984. They are reasons that influenced to a visible degree the formulation of the *"Initiative Zukunftstechnologien"*.

As an answer to the different local problems mentioned before, the Social-Democratic government in 1984 decided to start the *Initiative*. Its aim was to overcome social and economic problems by integrating a set of political measures tailored to the strategic needs of the federal state. This initiative covered four different parts:

- the programme Future Technologies, directed towards supporting the use of new technologies in eight areas: environment, energy, microelectronics, measures and regulations, ICT, biotechnology, human technologies and raw materials;
- the enforcement of technology transfer, especially tailored to the specific needs of small and medium-sized businesses in NRW;
- the technologically leading research in the eight areas mentioned; and finally,
- the programme Man and Technology – Socially Acceptable Shaping of Technology (MWF 1986:5, 16; Schatz-Bergfeld 1988:233-235).

The Initiative had an overall financial backing of 400 million DM for the period from 1985 to 1988, of which the SoTech programme obtained 15%. This meant approximately 15 million DM a year for the SoTech programme over this four-year period. It is easy to see that the initiative is a mixture of programmes working on several of the problems mentioned; but it is a remarkable point that scientific research, transfer, implementation to firms and humanization problems were considered together.

The main objectives of the SoTech programme were (and still are)

- gathering information about the impact of ICT in all areas of society (including, e.g. daily life, free time and the changing relation between citizen and state);
- developing and testing concepts – especially in the area of qualification, application of ICT in service and industry, design of workplace and organization of production processes in industrial production and services. Consequently, the projects include not only research but also consulting, reorganization and qualification in cooperation with users, employers and unions;
- creating chances for participation and co-determination offered by ITC through organizational and qualificational changes;
- disseminating and applying the results in a decentralized way – e.g. by cooperating with trade unions, the German Organization of Engineers (VDI),

local and governmental administrations, universities and research institutes, and employers' associations;
- defining criteria for the social acceptability of specific technologies on an operational level – as well as improving the discussion process by mobilizing, integrating and qualifying the groups involved (von Alemann et al. 1985; von Alemann/Schatz 1986; Schatz-Bergfeld 1988).

These programme incentives were conducted by research groups all over Germany from universities, research institutes, unions, and consultants; but they were obliged to at least cooperate with a partner in NRW. All projects were obliged to organize at least two workshops with other scientists and persons involved in the project, in order to present and to discuss results. This was an additional task for all projects, even for those forming qualification or consulting groups. Several congresses and conferences took place. The most relevant and interesting results from scientific discussions in the social sciences have been published in a series by the *Westdeutscher Verlag*. Within a four-year period, 110 projects were planned and completed by research institutions. The average financial volume of the projects was 470,000 DM; and the average duration was two years.

4.1 Patchwork or Integration?

First, we will discuss how far-reaching the integrational effort of the Initiative was, or if the initiative was rather a patchwork of programmes. This can be analyzed by asking whether the government intended to organize a cross-sectional integration of economic and social aims within the Initiative as a whole. In this particular case, social aims would have meant

- aspects of security and work satisfaction as one crucial point of the programme philosophy; and
- expanding discussion structures or institutional contacts for the social groups concerned by the implementation of these technologies. There, they could discuss the worthness or risks of the development, implementation and use of the technologies.

By referring to a statement made by Minister President Johannes Rau in 1985 concerning the social intentions of the Initiative, the aim of all the Initiative programmes can be so illustrated: "We will have to reach a consensus within our society concerning the question of which kinds of new technologies we want to see realized and which we do not, which kinds of technologies we are going to promote and which we do not. Not everything that is technically feasible is at the same time socially desirable" (translation cited according to Riepert 1988:2).

Accordingly, the SoTech programme stressed societal discussions on technology. Seen as a "cornerstone" (Schatz-Bergfeld 1988:234) of the Initiative, SoTech was meant to provide "an autonomous and socially-committed profile" for the whole approach. This statement illustrates the situation at the beginning of the Initiative's implementation. The SoTech programme – organized by the *Ministerium für Arbeit, Gesundheit und Soziales* (Ministry of Labour, Health and Social Affairs of NRW) – was meant to take an innovative role in both technology and policy (Schatz-Bergfeld 1988:234). The other programmes of the Initiative, however, remained conventional science- or market-oriented, technology-advancement programmes – leading to a fade-out of social discussion processes concerning the pros and cons of the technologies supported. This is specifically evident for the research programme of the Initiative (MWF 1986).

If one reflects upon the options of the Labour Ministry to bring forward the role of social aims in other programmes of the Initiative, it is necessary to take several factors into account:

- This kind of technology policy was a new task for the ministry, enlarging former responsibility for vocational training, labour market control and the control of work safety; therefore, it was also an integrational effort within the ministry.
- An appropriate administrative structure within the ministry to manage the programme had first to be acquired – before influences on other programmes of the Initiative could have any effects.
- The position and influence of the Ministry of Labour within the government was not too far-reaching.
- Only a small part of financial resources went to the Ministry of Labour. Hence, the internal 'weight' of the Ministry of Labour within the Initiative was small.

As result of all the above, a more active role of the Ministry of Labour within the Initiative was unlikely.

The scientific part of the Initiative was carried out and managed by the *Ministerium für Wissenschaft und Forschung* (Science and Research Ministry of NRW). The economic and transfer aspects of the Initiative were managed by the *Ministerium für Wirtschaft, Mittelstand und Technologie*[7] (Ministry of Economy, Medium-Sized Industry and Technology). There was continuous information between these ministries at the upper levels of administration; but the decision structures for the allocation of projects and for the hierachies of aims remained strictly separated along areas of administrative competence. Consequently, a cross-sectional impact of social aims was made impossible; and neither of the ministries fostered this point on their own initiative (MWF 1986). All the Initiative programmes had similar discussion and decision-making structures (e.g. consulting committees with unions, employers' organizations, scientists, scientific supervisors and consultants). In every committee, officials from other ministries were members.

Cooperation within the Initiative could be characterized by the image of four parallel pipelines of information, where each pipeline stands for one programme. A 'classical' way of integration (a "cross-over" between lines) was not intended, it seems. Only coordination at top governmental levels was implemented. As an administrative structure for the control of the Initiative, state government intitutionalized the *Interministerielle Ausschuß* (IMA), a cross-sectional committee of the ministries concerned with the Initiative in 1984. A second step was the *Technologiekabinett* (Government Subcommittee on Technology Affairs), initiated in 1985. This was chaired by the Ministry of Economy. Looking at the project level and based on our knowledge, we can at least say that between SoTech-financed projects and the projects or activities of other parts of the Initiative, there was no coordination at all.

This means that the integration of the Initiative's aims was very poorly implemented in four parts of the Initiative – or more precisely, its social aspects and discussion processes were functionally formulated by government and administration merely as a task of only *one* of the four programmes. Therefore, legitimate problems arising from the potentially-controversial aspects of the whole Initiative, were more or less concentrated in the SoTech programme.

Beyond this conclusion, one can easily see a difference between SoTech's technological focus and the scope of research areas on other parts of the Initiative. While the Initiative had to work on eight technology areas, SoTech only concentrated on ICT and humanization problems. The other programmes focussed only on research, industry, and administration; but SoTech projects had to cope with problems dealing with the world of work and daily life too, thus aiming at a comprehensive overview of the impacts of ICTs differentiated to the problems of specific groups (e.g. children and elderly people) (Schatz-Bergfeld 1988:235-237). This is to be seen as the second obstacle to integrating social and economic aims in the Initiative.

4.2 Organizational Aspects of Implementation

The SoTech programme was not a task with political priority in the Ministry of Labour; it was only one task among many. It fell under the jurisdiction of a subdivision of the Ministry of Labour. If the programme had been one of high priority within the Ministry, it would have been assigned to the staff of either the Minister or the Secretary of State. At the beginning of the Initiative, SoTech was seen as a regular task of the ministry, but limited to four years. Therefore, it was evident that it was scheduled to be terminated at the end of this period. The ministry did not intend to build up a continuing organization for the programme in 1984.

A second problem in implementing the programme was that the programme was not only research-oriented but *"gestaltungsorientiert"* ("shaping"- or action-research-oriented). This meant that the programme did not conduct basic research but rather research intended to cope with practical problems ('application-oriented'). Research funding and financing was the aim of another programme within the Initiative, and therefore the task of another ministry. The SoTech definition of tasks led to a contradictory concept regarding intervention in practical problems: research on practical problems should be transacted, but no financial resources were available for either the technology subsidies or investment goods needed to take the necessary second step: the design and allocation of specifically-designed technology. Hence, only "soft" forms of intervention were possible, such as consulting and qualification. This was how the development of technology ("hardware") faded out of the SoTech programme. This contradictory implementation concept was a response to changing political intentions. In the beginning, SoTech was seen as a kind of socially-oriented technology-assessment programme, but it turned out to be an action-research-oriented programme closely connected to the concepts of the union's programme to 'shape' technology.

Both the project tasks and the individual project funding were defined on the basis of a scientific overview of problems having to do with ICTs (von Alemann/Schatz 1986). They became more concrete through controversial discussions in the Consulting Committee of the SoTech programme. These discussions were backed up by the scientific support of the *Projektträger*, a group of eight scientists supervising and reporting as experts on the tasks and work carried out in the projects. The 110 projects that the SoTech programme financed, started in three intervals: between 1985 and 1987, according to the availability of financial resources.

Publication of the results and research findings is a central aspect for a discussion-oriented programme such as SoTech. Therefore, unlike other Initiative programmes, SoTech had the obligation to publish all its results. To optimize distribution of the results published, the Ministry of Labour decided to develop a specific organization for this task. Initially, this was done by the *Vermittlungsbüro* (distribution agency) of the SoTech programme. Today, the NRW office of the *Rationalisierungskuratorium der Wirtschaft* – a cooperative organization of industry, employers' organizations and the unions at Düsseldorf – distributes its research memoranda, leaflets and informational letters. This service is financed by the ministry. The books are available at book stores. Remembering that improving transfer was one of the main tasks of the Initiative, one thing is remarkable: not even in the distribution channels did coordination between the programmes of the Initiative take place in the first few years.

4.3 Changes during Programme Implementation

In the following section, we will provide some information on changes within the Initiative and SoTech during the implementation period. In our view, these changes show the increasing impact of SoTech on both administrative and ICT-oriented discussions.

To see how the implementation process of the SoTech programme and other parts of the Initiative changed, we will first concentrate on the changes in administrative organization during the implementation period. As a reaction to criticism resulting from its lack of integration, especially from the NRW unions, the *Interministerielle Ausschuß* was charged with the further development of the Initiative through cooperatively-financed projects. With this step, a specific mode for collective administrative decisions was founded; and several projects were financed by these resources. However, this does not at all indicate an integration on the project level of the different aims of the Initiative: there was no decision committee for *all* projects of all programmes, but only for those few commonly-financed projects. Decision and implementation structures remained separate.

As mentioned before, SoTech was from the very beginning the task of an administrative subdepartment of the Ministry of Labour. As a part of an internal reorganization within the ministry, the SoTech administration became a central aspect of the Labour Ministry's subdivision for structural problems in NRW and technology. Thus, within the ministry, the internal position of technology policy was strengthened. As a side-effect of the government's decision to continue with Initiative programmes, SoTech's administration became a continous task of the ministry. This institutionalization is one main indicator of its increasing impact.

On the other hand, financial resources for the programme were reduced to one quarter. This cut-back was due to an intention to evaluate the results of the projects and to set up new – or, at least, modified – aims for the ongoing programme. With respect to changes in German society and in the political system, this might be very difficult in the next few years.

Keeping in mind the government's intention to broaden social discussion processes on the problems of ICT use and implementation, the successive development of cooperative networks between researchers – initiated by the SoTech project workshops – is a point of success. Increased contacts among researchers from different scientific fields all over Germany – as well as the cooperation with employers, workers and employees – is one of the most impressive effects of SoTech. The publications, congresses and workshops carried out between 1986 and 1990 stress this feature. Seen from the point of view expressed by Minister President Rau in 1985, we can emphasize that the discussion-oriented approach of the SoTech programme was successful in many of the projects. Participation among the groups concerned worked out well over time.

On the other hand, the interventional aspect of discussions – as contradictory as they were – is scarcely measurable. Even they became an administrative problem when social actors and pressure groups began to oppose or criticize the results of single projects. One requirement is clear to social scientists: a social consensus needs at least two different positions and interest groups with a power base – and, secondly, a process of negotiations or conflict resolution among them. However, this fact imposes legitimate problems for programme administrators, who are used to routine tasks and to presenting results without controversial political implications.

Let us finally make a remark about cooperation and integration at the project level. Cooperation between projects from the different programmes was minimal; however, with respect to administrative coordination problems and to the legitimate difficulties of programmes enacted to broaden societal discussions, perhaps the only successful way to reach social aims might be to continue with separate programmes and projects. Coordination processes within the administration could more or less ameliorate aims and tasks of the single projects for a governmentally-defined compromise. This approach could bear the danger that projects could not produce a clear analysis of the restrictions of a socially-compatible implementation of ICT: they would be forced to stress other aspects in project design, too – thereby shifting the research focus.

5 Outcome and Impact – Learning Effects on Different Levels

Even taking into account all the critical points mentioned previously, there are several aspects of the SoTech programme and the Initiative that indicate an – at least gradual – change in technology policy in NRW. Some are worth mentioning here. However, this anamnesis is contradictory at some levels.

- At an administrative level, the first aspect is that the SoTech programme is no longer a single initiative spanning a short period of time, but rather a continuing task of the government and administration. After two years of scarce financing, both the evaluation and transfer of results has now brought financial backing from the Ministry for SoTech up to 4 million DM again. This still is not the level the programme received in the beginning; but due to the aforementioned financial restrictions of the federal states, it is remarkable. On the other hand, programme management within the ministry has still to struggle to retain this amount of money. Apparently, after almost six years, this kind of endeavour towards technology policy within the NRW government is not stable at all.
- Presently, there are other activities within technology policy in NRW that are affected by the SoTech programme and its socially-oriented strategies. The parallel structure to the administrative programmes in NRW's Parliament was the *Kommission Mensch und Technik* (the Commission for Man and Technology), established both to supervise the Initiative and to draw conclusions for Parliament. The commission, initially established for only one legislative period, has now been transformed into a continuing parliamentary committee with tasks comparable to the commission. This is unique in Germany; and as far as we now know, in Europe too. As a second activity, the *Institut Arbeit und Technik* (Institute for Work and Technology) in Gelsenkirchen was founded in

1988 as part of the NRW Science Centre. Besides the institute's own research tasks, installed there were the scientific supervision and administrative support of the SoTech programme.

- Examining scientific effects, one can perceive two different points having to do with the continuity of research groups (short-term perspective) and with scientific orientation (long-term perspective).

  Some of the scientific work groups were able to raise additional funds at the end of the two years, lending continuity to several SoTech projects. Some of these groups still work as scientific consultants for ICTs' implementation problems. Others managed to start new projects. But not all of the scientific groups financed by SoTech are able to continue their work. The research network financed by the programme is collapsing because funding, even by other means, is becoming increasingly difficult to obtain.

  From our perspective, another remarkable point is the qualificational aspect for scientists. In the SoTech projects, some 300 scientists in social and technical sciences worked for two years on problems regarding a socially-acceptable way to implement and use ICTs. This means, especially for younger scientists, a (re)-orientation concerning their work – and induced by this fact, for part of the sciences too. This effect is not measurable now, but there probably will be continuing and long-lasting effects within the sciences.
- Looking at the impact on working life, there were 'interventions' based on the projects in the work situation at the shop floor level. This work can be characterized as developing and testing methods for technology implementation. But as in other action-research-oriented programmes, there were problems transferring the new ideas and methods developed by the projects. The danger still exists that the models will become islands (i.e. isolated). Continuity and further development in several projects of the SoTech programme, is one way to increase the diffusion of results. The other way is to fund projects considering transfer problems in project design.
- Looking at the conceptual effects of the SoTech programme, we can emphasize that a socially-compatible way of implementing and using ICTs has been worked out and established as a *"Leitbild"* (model), drawing upon the results of the projects (von Alemann et al. 1992). They are now diffusing through the discussions on ICTs and will thereby certainly have an effect. On the other hand, if one looks back on a history of nearly 20 years of discussions on the humanization of working life in Germany, one has to recognize that this kind of innovation needs a long time to become established and to produce societal impacts. Maybe a comparable estimation can be made with SoTech: short-term effects, which are few, are overestimated in political discussions on the programme; while long-term impact is underestimated.
- Regarding the project level, the SoTech programme is now focusing on transfer aspects – on the one hand – and is trying to improve the integration of industrial management and administration on the other. This second point, which is a

necessity for the implementation as well as for legitimation of the programme, could contribute to a step backwards to the state of conceptual discussion. This could happen if industrial projects within firms or administrations become isolated and are not 'embedded' in activities working in the complex social environment of production and services. One of the central results of the SoTech programme was the experience that the main obstacles for an economically- and socially-compatible way of implementing and using ICTs, cannot be solved in a firm or on the shop floor level alone – especially not by single, two-year projects. These strategies need external backing and continuing support, e.g. by consultants or scientists and external qualification programmes. If this support is not available or is not organized according to the needs of the projects, a failure – or at least a partial one – is easy to predict.

Besides this aspect, though, we should mention that the concepts of discussion and public debate, as well as the development of methods and institutions for this purpose, were carried on in the ongoing projects of SoTech – as well as in other programmes such as the Bremen Work and Technology programme (*Sachverständigenkommission Arbeit und Technik* 1988) and the federal *Arbeit und Technik* (Work and Technology) programme, which succeeded the HdA programme.

6 Conclusions

Taking into account all the points mentioned before, it seems evident that a friction-free integration of social and economic aims failed. But in our view, the central point of the SoTech programme is the fact that problems related to social organization and production regulation by participatory means can no longer be ignored in the process of ICT implementation and use. The integration of both social and economic aspects is necessary; and this will be a continuing task for industry, unions, workers and administration in the future. The reason for increasing efforts and investing research into this aspect, is a normative one, i.e. the increase of security and satisfaction in the work place. It is also a functional and economic one, i.e. the necessity to increase worker involvement in the production process at the firm level in order to obtain economic success. These social and economic aims are still different and cannot be integrated into one aspect.

To further increase the complexity of our argument, it is necessary to see that the economic success of socially-compatible solutions – which in our view, would be optimal in the integration of both aims – is influenced by macro-political as well as by micro-political processes.

At the macro-level, political regulations define and influence the frame of economic success in production processes. Even a socially-acceptable but uneconomic way of production, or new technological appliance, can become the economic- and success-wise preferable for a company if the formerly-economic solution is levied with taxes or higher insurance fees. However, this way of internalizing external costs to achieve socially-acceptable production concepts, cannot be implemented without political conflicts and without the risk of missing economic aims.

At the micro-level, the results of several SoTech projects suggest that only in the concrete process of implementation and 'shaping' of technologies and by user participation, will a solution emerge and develop that ex post might be considered an economic one. The degree of "participation" or "involvement" (Thorsrud 1976), control or acceptance, use or misuse of ICTs cannot be correctly predicted. This is due to a continuing reformulation of the aims of implementation in the ongoing discussion and decision processes within firms (Becker et al. 1990; Behr et al. 1991; Manz 1990). Any technology programme can only support these processes – thereby attempting to give scientifically-based consulting, information and backing support to the different groups of actors. Socially-oriented technology programmes (as do others) act on social processes which find their manifestation in controversial debates and discussions. These discussions may lead to the "consensus" mentioned; but primarily, they mean conflicts of structurally-different interests on the shop-floor as well as within the economic and political environment.

Efforts to support these processes lead to administrative problems – they prototypically emerged in the implementation of the SoTech programme. Besides the structural point mentioned, it seems to us as if up to now it has been impossible for the administration to merely initiate and regulate societal discussion processes without being able to control the results – even if the administration is not able to otherwise execute more than a symbolic kind of policy. Legitimate reasons (e.g. the need for demonstrable solutions in firms), together with the decreasing amount of money spent on SoTech, might lead to a situation of a conceptual step backwards. If a social movement concerned with ICTs had emerged, the situation would be totally different.

Considering the political and administrative conflicts which socially-oriented technology programmes often produce (and seem to produce more often than other policies, because of efforts to integrate contradictory societal interests), it would not be astonishing that in Germany financial ressources and administrative staff be reduced in the coming years – because of financial restrictions due to the German unification or due to the changes in the international situation. Furthermore, as long as the labour market effects of these programmes are few (as they still are), there are other urgent problems on the political agenda to be solved prior to the socially-oriented 'shaping' of technologies – at least from an administrative point of view.

As can be implicitly realized, environmental aspects did not play a strong role in SoTech – neither at the level of programme formulation nor at the level of administrative organization nor in the implementation of most of the projects. As far as we can see at present, there will not be any further integration of environmental and social aspects in the programmes of the "Initiative" – at least not as cross-sectional tasks in all programmes. At present, the administrative view of environmental problems in NRW is focussed on "traffic break-down" in the Ruhr region and on the "waste-management crisis". An administrative structure which could import socially-perceived environmental dangers, does not yet exist; and therefore, it is easy to predict that the integration of environmental affairs in other programmes will remain close to existing administrative perceptions and legally-defined tasks (Luhmann 1986). We suspect that here, as in the integration of social and economic aims, it will be only through a furtherance of social discussion processes and public debates, and by an evolution of corresponding administrative structures, that an integration of the three groups of aims can be fostered in the long run.

Notes

[1] "Shaping" is not an accurate translation of the German expression *"gestalten"* but will be used here.

[2] The amount of money for computing and data-processing subsidies varied between 4% and 11% of the financial ressources spent by the Federal Ministry of Research and Technology between 1967 and 1979 (Beckerhoff/Sommerlatte 1982:114).

[3] One of the reasons for the possibility of this kind of policy, is that the legal authority for the universities is linked to the federal states. The establishment and the work of university departments is financed by the federal states.

[4] For coal mining, 1974 – 1984: 620 million DM; for steel, 1980 – 1990: 370 million DM; for non-nuclear energy, 1974 – 1984: 697 million DM; for nuclear energy (same period): 964 million DM (Vitols 1991:22).

[5] At the end of 1989, there were 14 science & technology parks in NRW focusing on different technologies; 10 of them were founded between 1985 and the middle of 1987. Only the parks in Gronau, Kleve, Gelsenkirchen and Schwerte are not in cities with universities, but they are all closely connected with universites or research institutes.

[6] However, this conjecture could not be verified by the different polls and surveys carried out in the years following (Hoschka/Lange 1983; Müller-Böling 1983; Jaufmann/Kistler 1986, 1988; Böltken 1988). Spectacular political actions concerning ICTs did not occur in the following years.

[7] The full title of the ministry is *"Ministerium für Wirtschaft, Mittelstand und Technologie des Landes Nordrhein – Westfalen"*. *"Technologie"* was added to the title in 1984 and indicates the increasing role of technology in industrial restructuring at that time.

References

Alemann, U. von, H. Schatz, G. Simonis, E. Latniak, J. Liesenfeld, U. Loss, B. Stark, W. Weiß (1992) *Leitbilder sozialverträglicher Technikgestaltung. Ergebnisbericht des Projektträgers zum NRW Landesprogramm 'Mensch und Technik – Sozialverträgliche Technikgestaltung'*, Opladen: Westdeutscher Verlag.

Alemann, U. von, H. Schatz, D. Viefhues (1985) *Zielsetzungen und Handlungsfelder des Programms Mensch und Technik*, Düsseldorf: Ministerium für Arbeit, Gesundheit und Soziales.

Alemann, U. von, H. Schatz (1986) *Mensch und Technik. Grundlagen und Perspektiven einer sozialverträglichen Technikgestaltung*, Opladen: Westdeutscher Verlag.

Becker, A., G. Ortmann, H. Schulz, A. Windeler (1990) *Computer und Macht in Organisationen. Mikropolitische Analysen*, Opladen: Westdeutscher Verlag.

Beckerhoff, D., T. Sommerlatte (1982) *Die Entwicklung der Datenverarbeitung in der Bundesrepublik Deutschland – Programmbewertung der DV-Förderung des BMFT 1967-1979*, Bonn: Bundesministerium für Forschung und Technologie.

Behr, M., M. Heidenreich, G. Schmidt, H.A. Graf v. Schwerin (1991) *Neue Technologien in der Industrieverwaltung – Optionen veränderten Arbeitskräfteeinsatzes*, Opladen: Westdeutscher Verlag.

BMFT (1984) *Informationstechnik. Konzeption der Bundesregierung zur Entwicklung der Mikroelektronik, der Informations- und Kommunikationstechniken*, Bonn: Der Bundesminister für Forschung und Technologie.

Böckler, M. (1990) *Sozialverträgliche Technikgestaltung. Zum Verhältnis von technologischer Reform- zu sozialwissenschaftlicher Praxiswende*, Bochum: Schallwig.

Böltken, F. (1988) Einstellungen gegenüber 'neuen Techniken': Technikakzeptanz im regionalen Vergleich, in: *ZA-Information* 22: 107-113.

Bruder, W. (ed.) (1986) *Forschungs- und Technologiepolitik in der Bundesrepublik Deutschland*, Opladen: Westdeutscher Verlag.

Bruder, W., N. Dose (1986) *Forschungs- und Technologiepolitik in der Bundesrepublik Deutschland*, in: W. Bruder (ed.) *Forschungs- und Technologiepolitik in der Bundesrepublik Deutschland*, Opladen: Westdeutscher Verlag, 11-75.

Dierkes, M., Th. Petermann, V. von Thienen (eds.) (1986) *Technik und Parlament. Technikfolgenabschätzung: Konzepte, Erfahrungen, Chancen*, Berlin: Edition Sigma.

Dose, N., A. Drexler (eds.) (1988) *Technologieparks. Voraussetzungen, Bestandsaufnahme und Kritik*, Opladen: Westdeutscher Verlag.

Ellwein, Th., W. Bruder (1982) *Innovationsorientierte Regionalpolitik*, Opladen: Westdeutscher Verlag.

GEWOS, GfAH, WSI (eds.) (1988) Strukturwandel und Beschäftigungsperspektiven der Metallindustrie an der Ruhr, Ergebnisbericht, Hamburg: GEWOS.

Grabow, B., H. Heuer, G. Kühn (1990) *Lokale Innovations- und Technologiepolitik. Ergebnisse einer bundesweiten Befragung*, Berlin: DIFU.

HdA (1982) *Ein Programm und seine Wirkungen: Analyse von Zielen und Aspekten zur Forschung 'Humanisierung des Arbeitslebens'*, Frankfurt/Main, New York: Campus.

HdA-Projektträger (1981) *Das Programm 'Forschungen zur Humanisierung des Arbeitslebens' – Ergebnisse und Erfahrungen arbeitsorientierter Forschung 1974- 1980*, Frankfurt/Main, New York: Campus.

Hoschka, P., K. Lange (1983) Einstellungen der Bevölkerung zum Computer – Ergebnisse empirischer Untersuchungen, in: *GMD-Spiegel* 3/4: 30-35.

Hucke, J., H. Wollmann (eds.) (1989) *Dezentrale Technologiepolitik? Technikförderung durch Bundesländer und Kommunen*, Basel, Boston, Berlin: Birkhäuser.

Jaufmann, D., E. Kistler (1986) Technikfreundlich? – Technikfeindlich? Empirische Ergebnisse im nationalen und internationalen Vergleich, in: *Aus Politik und Zeitgeschichte* 48: 35-53.

Jaufmann, D., E. Kistler (1988) 'Un-Möglichkeiten' eines nationalen und internationalen Vergleichs demoskopischer Daten – Einige Erfahrungen und Einsichten aus der Suche nach dem 'technikfeindlichen Deutschen', in: *ZA-Information* 22: 45-62.

Junne, G. (1989) Competitiveness and the Impact of Change: Applications of 'High Technologies', in: P.J. Katzenstein (ed.) *Industry and Politics in West Germany*, Ithaca, London: Cornell University Press, 249-274.

Katzenstein, P.J. (1989a) Industry in a Changing West Germany, in: P.J. Katzenstein (ed.) *Industry and Politics in West Germany*, Ithaca, London: Cornell University Press, 3-29.

Katzenstein, P.J. (1989b) Stability and Change in the Emerging Third Republic, in: P.J. Katzenstein (ed.) *Industry and Politics in West Germany*, Ithaca, London: Cornell University Press, 307-353.

Kitschelt, H. (1980) *Kernenergiepolitik. Arena eines gesellschaftlichen Konflikts*, Frankfurt/Main, New York: Campus.

Luhmann, N. (1986) *Ökologische Kommunikation*, Opladen: Westdeutscher Verlag.

Manz, Th. (1990) *Innovationsprozesse in Klein- und Mittelbetrieben. Soziale Prozesse bei der Einführung neuer Technologien*, Opladen: Westdeutscher Verlag.

Menneken, J.B. (1986) Die Forschungs- und Technologiepolitik der Bundesregierung, in: W. Bruder (ed.) *Forschungs- und Technologiepolitik in der Bundesrepublik Deutschland*, Opladen: Westdeutscher Verlag, 76-104.

Müller-Böling, D. (1983) Informationstechnik aus der Sicht der Benutzer im Wandel des letzten Jahrzehnts, in: *GMD-Spiegel* 3/4: 35-41.

MWF (1986) *Nordrhein-Westfalen Initiative Zukunftstechnologien. Wissenschaftsteil – Zwischenbilanz*, Düsseldorf: MWF.

MWMT (1990) *Technologie in NRW. Handbuch '90*, Köln: Verlag Deutscher Wirtschaftsdienst.

Naschold, F. (1989) Technikentwicklung und gesellschaftliche Regulation – Das Projekt 'Forschungszentrum Ulm', in: *WSI-Mitteilungen* 3: 144-155.

Petzina, D. (1989) Wirtschaftspolitik, in: H. Boldt (ed.) *Nordrhein – Westfalen und der Bund*, Köln: Kohlhammer Verlag, 122-135.

Petzina, D. (1990) Wirtschaft und Arbeit 1945 – 1985, in: W. Köllmann, H. Korte, D. Petzina, W. Weber (eds.) *Das Ruhrgebiet im Industriezeitalter. Geschichte und Entwicklung*, vol. 1, Düsseldorf: Schwann-Patmos Verlag, 491-569.

Riepert, W. (1988) *The Special Way of Technology Policy in North Rhine-Westphalia*, paper presented at the 1st Conference of the European Chapter of the International Association for Impact Assessment, Leiden/Delft.

Sachverständigenkommission Arbeit und Technik (1988) *Arbeit und Technik. Ein Forschungs und Entwicklungsprogramm*, Bonn: Neue Gesellschaft.

Schatz-Bergfeld, M. (1988) The North Rhine-Westphalian Project 'Man and Technology', in: *The Information Society* (5): 233-244.

Simonis, G. (1989) Modernisierungspolitik in Nordrhein-Westfalen. Vollbeschäftigung als Vision, in: J. Hucke, H. Wollmann (eds.) *Dezentrale Technologiepolitik*, Basel, Boston, Berlin: Birkhäuser, 348-377.

Simonis, G., E. Latniak, U. Loss, W. Weiß (1990) Gesellschaftorientierte Technologiepolitik. Das Landesprogramm 'Mensch und Technik – Sozialverträgliche Technikgestaltung', in: W. Fricke (ed.) *Jahrbuch Arbeit und Technik 1990*, Bonn: Dietz, 149-162.

Späth, L. (1987) Wirtschaftspolitische Umbrüche: Der Weg zur Innovationspolitik in Baden-Württemberg, in: H. Buddenberg (ed.) *Umbrüche – Neue Strukturen in Wirtschaft und Gesellschaft*, Herford: Busse & Seewald, 354-380.

Staatsministerium (ed.) (1983) *Zukunftschancen eines Industrielandes. Kongreß der Landesregierung Baden-Württemberg am 13./14. 12. 1983. Dokumentation*, no place: Staatsministerium Baden-Württemberg.

Thorsrud, E. (1976) Democratization of Work as a Process of Change towards Non-Bureaucratic Types of Organization, in: G. Hofstede, M.S. Kassem (eds.) *European Contributions to Organization Theory*, Assen-Amsterdam: van Gorcum, 244-271.

Väth, W. (1986) Konservative Modernisierungspolitik – Ein Widerspruch in sich? Zur Neuausrichtung der Forschungs- und Technologiepolitik der Bundesregierung, in: *PROKLA* (56): 83-103.

Vitols, S. (1991) Economic Restructuring in the Ruhrgebiet: Challenge for Trade Unions and Regional Policies, unpublished paper, Berlin: WZB.

Weißbach, H.J., J. Niebur (1986) Maschinenbau des Ruhrgebiets im Umbruch, in: W. Fricke, K. Johannson, K. Krahn, W. Kruse, G. Peter, V. Volkholz (eds.) *Jahrbuch Arbeit und Technik in Nordrhein Westfalen 1986*, Bonn: Neue Gesellschaft, 337-350.

Wildenmann, R. (ed.) (1985) *Umwelt, Wirtschaft, Gesellschaft – Wege zu einem neuen Grundverständnis. Kongreß der Landesregierung 'Zukunftschancen eines Industrielandes'. Dezember 1985*, Gerlingen: Staatsministerium Baden-Württemberg.

National Policies Devoted to Technology and the Environment in France: Towards an Integrative Approach?

Lahsen Abdelmalki and Thierry Kirat

By stating that "environmental protection cannot be regarded as a separate sector but must form part of all economic policies and social activities" (UN/ECE 1988:185), during the 43rd session members of the UN's Economic Commission for Europe clearly highlight the present and future stakes in national policies aimed at protecting the environment and safeguarding our natural heritage. The interdependence of the different factors involved make this a highly complex issue of particular significance since environmental protection has technical, political, social, cultural and ethical ramifications. Moreover, this issue inevitably has a worldwide impact as regards sovereignty and the coordination of national policies. This article does not seek to analyze and identify the various less damaging forms of economic development but will focus on the need for technology and innovation arising from the protection of our natural heritage.

The steadily growing volume of academic and applied research into environmental protection seems impervious to recent efforts deployed in the field of technological research. It is clear, however, that rapid technological progress must be made to quickly reduce CO_2 and NO_x emissions, the underlying causes of the greenhouse effect, eliminate phosphates contained in water and treat industrial and domestic waste.

This progress is spurred on by science, diagnosis techniques, different measures and the development of less polluting processes and products. It is, however, doubtful whether environmental goals are integrated into the technology policy of governments. This raises various problems, first and foremost the status of environmental policy within the general framework of national policies, particularly technology policy. Consideration must therefore be given to the rationale behind choosing the ambitions of public and private operators and the ways they fulfil them; possible rivalries between different power-vested operators (or groups of operators) and their inbred philosophies must also be considered. The protection and conservation of the environment is much more than a technical issue; it assumes an economic, social and political dimension. These matter a continuous undercurrent to the present article.

The first part of this article will analyze the features and trends of technology policy in France under the Fifth Republic since 1958, with emphasis on the implicit nature of the innovation process underlying policies benefiting science and technology. Attention will then turn to the constraints of environmental protection in technology policy. We shall show that while French technology policy does not set final objectives in environmental technology, such objectives are not overlooked. In particular, technology policy is more geared to the creation of institutions aimed at developing certain areas of technological research initiated by industrial firms than on shaping definitive national research programmes, except for a few interdisciplinary research programmes initiated by the CNRS. Recent trends in this area will also be discussed.

This article will show that environmental policy differs from technology policy in three respects: its theoretical premises, ways and means and objectives. The former draw on a welfare-oriented vision rooted in the conviction that since polluters pay for the damage they cause, levels of pollution in society are poorly regulated. We will try to understand the deep disparity between the ways and means and objectives. It is clear that public bodies and specialized organizations make little effort to finance environmental research. However, a recent reform of the French environmental policy, launched in connection to the implementation of a National Plan for Environment (NPE) has engendered a radical shift in this field and has led to the development of environmentally-compatible technologies.

This article will also examine how the environment can be protected through the discipline of political economy. This is a particularly important issue since economic analysis is not merely a tool with which to examine analytical and scientific neutrality (cf. Abdelmalki/Kirat 1991). Besides its implicit normative dimension, the application of economic analysis to solve environmental problems can serve some social groups eager to legitimize their own interests. It may also provide a reference framework within which to set goals and ways to implement a national environmental policy. It therefore seems crucial to investigate the very nature of politico-economic models, their heuristic value and their instrumentalization in society. This will be discussed in the latter part of this article which will show the two-fold limitation (heuristic and instrumental) to economic analysis based on a neoclassical approach.

This will be the first step towards recognizing an evolutionary and institutionalist approach to the question. This theoretical shift will be especially significant as analysis will concentrate on national policy, which requires a social and institutional strategy to tackle environmental protection. Finally, the glaring dichotomy between environment and technology policy will be discussed. This article will show that this dichotomy seems on the point of being removed by the emergence of new forms of organization and cooperation between different operators involved in protecting the environment.

1 National Technology Policy and General Innovation-Supporting Mechanisms

The effective protection of our natural heritage should be matched by upgrading technologies, i.e. a major effort must be made to develop 'clean' technologies in terms of prevention and repair. Under certain conditions, an analysis must be conducted from the perspective of technological creation. The major contribution made by the evolutionary theory of technological change (Dosi 1988) has been to show that the creation of technologies involves a collective and cooperative process. It therefore presupposes certain forms of research structures related to innovation which can be converted into relevant institutional measures. These in turn provide a framework for relations and cooperation between the different players involved in innovative processes. Technological creation then becomes a technical as well as an organized process; and ultimately, institutionalized. This raises the question of the role of a national policy in favour of science and technology.

1.1 Theory and Practice of National Policy in Favour of Science and Technology

As Andrew Jamison (1989) rightly pointed out, national policies dealing with science and technology develop according to ways in which the nature of the innovation process is conceived. In the industrialized countries, the goals of such policies are constantly shifting. The seminal shaping of policies supporting scientific and technical progress was based on the principle that major science is to tackle major problems. In the 1960s and 1970s, scientific policies were heavily influenced by a welfare-oriented view of the production of scientific knowledge able to support technological progress. According to this view, the emphasis was on market failures in order to allocate scarce resources to innovation as effectively as possible. Since information was supposed to be cheap and not fully appropriable, the production of scientific knowledge was deemed to produce public goods (Nelson 1959). Consequently, the further back one went into the sequence of fundamental research -> development -> industrialization -> production -> market, the more knowledge was supposed to be inappropriable. This reference framework made it necessary for scientific policies to compensate for the deficiencies of private firms in the production of scientific knowledge.

Recent studies have altered this view. Kline and Rosenberg (1986) showed that innovation is far more complex and systemic than traditional models usually suggest. Nelson and Winter (1982) and Dosi (1988), all three evolutionist theorists, have highlighted the sequential and cumulative nature of technological innovation. Jamison (1989:505) rightly pointed out that innovation is today being reconsidered as "a wide-ranging and multifaceted social activity" deserving an analysis based on the "nature and intensity of various categories of economic operators involved in its conception and diffusion" (Dufourt 1991:366). Such an analysis could throw light on the fact that national action in favour of technological progress is less a question of how many resources are allocated to the project than it is of creating and developing institutions responsible for implementing innovative activities.

1.2 From Scientific Policy to Technology Policy

In France, as Dufourt (1991) pointed out, national policy aimed at promoting scientific and technological progress has undergone three main phases of development since 1958. This is true for both government objectives and the related institutional framework and legislation. A logic of institutional change, punctuating national policies in science and technology, is matched by the search to adapt to structural changes that affect the industrial system. These changes may be interpreted as technological mutations that cross different sectors coupled with the emergence of new generic technologies in the areas of information technology, new materials and biotechnologies.

The three phases of development implemented in France over the past thirty years are as follows:

1. *From 1958 to 1970:* Policy concerning science and technology over that period of time could be called, to quote Henry Ergas' (1987) phrase, "mission-oriented". The emphasis at that time was placed on the production of major innovations in a few sectors considered as having national and strategic importance. Therefore, the supply of public goods related to innovation and their distribution in the industrial system were given a secondary role. As illustrated by the Plan Calcul in the computer industry or by the Concorde programme in the field of aeronautics, the typical public approach was at that time to launch large-scale programmes and to give absolute priority to fundamental research. Among the dedicated fundamental research institutions created over that period, the National Agency for the Promotion of Research (ANVAR) was devoted to increasing the vertical transfer of knowledge that can be described as follows: fundamental research -> applied research -> industrial development.

2. *From 1971 to 1978:* The priority was to help research programmes with clearly defined industrial goals. Privileged means of intervention were implemented in two fields; on the one hand, the environment of firms and, on the other hand, obstacles hindering innovation. This new national policy clearly illustrates the shift from an upstream form of management ('large-scale programmes') to a downstream one, which is mainly in the hand of firms themselves. From an institutional point of view, this evolution led in 1971 to the creation of the appointment of an "industrial relations director". At the same time, the National Bureau of Scientific and Technical Information (BNIST), the Regional Agencies of Scientific

and Technical Information (ARIST) working under the authority of the Chambers of Commerce and Industry, a National Bureau of Metrology (BNM), were set up.

3. *Since 1979:* National policy became marked by the need to find a new balance between fundamental research and industrial development. The emphasis shifted to the circulation of research results particularly national research results within the industrial system. In order to achieve this goal, the ANVAR was revamped several times between 1979 and 1984. Initially, it was made responsible for the management of financial aid and subsidies. Then in 1982, it monitored loans for industrial renovation of the Industrial Fund for Modernization (FIM). The legal status of the agency was also modified. It became an Industrial and Commercial Public Agency (EPIC) with three issues: promote the diffusion of research results, develop innovation and contribute to industrial modernization (Dufourt 1991).

1.3 Technology, Institutional Change and National Policy

It is important to point out that the French technology policy reveals real awareness of the pre-requisites needed to implement such a national policy. These pre-requisites are: the role of information in innovation processes; the complexity of relations between the different institutions involved; the form of organization between the different institutional partners whose co-operation is a key factor in the success of any innovation strategy. Innovation as a process of creation of technology is a collective, cumulative and interactive phenomenon, whose efficiency fully depends on the existence of a form of organization compatible with these characteristics (Allen 1983; Dosi 1988; von Hippel 1988).

Nevertheless, it is possible to analyze the institutional action plans launched to pursue technology policy. The same is equally true for the tools used to promote the practice of collective forms of technical research within the industrial system. Therefore, technology policy necessarily implies a transformation of institutional structures. It will guarantee optimal use of the technological resources available to the parties concerned since technological progress cuts across the boundaries of firms, industries and sectors. The very nature of modern technology, its trans-sectoral and generic character, underlies the workings of technology policy in France. As the following points out, "In France, particularly during the eighties, the search for stable patterns of partnership conducive to the creation of technology, whether this implies cooperation between firms or between firms and other institutions, as well as major efforts to develop fundamental research, are the two main trends in national policy" (Dufourt 1991:366).

This transformation of institutional structures can best be seen as a move towards the promotion of a mixed economy. This has been patently obvious since 1958. Change in legislation and the legal status of different forms of association between partners such as public bodies, also follow this trend. Since 1982, a great variety of such different bodies have been created, e.g. the Public Administrative Agency (EPA), the Scientific and Technical Agency (EPST) and the Public Industrial and Commercial Agency (EPIC). All are governed by regulations which provide a framework and a programme for governmental action in the field of research and technological development. According to Dufourt, this has been prompted by a "threefold desire: firstly to develop public research, liberalize management methods and finally promote collaboration with the private sector through the creation of subsidiaries" (Dufourt 1991:366).

This latter point reveals a certain consistency in French policy, since it demonstrates the effort undertaken since 1958 to promote collaboration between public and private actors in technological research. This collaboration is conducted within a working institutional framework. It is therefore clear that French technology policy is not merely based on mechanisms designed to correct the shortcomings of research in terms of funding research activities. This would be a rather poor way of solving the problem. On the contrary, it shows a certain clear-sightedness as to the nature and requirements of innovation processes. All this is reflected in the creation of institutional and legal conditions designed to facilitate public and private partnerships around the research projects. It is quite obvious, as we shall see later, that technological research related to the environment may here find a suitable conducive framework, whereas for a long time it seemed unable to find it in the institutionalized measures undertaken to pursue environmental policy.

2 French Environmental Policy, Past and Present

In reviewing French environment policy we must not only describe related trends, strategies and decisions but also cast a critical eye to identify strengths and weaknesses. It is obvious that merely describing any environmental policy is quite unsatisfactory; this should be backed by emphasis on the foundations of such policy.

2.1 Environment and Environmental Policy: An Introduction

While one may agree on a suitable action or policy to treat and protect the environment, there is some doubt as to the purpose of such a policy.

2.1.1 Definition and Scope of the Environmental Issue

Here, one faces a tricky, almost insoluble problem, given the absence of any objective and accurate references. Has the word "environment" anything to do with Nature, or should it be understood in terms of surroundings? Is it a global problem which could be equated with our civilization, or should it be boxed into economic, technical and cultural sectors? Whatever the answers to these questions, how can we qualify an analysis of the environment?

One could wave this embarrassing problem of definition aside and either argue that the environment is constantly changing, that it is a matter of intellectual speculation or to say that any definition, however accurate, is of little interest *per se*. This stance may lead one to focus on our living habitat and our natural heritage in order to assess these conditions or their scope. In the final analysis, one can fathom how difficult this task is. Environmental issues have suddenly been taken up by the economist. In many respects, economists feel ill-equipped to tackle these issues with only the traditional means at their disposal: they may opt for a globalist approach but run the risk of being overwhelmed, or they may choose a more fragmented approach which will ultimately leave them unable to grasp the numerous parameters that characterize the human environment and its evolution. As was suggested by Roqueplo (1990), there is a middle course of "interdisciplinary knowledge". It represents a qualitative leap between a multi-disciplinary approach and an immediately global one. Roqueplo himself admits that his solution clashes with what can be called three epistemological obstacles:

- Firstly, any knowledge in terms of environment supposes experimental and theoretical procedures which would give it a degree of plausibility. This only becomes possible if preliminary procedures already exist in order to settle disagreements or validate truths.
- Secondly, in any diagnosis on the environment, all laws of nature should be perfectly known and accounted for.
- Last but not least, government legislation on the environment may be assumed to be limitless. The links between these laws are probably very complex, and this complexity is compounded by a number of uncertainties: physical limitations, adaptability and the responsiveness of human beings.

In conclusion, we may say that environmental knowledge carries a risk in that it only derives its legitimacy from agreements that generate reasonable and objectively-based knowledge within a set and temporary time frame.

2.1.2 Trends and Practice in Environmental Policy

Most OECD countries have begun implementing environmental policies in the last 20 or 30 years. The general framework of the operational tools developed consists in the creation of economic instruments aimed at encouraging polluters to lower levels of pollutant emissions and stimulate clean technologies. These tools were mostly based on the premise that pollution is caused by external economies of consumption and production.

Two approaches corresponding to distinct types of environmental policy instruments can be followed in an environmental policy; the first can be termed a regulatory one, owing to the fact that it involves setting up institutional measures aiming at directly "influencing the environmental performance of polluters by regulating processes or products used, by abandoning or limiting the release of certain pollutants, and/or by restricting activities to certain times and areas through licensing, standard setting and zoning" (Opschoor/Vos 1989:12). The second approach is less constraining, since its aim is to provide economic and monetary incentives for polluters to behave in a more environmentally-compatible way. It entails the implementation of several kinds of economic instruments such as charges, subsidies and deposit-refund systems.

A new economic sector might be created that would integrate the biological, aesthetic and scientific value of Nature to move towards optimum growth in terms of economic return on investment and development. At the same time, such integration raises the acute problem of the type of approach that would be used: thus, the attacks on forest land and noise in urban areas for the 'intersection' between the economy and ecosystems. Do they exclusively belong to the realm of natural sciences, or are they merely economic problems; in that case, how are they to be analyzed? W. Kapp, Kneese and Ayres provided an original answer, but unfortunately its appeal was short-lived. They thought the identification and quantification of physical flows from the point of view of extraction and waste should precede monetary analysis (see Bourguinat 1973).

Under these conditions, what could an environmental policy be, and what should be its aim? Again, one inevitably hesitates when choosing a precise course of action. To find a way out of this dilemma, one may choose between the satisfaction of having raised a practical problem or take a negative attitude which involves shaping environmental policy in terms of what it should *not* be, i.e.:

- *a short-term policy*: indeed, any radical change in soil management, production plants that cause pollution and forecasting methods require time and money to such an extent that the greater need for material resources places constant pressure on the use of space and non-renewable resources;
- *a curative policy only concerned with symptoms*: the risk here would be to deal only with well-known consequences of pollution such as mental fatigue, nervous tension and various frustrations, only detectable after a long period of accumulation;
- *a fragmented policy, either in terms of speciality or in terms of degree of gravity*: specific corrective actions may not only be ineffectual but they could also defeat the original purpose. For example, removing pollution from the air could pollute water; burning industrial waste might affect the purity of air. While focusing one's attention on industrial processes, one may overlook different products and the way they are used. The emphasis below will be on the need to distinguish between the so-called end-of-pipe technologies and genuine 'environmental technologies'.

2.2 Features of French Environmental Policy

For a long time, national policies in environment, technology, industry or energy assumed that the industrial system of production was conceived as a conservative system, and not in terms of a dissipative one (Ayres 1989). The environment has remained a kind of 'alien' factor; to integrate it was considered almost as an expensive form of regulation.

The environmental policy implemented in France currently undergoes dramatic changes related to both its institutional architecture and its aims and scope. These changes will be analyzed after summarizing the main characteristics of the policy which have prevailed until the French government adopted the National Plan for the Environment in December 1990.

2.2.1 A Belated, Flexible and Centralized Environmental Policy

In France, environmental policy is both overdue and conservative. This delay can be observed at two levels. Firstly, the Ministry of Environment was created in 1990 (admittedly it then replaced the Secretary of State for the environment created in 1974). Furthermore, paltry funds were allocated, only 0.07% of the total budget allocations to all ministries in the 1990 budget (in 1989, the environmental budget was only 0.06% of the total allocations). While this small allocation for the environment can explain the difficulties in coordinating the different actions, it also highlights the very cautious attitude of the government towards the environment.

The hallmark of state action is its lack of interest in large-scale reforms, probably because it has too much to deal with and this would cause major disruptions; the costs would be high and the benefits may not be clearly discernable. Therefore, the state has merely a regulating action with very unequal constraints. It is there to compensate for the very low commitment of both companies and households. In 1987, the state expenditure in national environment programmes accounted came to 53.8%, companies 26.4% and households only 19.8%. About 66% of that money paid for running costs, while 34% was used to finance investment projects. Today these two budgets respectively absorb 68% and 32% of the total amount allocated to the public administrations (see Tables 1 and 2).

The relative weakness of resources allocated to investment, which aims to create a suitable technological and institutional framework likely to influence the programming and implementation of national decisions in environmental terms, has also been regularly decreasing over time. In 1967, it represented 60% of the environmental budget; in 1975 the figure had sunk to 50%. In 1987, the share of households and companies in investment represented respectively 10.2% for the former and 6.8% for the latter in terms of national environmental outlay (Secretary of State for the Environment 1989). The greatest challenge facing environmental policy in France is therefore to abandon this neutrality over the rules of the market, as such neutrality exerts a constraint. This conservative attitude may ultimately

Table 1: Structure of the national environmental outlay in 1987 (millions of Francs in 1988 prices)

	Public administrations		Industries		Households		Total	
	Amount	%	Amount	%	Amount	%	Amount	%
Water	19,243	26.7	6,697	9.3	3,702	5.1	29,642	41.1
Air	–	–	5,862	8.1	562	0.8	6,424	8.9
Noise	544	0.7	729	1.0	1,073	1.5	2,346	3.2
Waste	11,234	15.6	5,740	8.0	1,289	1.8	18,263	25.4
Ecological patrimony	485	0.7	–	–	–	–	485	0.7
Environment	7,267	10.1	–	–	7,620	10.6	14,887	20.6
Total	38,773	53.8	19,028	26.4	14,425	19.8	72,047	100.0

Source: Secrétariat d'Etat chargé de l'Environnement (1989:39)

Table 2: Environmental outlay per agent and per type in 1987 (millions of Francs in 1988 prices)

	Investment		Functioning		Total	
	Amount	%	Amount	%	Amount	%
Public Administrations	12,277	31.7	26,496	68.3	38,773	100
Industries	4,898	25.7	14,130	74.3	19,028	100
Households	7,357	51.6	6,889	48.4	14,246	100
Total	24,539	34.0	47,515	66.0	72,047	100

Source: Secrétariat d'Etat chargé de l'Environnement (1989:39)

cancel any effort to change attitudes to waste and pollution. For example, only lately did some people insist on the possibility of upgrading techniques to improve energy efficiency by integrating within the latest technology in various materials.

Although it has been regulatory, French environmental policy has remained flexible to accommodate the difficulties of integrating the social cost of economic activities into economic calculations or of devising a large-scale environment scheme such as in other economic or social sectors. Another obstacle has been the delays and hesitations in translating the identified social costs of public decisions into its inherent or civilian structures.

2.2.2 The Tools of the Environmental Policy

As far as France is concerned, one can distinguish two courses of action in environmental policy: the first relates to operational indirect regulatory instruments, in connection with public management of the natural environment. The second, more recent than the first, includes research programmes in environmental fields, aiming at producing advanced knowledge in the functioning of ecosystems and pollution-producing processes.

Concerning the former, except in a few, well-defined fields such as water, air and waste, the role of the state has been overcentralized until recently; but as we shall see, this did not clash with its segmented, sector-specific character. The major decisions and major resources (however limited) were controlled by the central public bodies. While this centralized form of management is needed to offer a regulatory and uniform framework for environmental actions, it soon becomes sterile if it is not completed by suitable local measures able to absorb the many situations that arise in the field. A few examples will show the methods and tools used, as well as their limitations. We shall see that these measures are either ineffectual – this is the case for the management of social costs in public transport – or that they are misapplied. This is especially true of the management of water and waste.

The management of water shows that even when the public system is organized and decentralized, it is nevertheless affected by certain factors which eventually cause a relative inefficiency of its methods of management. In 1964, the Financial Water Agencies (AFB) promoters wanted the water supply to be managed like many other commodities, i.e. in accordance with prices and demand. For this system to work, it was necessary to make consumers feel responsible through an 'internalization' of the global cost of the service which would take into account the cost of supplying water as well as the cost of treating liquid waste. It was then necessary to work out an accurate system of water charges to charge each individual consumer only for water actually used and at a rate that truly reflected consumption (i.e. time of year, quantity, waste, quality of the site). But owing to the pressure of industrial firms and local authorities, the whole organization drifted into the mutualization of costs which in turn blunted the responsibility of members, whereas taxes were originally there to remind them of their responsibilities (Martin 1988).

Similarly, local authorities are in charge of waste collection. They are therefore at the heart of a specific organization of actors, and because of this position, they have a wide choice in the way they run this service. They generally subcontract these responsibilities to private companies, using the legal framework of a subcontracting kind of executive management (known as "gestion déléguée"). These private firms are generally large diversified industrial groups such as the "Générale des Eaux" or the "Lyonnaise des Eaux". Their experience in the water and waste business means that they are well qualified to deal with environmental problems, especially in urban areas. Because of the conjunction of several phenomena, such as the considerable aging of the urban technical networks, they have to run an ever-increasing number of different forms of pollution and many of these firms have accepted to make considerable efforts.

The way in which road tax and freight transport are worked out in financial terms show how difficult it is to move from global social costs to the implementation of a market price. Indeed, several studies (Bernadet/Lasserre 1985; Quinet 1990) have proved that whatever the options, if external social costs are not taken into account it will lead to the wrong choice being made. Long-haul freight transport offers a range of alternatives: railways, roads or waterway. From a collective point of view, when social costs are taken into account, rail proves cheaper than road; whereas in strict market terms, the contrary is true. In the same way, silent, non polluting cars would cost more to start with, but they would spare local authorities from footing the environmental bills for noise and air pollution.

The second thrust of the environmental policy, less important than the operational-regulatory trend, relates to research programmes. It has been implemented only relatively recently, since no support of environmentally-oriented academic research had been provided until the creation in 1979 of the Interdisciplinary Research Programme into the Natural Environment (PIREN). Following an interdisciplinary approach in the field of fundamental research, the research programmes supported by the PIREN were implemented by public research bodies such as the CNRS, the universities, the Secretary of State for the Environment and various specialized public institutions. The scope of research was rather large, since it included eight themes:

- continental aquatic systems,
- rural environment,
- research into the atmosphere,
- coastal regions,
- ecotoxicology,
- environment and health care,
- legal, sociological and economic aspects of environmental protection, and
- history of the environment.

Research programmes undertaken under the PIREN framework were essentially academic and fundamental; no support was granted to applied technological research, whether solely public or in collaboration with the industries potentially or effectively involved. The only institutional framework devoted to this comprised research programmes designed and coordinated by the Department of Research, Studies, Treatment of Information on Environment (SRETIE) of the Secretary State of Environment. It should be noted that this ministry has not been provided with a body responsible for the technological and scientific aspects of environmental protection. Before the creation of the SRETIE in 1984, the predecessor was the Mission des Etudes et de la Recherche which had neither resources nor power. In 1989, the SRETIE has began implementing seven multi-disciplinary research programmes in cooperation with the Research and Technology Ministry (*Le courrier du CNRS* 72, May, 1989):

- ECLAT (Evolution of climate and atmosphere),
- "toxic waste",
- "large rivers",
- "ecologic consequences of agricultural crisis",
- "spread of coastal seaweed",
- "genetics and environment", and
- "eco-products and clean technologies".

The implementation of these programmes may actually be considered as a first step towards broadening the scope of environmental policy, which implies inter-ministerial cooperation.

2.2.3 The Institutional Infrastructure of Environmental Policy and its Recent Evolution

In the two last years, several major shifts in the goals, scope and practices of environmental policy have occurred; they reveal an unprecedented awareness of the technological challenges linked with the search for new less environmentally damaging actions. From an institutional point of view, important changes should be underlined; the most important is the reinforcing of the interlinks between environmental and technology policies. The National Plan for the Environment states that: "the machinery (of environmental policy) are ill-adapted. Neither the law, poorly applied, nor the economic incentives, too low, have succeeded in providing a durable and deep integration of environmental issues in the society" (quoted by F. Billaudot 1991:334). This National Plan, adopted by the government in December 1990, has been jointly developed by the Ministries of Environment and of Research and Technology. It aims to largely increase research in the field of environment, which is expected to represent 5% of the national research effort in 1995, so as to catch up with the German level.

The National Plan for the Environment sets the following priorities in that area:

- *Rational coordination of public support for environmental research*: It is now the responsibility of the Coordination Committee of Public Research Concerning the Environment (CCRPE) to fulfil this function. The Committee will also be responsible for studying measures to improve the efficiency and the coordination of research programmes. It will have to control the coherence of these programmes with EEC programs such as STEP/EPOCH.
- *Stimulating technological research as far as "clean technologies" are concerned*: For this priority the Environment Ministry has just decided to set up an institutional framework for technology policy to achieve this. In this regard,

the recent creation of two *Groupements d'Intérêt Public (GIP)* in the fields of water and forest ecosystems may be considered as precursors of a close link between environmental and technology policies owing to the constitution of stable, cooperative patterns of organization of research which implies public research bodies as well as private firms.

At the administrative level, the National Plan for the Environment initiates both the reform of existing institutions and the creation of new ones. Concerning the Ministry of Environment itself, it was highly unstable before 1991, owing to frequent changes which affected its attachment to other ministries. Moreover, its means of intervention were scattered and it had no external services. The actions undertaken by this ministry 'in the field' required the involvement of decentralized departments of other ministries (regional public works authorities, departmental agriculture authorities). This lack of locally based bodies has been overcome by setting up the Regional Environment Authorities (*Directions Régionales de l'Environnement, DIREN*) and by converting the *Directions Régionales de l'Industrie et de la Recherche into Directions Régionales de l'Industrie, de la Recherche et de l'Environnement (DRIRE)*. 22 metropolitan DIREN's have been created by amalgamating pre-existing specialized departments. The role of the latter is:

- to fulfil all missions attributed by the Ministry of Environment, in the following fields: water, protection of natural areas and the landscape, enhancement of the urban and architectural heritage;
- to coordinate, process and circulate data linked with the environment and contributing with new institutions such as ADEME, IFEN, INERIS to upgrading the quality of scientific knowledge, the design of methods of investigation, evaluation, management and protection of the environment;
- to contribute to informing and educating people about the need to protect the environment.

As already mentioned, the regional industry and research authorities are today referred to as the *Directions Régionales de l'Industrie, de la Recherche et de l'Environnement (DRIRE).* Within each DRIRE, a regional department is set up to manage the industrial environment.

Three new public bodies, whose rationale is to provide adapted and relevant scientific bases and to nurture a new framework of the environmental policy, have been set up: the Agency for the Defense of the Environment and Energy-Saving (ADEME), the French Institute for the Environment (IFEN) and the National Institute for Industrial Environment and Risks (INERIS). These institutional structures were given credence by the following statements made in the text of the National Plan for the Environment: "The way the environment will be integrated into scientific and technical progress will play a crucial role in the near future" (Billaudot 1991:345). "In France, a large gap prevails in the field of environmental databases owing to poor use of existing information and the little effort made to conducting research in this field" (Billaudot 1991:345). The ADEME is the result of the amalgamation of the Agency for Air Quality (AQA), the National Agency for Waste-Recycling (ANRED) and the French Agency for Energy-Saving (AFME). Its creation reflects an institutional shift that underlies the need to express in institutional form the interdependence of different issues which were previously handled by specific and dedicated agencies. The ADEME, which has the status of a Public Industrial and Commercial Agency (EPIC), is responsible for combating air pollution, monitoring waste recovery and recycling, energy conservation, reducing harmful noise levels and supporting the development of clean technologies and technological research. It draws its financial resources from a system of taxation and licence fees.

The IFEN was conceived as a technical instrument for the Ministry of Environment; it should quickly become an equivalent of the INSEE as an environment database. The missions of this Public Administration Agency (EPA) are in environment connected scientific and technical information, coordination of studies implemented in the field of observation, measurement techniques, data processing and the supply of expertise. The IFEN is also able to undertake research or expertise activities in a contractual framework. Its missions relate to soil use, natural resources, air and water quality, waste management, urban environment, technological and natural risks and the sociology of social behaviour in the field of the environment.

Finally, the INERIS has been set up in order to provide a body devoted to applied research activities in an institutional framework connected with environmental policy. The INERIS is able to undertake various activities connected with scientific research, tests and experimentation protocols and detection methods. It also contributes to setting standards and formulating technical regulations. In addition, the INERIS provides technical assistance to the DRIRE, financial water agencies, ADEME, public authorities and local authorities. It is worth noting that the INERIS should also participate in cooperative research or expertise programmes, conceived in terms of partnership, with specialized bodies such as the CNRS or INRA research teams, the universities in the identification and measurement of pollutants, analyses of the ecological impact of various pollutants, fabrication tests, inspection and certification.

We must here refer to the recent renewal of both the PIREN, today called the CNRS Environment Programme, and the Interdisciplinary Research Programme into Energy and Materials (PIRSEM); the latter has seen its environmental component straightened. One of its most important objectives is to promote the development and diffusion of environmentally-compatible technologies. The former aims at stimulating the inter-disciplinary, inter-institutional and international dimensions of research programmes undertaken within its terms of reference. The PIRSEM is focussed on the synergy which ties together process optimization and pollution abatement. As C. Claverie, Director of the PIRSEM, says: "In 1991, one cannot but ascertain that the industrial's demand for new, less-polluting technologies, or able to provide a better use of fossil resources, becomes more important than the need for simply energy-saving technologies". Four priorities have been set within this programme: energy production and conversion, industrial processes, transport and urban planning. Moreover, PIRSEM's strategies derive from an awareness of the inseparability of research connected with energy savings and development of clean technologies.

3 The Technological Dimensions of Environmental Protection and their Impact on National Policy

As we have seen, the technological research and organization of innovation production connected with the protection of our natural heritage is not a matter for an environmental policy setting strict goals and matching resources in financial, human and physical terms. French environmental policy is mainly about regulating, redistributive and allocative so it should not be seen as a technology policy. In addition, technology policy is very similar since it does not specifically privilege any technological creation connected with the environment. Two questions therefore arise: firstly, would it be a good idea for the state to mobilize public resources for finalized programmes to compensate for the gap in technological research in this field? Can one reasonably consider that French technological policy be disconnected from environmental problems?

Both questions form the backbone of the following argument which will fall in three parts. Firstly, we shall show that any effort in terms of technological research into environmental protection is not so much a quantitative question of volume of resources as a qualitative question of relevant institutional structures and arrangements related to the characteristics of technological creation. Thus, the various measures taken in the context of French technology policy, and especially in the field of legal provisions governing various forms of partnership, are clearly moving in that direction. In particular, the use made of different structural options within the institutional framework, as defined by technology policy, stems from the deliberate behaviour of operators; we shall then see that the 'dovetailing' between technology and environment policies occurs at the grassroots level. We shall illustrate that thesis before focusing on the limitations of current forms of organization which we shall then correlate with the relevant industrial strategies of firms and economic sectors.

3.1 Technological Creation and Environment: From Knowledge to Innovation

A useful approach would seem to be to look at this question in the context of a fundamental coherence which governs the organization of technological research. This coherence is based on the compatibility between the structural evolution of industries and technologies and institutional structures grouping different operators involved in the creation of technologies.

In the field of technological research connected with the environment, three main dimensions appear, i.e. i) the strengthening of scientific knowledge into factors related to the creation of pollution; ii) the development of measuring devices and of a metrology which could allow a finer characterization of pollution levels in order to find suitable technical solutions; *iii*) the need for a cooperative and synergistic research structure. On the first point, the available knowledge seems lacking in a number of disciplines. It is still impossible to detect certain pollutants used in agriculture, e.g. diazins in drinking water. No one is quite sure about the side effects of certain water purification processes. The aluminium salts used in these processes are even claimed to be a possible cause of Alzheimer's disease (Mallevialle/Chambolle 1990).

The most acute problem regarding air pollution is the disastrous effect of CFC on the ozone layer. Roqueplo (1990) has clearly pointed out the challenges faced by scientific knowledge. All factors combine in such a complexity that they can only be scientifically tackled by interdisciplinary research. But given the French structure of scientific research, this effort clashes with the institutionalized partitioning of the different disciplines. Hence, this very need calls into question the role of the state, since it requires a re-organization of the institutional pattern of public scientific research.

The setting of normalized infra-technologies regarding measurement, metrology and classification methods seems a key prerequisite for any efficient technological research as regards water and fumes analysis, urban hydrology and drainage (Kirat 1990). Any research concerning efficient technological processes entails a capacity to characterize various forms of pollution and proper instrumentation. Yet many experts agree that these infra-technologies are still underdeveloped (Enel 1987).

3.2 Trends in the Organization and Nature of Technologies in the Field of Environmental Protection

In France, a shift in policy affects the management of water and purification networks in urban areas, as well as waste treatment. The behaviour of the operators involved in those activities is more geared to the reinforcement of research and technological innovation programmes. It must be said that these changes stem from a growing convergence among actors: mainly industrial firms, local authorities and public research institutions.

The types of technologies explored and the forms of organization of technological research seem worthy of interest. New membrane techniques used in the treatment and purification of water were adapted from an original technology that was first used by the food industry. In the same way, real-time management of such urban technical networks as water, drainage and electricity implies the adaptation and the setting of devices based on information and communication technologies. Générale des Eaux has adapted devices designed to measure vibrations to detect leaks in pipework. These few examples mean to illustrate the idea whereby new technological solutions implemented in the various sectors of environmental technologies have a trans-sectorial character and therefore require cooperative structures in the organization of research.

Such structures can be observed in bodies such as NANCIE and GEMCEA which group industrial firms, local councils, public administrations and public research centres connected to the CNRS or to the universities. Such groupings were made possible by a set of measures defined within the framework of technology policy, especially in terms of legal provisions dealing with these forms of association. NANCIE is a Public Administrative Agency (EPA) which includes within the same cooperative structure the Urban District of Nancy, three universities, six schools and four private laboratories. The structure is successfully conducting research into the formation of a biofilm, i.e. it is finding ways to destroy bacteria living inside the pipework which are supposed to protect pathological germs that might accidentally enter the network, thus affecting the quality of the water supplied by the network. The research programme also includes the industrial firm Pont à Mousson, which produces the pipes. The same structure may be observed in the case of GEMCEA, whose status is that of a GIP that groups NANCIE, Pont à Mousson and the Générale des Eaux in order to research into the fields of normalization of measures, material resistance and the quality of pipes (Cans 1990). All this tends to show that the design of technological innovations, as far as environmental protection goes, requires cooperation between the different players within a research framework where functional synergies and interdependences are important. Moreover, some water specialists have come to realize that the quality of drinking water will in future depend on three main factors of evolution: i) the standard of quality must be progressively improved on a European scale; ii) the increasing reduction, both in qualitative and quantitative terms, of water resources, will require the development of techniques to recycle liquid waste; iii) all will depend on the technological progress in the fields of biotechnologies, polymers and membrane techniques.

In an urban environment, the same phenomena can be observed. Indeed, large metropolises demand greater efficiency of urban services as well as an instrumentalization to measure levels of pollution, mostly air pollution. As far as technologies are concerned, these requirements can be translated in terms of transversality;

and this is quite obvious in the field of communication technologies, but it may ultimately clash with sectorial logics (Drouet 1987).

At this stage, it must be noted that technological research projects dealing with the environment are initiated by actors who are responsible for environmental management problems. In addition, these innovating actions are often conducted outside the scope of the environmental policy which – except in a few cases, such as PIRSEM – does not interfere with technological research, even if some public bodies (e.g. AFB, INRA, BRGM) contribute to these actions. However, things are moving in the right direction. As stated above, the newly created environment-linked agency, the ADEME, unites three different bodies: the AFME, the ANRED and the AQA. It is worth noting that up till now neither the ANRED nor the AQA had research budgets to support technological research in their fields of action. This grouping also shows that the control of polluting emissions in the air should go hand in hand with a vision implying that such a control is not so much a question of saving energy as it was thought in the 1970s, as a way of integrating "all progress accomplished in the field of materials and information techniques" into new equipment (Chartrier 1990:12).

Nonetheless, both technology and environmental policies face a major challenge: according to Kemp/Soete (1990), the implementation of technological solutions to environmental issues can follow two approaches. One is to promote 'end-of-pipe' (or 'clean') technologies; the second is more radical in that it involves the creation and diffusion of 'cleaner technologies'. In the first case, the purpose of such technologies is to provide marginal corrections to well-established and irreversible technological strategies, whereas in the second case, what is at stake is the move towards new technological trajectories, more 'environment-friendly' than in the past. The possibility of such a dramatic change would presuppose a systematic integration of environmental constraints within the sets of criteria used in the early conceptual and design phases of technical artefacts, whether products or processes.

On this issue, a recent OECD report states that the environmental policies implemented in the OECD countries "have focused on regulating major sources of pollution, while mobilizing a system of regulations involving sanctions-associated constraints in order to impose the conception and use of pollution reducing techniques at the end of processes" (OECD 1992:211). In the case of France, the inducive character of environmental policy in this respect should be called into question. The other more directly-regulating aspect of the environmental policy consists in the promulgation of technical and environmental standards and is aimed at increasing the pace of technical progress. Its effects should also be questioned; indeed, it is more efficient in stimulating the adoption of "end-of-pipe" technologies than in providing inducement mechanisms able to sustain a systematic search for new and cleaner technologies. Another phenomenon also deserves attention: according to Kemp and Soete, these environmental technologies "differ from the traditional process of 'normal' technological change involving a succession of new and more efficient production techniques" (Kemp/Soete 1990:251). The sources of such differences arise from factors affecting both the supply and demand of environmental technologies: appropriability conditions, technological opportunities, market demand, problems related to knowledge and information, insecurity and uncertainty, lack of user-producer relationships, etc. (see Kemp/Soete 1990).

3.3 Do Environmental Technologies Reveal a Rift in the Industrial System?

The economic and ecological stakes connected to the environment seem to have a different meaning for two main categories of players. On the one hand, we find firms who are involved in the production of goods and services directly connected with the environment. On the other hand, there are traditional industrial firms. The former generally deal with water management, industrial waste and civil engineering, whereas the latter are mostly heavy industries (steel, chemicals, petrochemicals) or consumer good industries (cars, electrical appliances).

The former category reveals efforts to innovate and develop technologies that will preserve or improve the quality of the natural environment. It is, however, obvious that the most of the industrial system is reluctant to take part in the process of developing new technologies to protect the environment. The only strategies that are being developed by the latter category aim at producing eco-products such as phosphate-free washing powder, lead-free petrol and CFC free aerosols. However honourable their purpose, these products do not represent more than 1% of total household consumption.

Table 3: Main components of the environment industry, 1990 (estimated part of the value of production, in %)

	North America	Europe	Japan	Total OECD
Equipment	74	76	79	76
Water and effluents retreatment	24	34	22	29
Waste	25	15	22	21
Air purity control	12	17	25	15
Others (hearth preservation, noise reduction)	13	10	10	11
Services	26	24	21	24
Total	100	100	100	100

Source: OECD (1992:217)

According to OECD estimates on the main segments of the so-called environmental industries, three quarters of the industry's output include equipment produced to protect the environment, mainly "end-of-pipe" equipment (see Table 3). Moreover, the OECD distinguishes three methods of entry and three groups of firms in this area:

- the traditional insiders which produce equipment and goods designed for water and sewage effluents treatment;
- the newly-created and innovative small and medium-sized firms specialized in the supply of consulting or tailored equipment on the basis of specific know-how;
- other firms, mainly large companies, who integrate the production of environmental equipment or services in a global diversification strategy; this tends to be the case in the engineering, chemicals and electronics industries, and for activities such as toxic waste management, plastics recycling and waste treatment.

The OECD concludes that on the basis of past experience, one can expect that the most likely future trend will be the continuous improvement of existing technologies and the addition of "end-of-pipe" technologies. If this trend is confirmed as the prevailing trend in future technology, it is also plausible that a shift towards more drastic changes in processes and products may be prevented.

It is difficult to avoid mentioning various inertias stemming from the economic weight of well-known polluting sectors. They can be part of the general strategy of some industrial firms, such as PSA (Peugeot) which deliberately refused to extend the catalytic converters to the whole of the French automobile industry, or it may be the consequence of the irreversibility of previous technology choices (Hourcade 1989). One may interpret the opposition of a large part of the French automobile industry to catalytic converters as being a choice between the "ecological risk" and the "economic risk" or to quote Roqueplo (1988), between the "direct" risk (irreversible damage to the environment) and the "indirect" risk (the risk carried by industrial firms, derived from combatting this damage). In the case of France, Roqueplo says, PSA opposition can be explained by the fact that the German firm Bosch has an almost complete monopoly over the fabrication of the mechanical part of the injection system. Therefore, PSA fears that German industry might have an unfair edge over the French automobile industry (Roqueplo 1989:23).

Thus, what seems crucial today is that parallel to the ecological awareness that is dawning on industrialized countries, we see that environmental imperatives are giving way to economic ones: in other words, the latter may very well use the former to their own ends. This may have different impacts, only one of which would serve to protect our natural heritage: i) the constraints of economic competitiveness may push environmental protection into the background as we saw with PSA; ii) ecological arguments may be used *a contrario* to legitimize some technological choices[1]; iii) ecological qualities may be used as a starting point for new competitive industrial strategies.

The last proposition is the only one that fits environmental constraints but, from the point of view of international competition, it may very well be diverted from its original purpose and used as a first category option by non-innovating countries.

4 National Policies: Between Theoretical Views and Social Realities

For the last thirty years, the economics of the environment has appeared as being scientifically autonomous in view of its object, but orthodox because of its methodology. 'Environmental economics' are based on the use of neoclassical micro-economic analytical tools as regards environmental protection and preservation of our natural heritage.

4.1 The Instruments of Economic Analysis in Environmental Policy

The basic analytical and theoretical premises of conventional environmental economics are to be found in the acceptance of the divergence between the private and social impact of micro decisions, which economists call 'externalities' after Pigou. The apparition of such negative technological externalities as pollution could jeopardize the efficiency of market mechanisms as far as the achievement of a social optimum is concerned. If the market fails, it is up to public authorities to make up for this deficiency to stimulate the market. Setting a shadow price to external phenomena boils down to allowing operators to integrate the external impact of their decision into their economic calculations. Procedures aiming at taxing polluting emissions then becomes akin to behaviour-rationalizing tools and to offer a solution to solve the problem of incompatible use of property rights. The Polluter-Pays Principle (PPP) is one of the two cornerstones of conventional environmental analysis, the second being axiomatic of rational choice in a world of certainty and perfect information, based on methodological individualism and on the substantive rationality postulate of operators, which is more than obvious in the cost/benefit analysis. Moreover, neoclassical axiomatics derive from an "economistic reduction" of its objects. Marriage, crime, religious practice or natural environment are all considered from an unidimensional point of view, i.e. in terms of advantages versus costs, opportunity cost and maximization under constraint. Hence, all phenomena are seen in terms of price and cost, which reduces any human behaviour to that of the *homo oeconomicus* (Bourguinat 1973; Passet 1983).[2]

Yet between fundamental principles entirely based on theory and their transcription into the reality of environment policies, there are discrepancies due to estrangement as well as an essential transformation relation. One is only to take one example to confirm this analysis: the PPP which underlies the tax and fees attitude towards polluting emission. In France, it is the preferred tool of the environmental policy as far as water and waste are concerned. To AFB, the theoretical principle has obviously been corrupted by its implementation. Indeed, whereas the taxes collected by the Agencies were inspired by a philosophy based on internalization of externalities, the AFB have become some kind of mutual benefit societies, *i.e.* a system which allows the externalization of internal costs, "blunting the sense of personal responsibility of its members, whereas taxes were intended to remind each consumer of their responsibilities towards others" (Martin 1988:117). Opschoor and Vos's analysis (1989) confirms all this: indeed, while taxes have a two-fold content, both inducive and redistributive, they show that the latter prevail over the former: "One could consider the French approach as a pragmatic generalization of the Polluter-Pays Principle (PPP) in a situation of powerful interests: the polluters in general do raise the funds necessary for realizing environmental objectives and efficiently allocate funds, but the reallocation of these funds takes the form of financial assistance, as decided *by* the polluters themselves" (Opschoor/Vos 1989:29).

Then they point out that this redistributive aspect which does not fit in with the inducive nature of the polluter-pays theory, is fully applied in France as far as air and water are concerned. In this case, taxes collected by the AQA are too small to have an inducive effect, whereas 90% of their total is handed to the actual polluters as subsidies to buy monitoring equipment. Only 10% of the funds are allocated to help technological research (Opschoor/Vos 1989:35-36).

The appropriation of analysis tools developed within the neoclassical model by economic and political decision-makers, poses two problems: i) their status, in terms of a legitimizing tool that is being used and over-used; ii) their relevance. The first aspect has been very well described by Hourcade (1991) and Godard (1990) who rigorously showed that any economic calculation only makes sense in concrete conditions of negotiation and the behaviour of operators so that economic rationality then becomes the "tool of legitimacy and a weapon for strategic domination" (Godard 1990:222). The second aspect refers to the relevance of the analysis as a reference for the choice of both the objectives and instruments of the environmental policy, as well as its relevance in terms of a medium of information necessary for the decision-making process.

As regards this last point, safeguarding the conditions of optimality of our natural heritage is often taken for granted as soon as external factors have been corrected, i.e. as soon as polluters are paying. This principle, owing to its implicit moral aspect, may first appear very attractive; but it nevertheless poses many problems. Firstly, "the very notion of externality is ambiguous. It presupposes that one can, at the basic fundamental level, isolate what strictly belongs to the operator or the firm concerned, from what belongs to other operators, or other firms or groups. ... Is it not paradoxical in an increasingly interdependent world to claim to isolate factors belonging to a specific unity from exterior influences?" (Bourguinat 1973:1526). So what of the future environmental consequences of decisions that are being taken today? As a corollary to this, there remains the question of inter-temporal allocation, inasmuch as the preservation of our natural heritage may appear as an imperative for the transmission of natural assets from generation to generation. Thus, the basic neoclassical tools (such as price system, efficient allocation) cannot accept an inter-temporal dimension, an impossible matching between present supply and future demand to use and enjoy the natural environment (Henry 1990).

As far as the doctrinal inspiration of environmental policies is concerned, the neoclassical model[3] seems to be gaining in importance; this is the consequence of the shift in both means and objectives towards a market dimension in most industrialized countries. Often, the 'command-control' regulating attitude based on a set of legal measures is being replaced by liberal inducive, correcting mechanisms that are both allocative and market oriented. The archetype would be the "rights to pollute markets". This is disquieting since it assumes that it is possible to set efficient control mechanisms, that the levels of tolerance and the way pollution arises are well known. But all these conditions are far from being met.

The neoclassical approach can be characterized by its static reasoning, its allocative regulation and 'economistic reduction'. These elements can be discussed and used with a view to reformulating institutionalist thinking that can be discerned here and there (Söderbaum 1990a, 1990b). After underlining the methodological options involved, we shall show that this approach has a two-fold merit: from a heuristic point of view, i.e. understanding and knowledge of phenomena, and from the point of view of analyzing environmental policies.

4.2 Environmental Protection as Central to Social Attitudes and Technological Choices

The arguments can be summed up in four points:

1. The degradation of the environment and of natural ecosystems is mostly irreversible. Therefore, as Söderbaum rightly pointed out, the cost/benefit analysis becomes inappropriate for "such an analysis is based on the idea that processes are reversible and that everything can be traded off in monetary terms" (Söderbaum 1990b:3).

2. The environmental effects of industrial and technological choices are uncertain. From a political viewpoint, this implies the need to establish procedures to gain information and knowledge over a long period of time, to mobilize and create specific expertise capacities in order to take into account the complexity of phenomena involved in the formation of pollution (Roqueplo 1990).

3. Environmental problems are both multi-dimensional and complex. They cannot be reduced to merely problems of money, price and cost. This complexity is due to the interdependence of effects, to the accumulation of phenomena as well as to social psychology. For example, the treatment of industrial waste causes new forms of pollution, especially as regards the release of fumes into the atmosphere. It would be possible to reduce these emissions by organizing selective specialized procedures to recycle this waste, but this would imply selective collection of waste, therefore a change in the behaviour of operators, particularly households.

4. Solutions to preserve the environment also have a moral dimension inasmuch as they commit society as far as future generations are concerned.

Consequently, it would seem useful to place the emphasis on the fact that the preservation of the environment is a question of technology as well as of economics and social attitudes. The assumption contains the prerequisites to draw up an institutional concept which would meet three imperatives. The first has to do with taking into account of time and history. The second deals with institutional machinery, structures and legislation which requires study in terms of rationality, evolution and the issues at stake. The latter is concerned with social attitudes beyond their mere *homo oeconomicus* aspects. On this last point it is obvious that any meaningful analysis must be based on an endogenization of costs, values, institutions and technologies as well as on conflicts between different groups of players (Söderbaum 1990b).

This brings us back to the theoretical institutionalist framework whose major contribution is to express the relationship between the nature of the evolution of social institutions and industrial societies in terms of technologies and social behaviours. As Söderbaum rightly pointed out, the success and efficiency of any national policy depend on the legal provision and regulation imposed by the state and on the jurisdiction and behaviour of local councils, particularly the behaviour of private operators, whether firms or households. Thus, "the success or lack of success of some particular institutional arrangement will largely depend on the people concerned, their values, attitudes and behaviours" (Söderbaum 1990b:5).

From that point of view, preserving the environment and our natural heritage can be envisaged in an alternative way from the one favoured by both neoclassical economists and engineers. The former are essentially concerned with two things: firstly, to 'get the price right' and secondly, to establish a coherent structure for property rights (Swaney 1987). The latter are only interested in finding technical solutions to technical problems. While there is no denying that these two attitudes are important, one should nevertheless avoid the pitfalls of 'economistic' or 'technical' reductions. Of course, the preservation of the environment has both economic and technical dimensions; but it is also an ethical and social issue that challenges our modes of thinking, attitudes and social institutions. Society is not seen as a accumulation of operators, but rather as an organic body deserving holistic and historical analysis. This dimension is important inasmuch as it makes it possible to understand today's environmental problems in the light of past thinking and social concepts. To illustrate this argument, one could quote the case of pollution connected to abnormal rainfalls in France. Indeed, present drainage systems in France are based on an 'hygienic' approach expressed in the 1894 law on sewage. According to this view, drainage entails getting rid of liquid waste through a network of sewers and drains. Between 1918 and 1939 the first water purifying plants were created. To prevent these plants from being overloaded by rainfalls, rain outlets were installed in the primary network in order to discharge into lakes, rivers or the sea the rain-water/liquid-waste mixture with which the purifying plants could not cope. These rain outlets today seem to spread pollution since they diffuse pollutants contained in rain water. It is only recently that new techniques different from the 19th century hygienic approach have been investigated. Instead of draining water out of the urban environment, no one tries to slow down the flow (Desbordes et al. 1990).

Finally, as has been shown by many analysts (Henry 1990, Söderbaum 1990a, Swaney 1987), there is uncertainty as to the consequences arising from the ignorance of the ecological chain of cause and effect. There is still considerable uncertainty today. It implies that the underlying behaviour of operators cannot be anything other than procedural, and this also affects public decisions. As a result, government decisions are often short-sighted due to a lack of knowledge about what accelerates the process of degradation or how to help preserve our ecological heritage.

5 Conclusion

The preservation of the environment has both a paradoxical and an ambivalent relationship with social institutions and technology. If technology can be defined as "knowledge about how to do things" (Samuels 1977:872), then it is undeniable that "many environmental problems arise as unintended by-products of economic activity and therefore involves the application of knowledge to the material world" (Swaney 1987:1748-1749). On the one hand, the development of technology affords mankind more opportunities to damage its natural environment. On the other hand, preserving this environment requires the development of technology. This is only feasible and viable if there are social forces willing to cooperate and work towards this goal and if national policy measures contribute to an awareness of fundamental issues and institutional steps taken to give them tangible form. While institutionalist economists see the evolution of society as being the product of various tensions between technological and institutional progress, it must be admitted that the protection of our natural heritage calls for a rethinking of technologies and social institutions. It is true that the institutionalist theory marks a certain technological determinism (Samuels 1977). However, technology is unquestionably a social artefact and the technological options taken depend on factors themselves determined by power structures and the attitudes prevailing within society as a whole. If one shares this view, it then becomes possible to think in terms of dynamic adjustment between the evolution of society and that of institutions in a framework of reciprocity determined and structured by national policies.

We believe that such a perspective can serve to re-appraise the relationship between technology, social institutions and the preservation of our natural heritage. To monitor this imperative requires thorough knowledge of the problem, i.e. a full understanding of social phenomena and ecosystems with all their complexities.

Notes

[1] This is especially true of nuclear energy. Some political groups hurriedly justified it and excluded it from the public debate, simply because it caused less pollution than traditional plants.

[2] Neoclassical economics only deal in abstract concepts and is not interested in social facts as defined by Emile Durkheim (1983).

[3] A corollary to this, this model sees the environment limited in economic terms. The various elements of Nature are viewed as public goods, since supply regularly outstrips demand. It would seem more appropriate to talk in terms of "pseudo public goods" which, inasmuch as they are rare, have no market value and lead to incompatible demand.

References

Abdelmalki, L., Th. Kirat (1991) Les politiques de l'environnement au défi de la maîtrise sociale de la technologie, paper prepared for the symposium 'The Social Mastery of Technology: Issues and Methods', Lyon.

Allen, R.C. (1983) Collective Invention, in: *Journal of Economic Behavior and Organization* (4): 1-24.

Ayres, R.U. (1989) Le métabolisme industriel et les changements de l'environnement planétaire, in: *Revue Internationale des Sciences Sociales* 121: 401-412.

Bernadet, M., J.C. Lasserre (1985) *Le secteur des transports*, Paris: Economica.

Billaudot, F. (1991) Les mutations administratives de l'environnement: aspects de l'application du Plan National pour l'Environnement, in: *Revue Juridique de l'Environnement* 3: 333-353.

Bourguinat, H. (1973) L'économiste et l'environnement: propos liminaires, in: *Economies et Sociétés* F, 25: 1513-1537.

Cans, R. (1990) De mauvais tuyaux, in: *Le Monde*, 31 October.

Chartier, P. (1990) Recherche technologique, énergie et environnement, in: *Le Monde*, 17 October.

Desbordes, M., J.C. Deutsch, A. Frérot (1990) Les eaux de pluie dans les villes, in: *La Recherche* 221 (21): 582-589.

Dosi, G. (1988) Sources, Procedures, and Micro-economic Effects of Innovation, in: *Journal of Economic Literature* 3 (27): 1120-1171.

Drouet, D. (1987) Les acteurs du génie urbain en Europe: des structurations très contrastées, paper prepared for the 'Séminaire Européen L'innovation et la recherche en génie urbain', Lyon: Institut National du Génie Urbain.

Dufourt, D. (1991) Les politiques technologiques: une nouvelle rationalité de l'intervention de l'Etat dans le système productif?, in: J. De Bandt, D. Foray (eds.) *L'évaluation économique de la recherche et du changement technique*, Paris: Editions du CNRS, 351-380.

Durkheim, E. (1983) *Les règles de la méthode sociologique*, Paris: Presses Universitaires de France.

Enel, F. (1987) *La métrologie et la gestion des réseaux d'assainissement*, Paris: Ministère de la Recherche, Ministère de l'Equipement.

Ergas, H. (1987) The Importance of Technology Policy, in: P. Dasgupta, P. Stoneman (eds.) *Economic policy and technological performance*, Cambridge, MA: Cambridge University Press, 51-95.

Godard, O. (1990) Environnement, modes de coordination et systèmes de légitimité: analyse de la catégorie de patrimoine naturel, in: *Revue Economique* 2 (41): 215-242.

Henry, C. (1990) Efficacité économique et impératifs éthiques: l'environnement en copropriété, in: *Revue Economique* 2 (41): 195-214.

Hippel, E. von (1988) *The Sources of Innovation*, Oxford: Oxford University Press.

Hourcade, J.C.(1991) Calcul économique et construction sociale des irréversibilités: leçons de l'histoire énergétique, in: R. Boyer, C. Chavance, O. Godard (eds.) *Les figures de l'irréversibilité en économie*, Paris: Editions de l'EHESS, 279-310.

Jamison, A. (1989) Technology's Theorists: Conceptions of Innovation in Relation to Science and Technology Policy, in: *Technology and Culture* 3 (30): 505-533.

Kemp, R., L. Soete (1990) Inside the 'Green Box': On the Economics of Technological Change and the Environment, in: C. Freeman, L. Soete (eds.) *New Explorations in the Economics of Technological Change*, London, New York: Pinter, 245-257.

Kirat, Th. (1990) Les normes, instruments de gestion de l'innovation?, in: *Economies et Sociétés* SG, 15: 165-191.

Kline, S.J., N. Rosenberg (1986) An Overview of Innovation, in: R. Landau, N. Rosenberg (eds.) *The Positive Sum Strategy*, Washington D.C.: National Academy Press, 275-306.

Le courrier du CNRS (1989) 72.

Mallevialle, J., Th. Chambolle (1990) La qualité de l'eau, in: *La Recherche* 221 (21): 598-606.

Martin, Y. (1988) Quelques réflexions sur l'évolution des Agences de Bassin, in: *Annales des Mines* July-August: 117-119.

Nelson, R. (1959) The Simple Economics of Basic Scientific Research, in: *Journal of Political Economy* 3 (67): 297-306.

Nelson, R., S. Winter (1982) *An Evolutionary Theory of Economic Change*, Cambridge, MA: Belknap.

OECD (1992) *Technologie et Economie*, Paris: OECD.

Opschoor, J.B., H.B. Vos (1989) *Economic Instruments for Environmental Protection*, Paris: OECD.

Passet, R. (1983) *L'économique et le vivant*, Paris: Payot.

Quinet, E. (1990) *Analyse économique des transports*, Paris: Presses Universitaires de France.

Roqueplo, Ph. (1988) Les enjeux de la réglementation européenne pour les acteurs économiques: l'exemple de la pollution de l'air, paper at the 'Journées de l'environnement du CNRS, "Le Droit et l'environnement" '.

Roqueplo, Ph. (1989) L'écologie est-elle en train de rapprocher la France et la R.F.A. ou de les séparer?, Discussion paper at the Fondation Saint-Simon, Paris.

Roqueplo, Ph. (1990) Comment évaluer les recherches destinées à fonder la fiabilité d'un diagnostic? Exemple des problèmes globaux d'environnement, in: *Journées CNRS sur l'interdisciplinarité*, 12 February.

Samuels, W.J. (1977) Technology vis-à-vis Institutions in the JEI: A Suggested Interpretation, in: *Journal of Economic Issues* 4 (11): 871-895.

Secrétariat d'Etat chargé de l'Environnement (1989) *Données économiques de l'environnement*, Paris: La Documentation Française.

Söderbaum, P. (1990a) Neoclassical and Institutional Approaches to Environmental Economics, in: *Journal of Economic Issues* 2 (24): 481-492.

Söderbaum, P. (1990b) Neoclassical and Institutional Approaches to National and Private Environmental Policy, paper presented to the 'Annual Conference of the E.A.E.P.E.', Florence.

Swaney, J. (1987) Elements of a Neo-institutional Environmental Economics, in: *Journal of Economic Issues* 4 (21): 1739-1779.

UN/ECE (1988) Regional Strategy for Environmental Protection and Rational Use of Natural Resources in ECE Countries Covering the Period up to the Year 2000 and Beyond, in: *Environmental Policy and Law* 5 (18): 175-186.

The Historical Development of Japanese Science and Technology Policy in Conjunction with Socio-economic Policy

Masahiro Kawasaki

1 Introduction

Today, science and technology policy is highly ranked in the policy priority areas of both public and private sectors throughout the world. It is well recognized that science and technology is a driving force for economic and social development. It is further recognized that recent changes in the mode and mechanisms of science and technology development and innovation, as well as the globalization of technology, are having profound impacts on the policy-making environment. On the basis of these changes in science and technology, transformations are also occurring in East-West relations and the relationship between developed and developing countries. For Japan, the changing circumstances are forcing her science and technology policy to move toward new approaches which take into account her *global leadership-sharing* position. At this turning point in Japanese science and technology, and in order to learn from the past, this paper aims to review the historical development of that policy – particularly focusing on its relationship with social and economic changes. Several epoch-making recommendations of the Council for Science and Technology (CST) shall be reviewed since they proposed the overall S&T policy framework from time to time, reflecting upon the socio-economic goals of the time.

From the author's overall view, Japan gave high priority to the development of science and technology as an indispensable factor in her industrial and economic restoration following World War II. Then through the decade of the 1960s, the era of high growth in the Japanese economy, this recognition was strengthened. As such, it formed the basis of Japanese science and technology policy until 1984. During that period, Japanese S&T policy did not focus independently on science and technology development – in particular, science policy – but also on how S&T can contribute to solve economic and social problems. Thus, S&T policy worked as a tool for societal and economic policy. In 1984, however, the CST proposed a new direction for S&T policy that gave a high priority to the progress of science and technology within the overall policy framework. This was a dramatic change of S&T policy. In this direction, there has been a policy shift to strengthen basic research and to promote creative science and technology. This shift seems to bring about today's S&T policy issues, since they derive mainly from the frictions between new policy objectives and the traditional institutions.

Throughout this paper, the author will present what can be learned from past Japanese experience, in order to seek new directions for a new S&T policy framework.

2 S&T Institutional Framework and S&T Policy as Built into Industrial Policy (1945-1955)

During the period from 1945 to 1950, various governmental actions were taken. Together, these formed the institutional framework of science and technology policy for postwar Japan.

World War II completely destroyed living conditions, industry and economy in Japan. In the period from 1945 to the beginning of the 1950s, Japan was forced to build up both industrial and economic capabilities from a "zero" base of social and economic fundamentals – such as lack of raw materials, consumer goods, industrial production capacities, capital and foreign currency, etc. The demobilization of the military resulted in high unemployment. Under the guidance of the GHQ (General Headquarters of the Occupation Forces in Japan), policy priority was given to urgent counter-measures to overcome both the lack of basic materials and vicious inflation, as well as to reconstruct the base of economic activities. Along these lines, the major policy goal was to regulate the supply of food and industrial products (*Rinji Busshi Jukyuu-Chousei Hou* [Law of Provisional Measures of Coordination of Supply and Demand of Goods], 1945). The Top Priority Production Policy of Coal, Iron and Steel (*Sekitan, Tekkou wo Suuten to suru Keisha-seisan Seisaku*, Cabinet decision, February 1946) was adopted by the Cabinet, in order to restore the overall industrial base of Japan. In this way, the Japanese economy took a path of recovering its footing, coupled with the special procurement boom of the Korean War (1950-1953).

Likewise, the GHQ took several actions to eliminate military R&D and to shift science and technology structure towards peaceful purposes. In 1947, the *Gakujututaisei-Sasshin-Iinkai* (Advisory Committee for the Reform of Science and Technology Organizations) was established. In the next year, this committee recommended to the Prime Minister that (i) the government establishes the Science Council of Japan (JSC), of which all members are to be elected by scientists and engineers themselves; and (ii) the Science and Technology Administration Council (STAC) be organized as an administration body for science and technology policy. The chairman of STAC was the Prime Minister, and its members were constituted by the administrative Vice-Ministers of all ministries and agencies concerned – as well as several wise men appointed by the Prime Minister. STAC was obliged to (i) coordinate overall science and technology policies conducted by various ministries and agencies, (ii) promote international cooperation in the field of science and technology, and (iii) take any administrative actions required to implement JSC recommendations. Both the JSC and the STAC were established in 1949. At the same time, the Ministry of Commerce and Industry (the predecessor of the present MITI) established the Agency of Industrial Technology (the predecessor of the present AIST of MITI) based on the recommendation of GHQ, in order to promote industrial technology. In addition to these actions, there was a reform of other scientific institutions, mainly universities, under the National School Act; and statistics were introduced to R&D under the Statistics Act of 1947.

Over the period, it can be said that Japanese science and technology policy was firmly formulated as an important tool of Japanese industrial policy. In order to develop Japanese industrial structure both quickly and smoothly – a realization of the shift from light industry to more lucrative heavy and chemical industries – the government provided various industrial policy measures. First, in order to regulate both foreign exchange and trade and to guarantee foreign investment in Japan, two laws were respectively established: the Act for Control of Foreign Exchange and Trade in 1949, and the Act for Foreign Investment in 1950.

Both acts provided effective policy tools, not only for the improvement of the industrial structure, but also for the development of Japanese industrial technology. This lasted until 1967, when the government adopted the liberalization policy of trade and capital transaction (Japan became a member of OECD and of the article 8 of IMF in 1964). In 1949, the Act for Industrial Standardization came into force; and in 1952, the government provided the Act for Promotion of Industrial Rationalization and the Government Order of Tariff Exemption for Import of Important Equipments and Machines. All worked towards accelerating the modernization of manufacturing industries. Under the Foreign Investment Act, the government operated the strict screening of technology import with the promotion of domestic industry in mind. This screening adopted impact analysis for small and medium-sized enterprises, assessment of export-promotion and import-reduction potentials, and market analysis of products derived from imported technologies.

In addition, reflecting upon the speech on the peaceful use of nuclear energy presented by President Eisenhower of the United States at the General Assembly of the United Nations in December 1953, Japan set up a nuclear energy R&D budget for the fiscal year 1954 on the basis of a proposal made by the House of Representatives. In this connection, the JSC declared the three guiding principles of nuclear energy R&D – that is, *opening to the public, democratic and autonomous*, in order to ensure the peaceful utilization of nuclear energy. They were incorporated in the Basic Law For Atomic Energy and the Establishment Act of Atomic Energy Commission, enforced in 1956. Since then, these guiding principles were applied to various features of science and technology policy – in particular, to governmental *major scientific* programs, such as the space development program. It must be noted that these principles have mapped out the overall science and technology policy until today: science and technology activities must be directed towards peaceful purposes in Japan.

In 1953, after Japan's economic recovery from the destruction of war, Japan began to set policy targets focusing on the future progress of Japanese economy. The *Keizai Singikai* (the Council of Economy, the highest advisory board to the Prime Minister, established in 1952) recommended to the Prime Minister an integrated policy framework for building up a self-sustaining economy. In this recommendation, the Council proposed the following *three goals* and *four guiding principles* for economic independence:

Goals:
1. Normalization of trade, in particular promotion of export.
2. Improvement of self-supply; capability through development of domestic natural resources.
3. Building up economic capacity through capital accumulation.

Principles of Action:
1. Stabilization of the economy.
2. Self-support of main economic constituents.
3. Cooperation between capital and labor.
4. Promotion of science and technology.

In this context, the government formulated the Five-year Plan for Self-supporting Economy in December 1955. Part 2 of this plan stressed the promotion of science and technology as an important policy priority for attaining its goals. Behind this plan, a significant consensus was formed among leading economists and government officials: the *White Paper on Japanese Economy* of 1956 proclaimed, with an illustration of an analysis of Japanese economic development, that the increase of investment initiated by *technological innovation* was a key factor of economic growth (Economic Planning Agency 1956; Commission on the History of Science and Technology 1991:84). Also, the *White Paper* declared that the postwar period was over, since the Net National Income per capita in 1955 exceeded the peak of the prewar period.

During this period, the government also launched ambitious plans through public endeavours to improve socioeconomic infrastructure through such public entities as Nippon Telephone and Telegram (the predecessor of NTT) and Japan National Railway Corporation (divided into nine private companies in 1987). These plans presented stable markets to industry and played a significant role to stimulate industrial R&D in the respective field of new technologies, parallel with improvement plans for other infrastructures. NTT began to construct a microwave telecommunication network between Tokyo and Osaka in 1954 and completed the nation-wide network in 1957. Japan National Railway achieved the electrification of all lines by 1956 and in 1958 started the development program of the *Shinkansen* system (a very high-speed railway system), operated in 1964 between Tokyo and Osaka, as the New *Tokaido* Line (i.e. *Shinkansen*).

3 Toward the Integration of Science and Technology Policy under Economic Growth and Liberalization Policy (1956-1969)

Innovations during the 1950s and 1960s were mainly achieved by the intensive importation of advanced industrial technologies from Western countries, armed with the legal framework mentioned above. Together with those intensive imports, including technology management such as the QC, various sectors of the industry made enormous efforts to build up indigenous R&D capabilities (see Table 1 and Figure 1). Based on the good match of technology import and indigenous R&D efforts and guided by the government (for instance, with long-term low-interest loans from the Development Bank of Japan: see Table 2), Japanese industry rationalized and modernized its production facilities, thus building up the prerequisites for catching up with more advanced countries. Completion of catching-up was seen in the textile, camera, sawing machines, ship-building, iron and steel industries during the first half – of the 1960s; and signs of building catching-up potential were seen in the chemical, machine and electrical equipment industries. Furthermore, new industries were launched during this period: such as the petrochemical, synthetic fiber, and electronics industry, etc. For the machine industry, the government provided in 1956 the Act for the Promotion of the Machine Industry; and also in 1957, it enacted the Act for Promotion of the the Electronics Industry. This provided tax exemptions and financial support for joint R&D projects, such as for color TVs and computers, etc.

As for industrial R&D activities during the 1960s, it is noted that major companies began to establish central or corporate research institutes. Figure 2 shows the number of research laboratories established in the industry. The R&D expenditure of private industry increased remarkably, as is shown in Table 1. Under a policy of high economic growth during the 1960s, Japanese industry not only expanded its scale but also developed its own R&D constitution. It is commonly explained that high economic growth in 1960s was caused by the mechanism that *investment invites further investment*. This industrial investment was targeted not only to production capabilities but also to R&D. Gradually, indigenous R&D upgraded its potential to an extent sufficient for competing with more advanced countries. The electronics and communications industries particularly built up their potentials to develop and commercialize products such as the IC calculator in 1967, the laser in 1965, the LSI in 1969 and optical fibers in 1970.

Table 1: Trend of Japanese R&D expenditure (1948-1987, in 100 million yen)

	A – Total	A/D (%)	B – Government	B/A (%)	C – Industry	C/A (%)	D – National Income
1950*	147	(0.4)	n.a.		n.a.		33,815
1953**	466	(0.8)	n.a.		n.a.		58,517
1954**	531	(0.9)	n.a.		n.a.		64,877
1955**	562	(0.8)	n.a.		n.a.		71,818
1959***	1,489	(1.5)		(29.1)	1,056	(70.9)	107,498
1960***	1,844	(1.6)		(25.3)	1,377	(74.7)	130,091
1965***	4,258	(1.6)	1,311	(30.8)	2,947	(69.2)	263,608
1970***	11,953	(2.0)	3,014	(25.2)	8,939	(74.8)	610,297
1975***	26,218	(2.1)	7,208	(27.5)	18,710	(72.5)	1,124,716
1980***	46,838	(2.4)	12,096	(25.8)	34,742	(74.2)	1,993,352
1985***	81,164	(3.2)	15,740	(19.4)	65,424	(80.6)	2,545,192
1986***	84,150	(3.2)	16,517	(19.6)	67,633	(80.4)	2,644,507
1987***	90,162	(3.3)	17,983	(19.9)	72,179	(80.1)	2,740,767
1988***	97,752	(3.3)	18,014	(18.4)	79,655	(81.6)	2,919,421
1989***	106,276	(–)	21,178	(19.9)	85,098	(80.1)	(–)

Sources:
* Science and Technology Administration Council (1951:6-7)
** Statistics Bureau of the Prime Minister's Office (1953-55:2-3; 1956:97-120))
*** Statistics Bureau/Agency of Administrative Management (1961:19; 1976:22; 1984:29; 1990:29)

Table 2: The trend of loans for promotion of science and technology by the Development Bank of Japan (in 100 million yen)

Loan Items Year	1951	1952	1953	1954	1955	1956	1957	1958	1959	1960	1961	1962	1963	1964	1965	1966					
	1.8	2.3	5.9	0.5	0.7	4.3	4.5	6.9	5.4	12.2	4.0	5.5	20.5	16.3	12	12					
1. Commercialization of New Technology	1967	1968	1969	1970	1971	1972	1973	1974	1975	1976	1977	1978	1979	1980	1981	1982	1983	1984	1985	1986	1987
	17	75	102	122	168	145	199	207													
2. Establishment of Laboratories	4	11	5	10	15	10	2	2													
3. Development of Heavy Machinery	25	28	27	17	41	24	14	5													
4. Production of Synthetic Rubber	10	13																			
5. Development Of Computers	70	90	155	215	402*	150	215	325	574	449	385	553	471	554							
6. Ocean Development				3	2	13	18														
7. Electronic and Machinery Industry								85	91	105	83	78	102	145	120	134	168	240			
8. Development of Home-made Technologies									254	280	247	659	512	265							
9. Development of Information-oriented Systems															451	509	540	571	971	835	1104
10. Development of Industrial Technologies															390	461	574	750	663	635	816

* Amount of 1971 includes the loan for the software development (12 million yen).

Source: Commission on the History of Science and Technology Policy (1991:184)

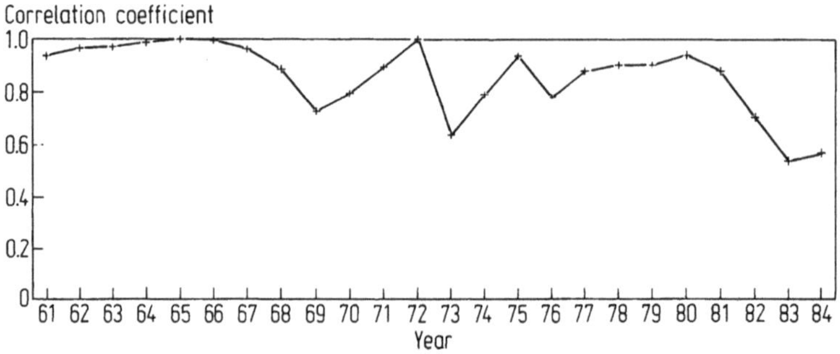

Figure 1: Correlation coefficients of technology import and R&D expenditure

In this figure, I used R^2 for the explanation of correlation between R&D expenditure and the payment of technology import instead of r usually used (refered to Kodama 1991: 58). The Range of R^2 is from 0 to 1.

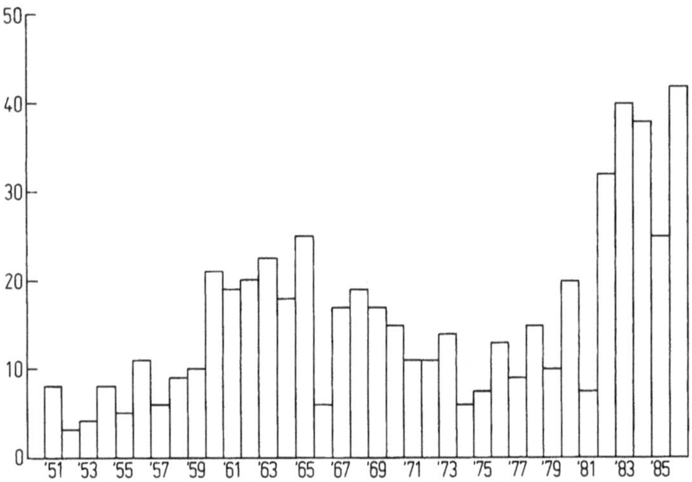

Figure 2: The trend of the annual number of establishment of corporate labs
Source: Commission on the History of Science and Technology Policy (1991:217)

At the government level, despite the STAC playing its own coordination role over the past period, voices of individual industry-related ministries were dominant in approval procedures for technology import, etc. In addition, there appeared a series of new S&T policy issues which required strong coordination at the national level: nuclear energy, aeronautical and material R&D. In cope with these urgent issues, serious discussions arose in Parliament, industry and government regarding *how government organization for science and technology should be*

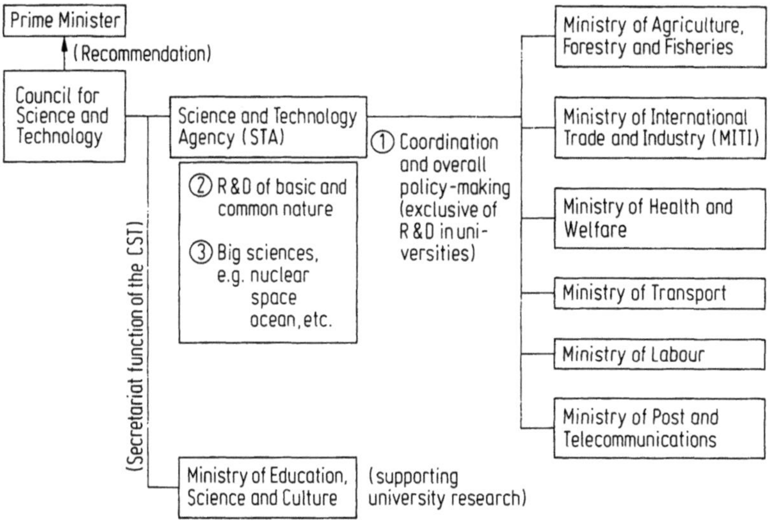

Figure 3: Administrative structure of S&T policy (functional chart)

organized. Accordingly, the government abrogated STAC and tried to integrate overall administrative activities for science and technology policy. As a result, the Science and Technology Agency (STA) was established in 1956. Also, new governmental R&D organizations for nuclear energy, material and aeronautical S&T were established under STA during the period from 1956 to 1959. Thus, today's administrative structure for science and technology policy was principally formed during this period (see Figure 3). In particular, the establishment of STA in 1956 was accepted as a clear policy sign that the promotion of science and technology had become one of the priority policy issues at the ministry level in Japan. Not surprisingly, there were serious conflicts among the existing ministries about what competences the STA should have. The responsibilities of the STA were defined as follows:

- to coordinate all governmental R&D undertaken by ministries and agencies, but not by universities (this was managed by the Ministry of Education);
- to formulate national S&T policy (except university policy); and
- to promote the development of major comprehensive science and technology – such as nuclear energy, aerospace, material science and processing.

Several R&D organizations were newly established under the STA, which include the

- National Aerospace Laboratory (NAL), in 1956;
- National Research Institute of Metal (NIRM), in 1956;

- Japan Atomic Energy Research Institute (JAERI), in 1956 (called "Tokushu-Hojin", a semi-governmental entity which shares most of its capital with the government);
- Nuclear Fuel Corporation (Tokushu-Hojin, the predecessor of the Power Reactor and Nuclear Fuel Development Corporation PNC), in 1956;
- National Institute of Radiological Science (NIRS), in 1957;
- Japan Information Center of Science and Technology (JICST, or Tokushu-Hojin), in 1957; and
- Institute of Physical and Chemical Research (RIKEN), in 1958.

In spite of the restructuring attempt, there has still remained from the viewpoint of efficiency of policy formulation, a question of how to coordinate and integrate university activities into national S&T policy. In response to this question, the Council for Science and Technology (CST) was established in 1959 as the highest advisory board of S&T policy. The role of the CST is to make recommendations on science and technology policy (including university research policy) at the national level, as well as to coordinate science and technology activities conducted by all ministries and agencies. The chairman of the CST is the Prime Minister; and its members are composed of the Minister for Science and Technology, Minister for Education, Finance Minister, Minister for Economic Planning, the President of the JSC and five experts designated by the Prime Minister. In addition, any minister could participate in the plenary meeting of the CST. Therefore, the CST's function can be recognized not only as a pure advisory board, but also as a Cabinet-like body on science and technology policy. The legal status and legitimate functions of STA and CST have not changed until today, although the CST has been required from time to time to strengthen its overall coordination function of S&T activities. Given the function of the CST, Japan's S&T policy-making structure has obtained integrity at the Cabinet level in conjunction with other policy issues, particularly industrial and economic policy.

Just after the establishment of the CST, an inquiry was conducted by the Prime Minister on a comprehensive S&T policy framework for the 1960s. At the same time, the Council of Economy prepared a recommendation to the Prime Minister on a new direction for economic policy, reflecting that Japanese economic and industrial performance had reached their goals in terms of annual growth rate of the GNP, development of industrial structure and improvement of balance of trade as settled by the New Long-term Economic Plan of 1957 (see Tables 3 and 4). In 1960, both councils presented recommendations on science and technology policy and on economic policy, respectively. The recommendation of the CST was recognized as *the first overall policy framework for science and technology.* Moreover, the economic policy framework called the National Income Doubling Plan presented a way to realize high economic growth. Both of them positively affirmed that science and technology were key factors in economic and industrial development and that S&T policy had a close linkage with economic policy. The structure of the recommendations are described in Table 5, showing that overall S&T policy was strongly consolidated in economic growth policy. Along these lines, the STA established the Research and Development Corporation of Japan (JRDC) in 1961, with the function of encouraging the transfer of R&D results from national research institutes or universities to industries.

Table 3: Annual growth rate of GNP (%)

	Nominal term	Actual term
1950	16.9	12.7
1951	37.9	13.6
1952	12.4	9.3
1953	15.8	8.2
1954	5.0	3.6
1955	10.3	9.6
1956	12.8	8.5
1957	8.4	7.0
1958	2.7	5.1
1959	20.8	17.3

Average annual rates:
1947-1952: 11.5%
1953-1958: 6.8%
1953-1959: 8.3%

Source: Okita (1960:6)

This industry-oriented nature of science and technology policy is also seen in the CST recommendations which followed, while the goal of policy was changed from economic growth to open-door economic policy. In 1966, the CST proposed a new policy framework to cope with the emerging open-door economic system. In particular, the CST focused on the international competitiveness issues of Japanese industry in compliance with the liberalization of technology import implemented in 1967. The proposal designated 102 R&D objectives for social and economic development, as well as 13 fields of science all requiring the government's support during the next 10 years. Also, it emphasized that the government should improve its R&D basis and the environment of national research institutes/universities – such as increase of the per capita research fund, modernization of R&D facilities and equipments, strengthening S&T information activities, etc. Other priorities were to provide tax exemptions for R&D expenditure in the private sector and to increase the number of scientists and engineers –

Table 4: Change of industrial structure (%)

	Primary Industry	Secondary Industry	Tertiary Industry
1. Added Value Base			
1934-1936	19.8	30.8	49.4
FY1951	25.0	32.4	42.7
FY1955	22.8	30.8	46.8
FY1956	19.4	33.4	47.6
FY1957	18.8	33.9	47.8
FY1958	18.4	33.2	48.8
FY1970	10.1	38.6	51.3
2. Employment Base			
FY1930	49.3	20.4	30.2
FY1940	44.1	26.0	29.9
FY1950	48.3	21.9	29.8
FY1955	41.2	23.8	35.0
FY1970	23.7	32.2	44.1

Source: Okita (1960:8)

a well as to improve post-graduate education. As for the activation of national research institutes, the government approved the construction plans for Tsukuba Science City in 1963, in order to improve their research facilities and environment. The initial construction plan – that 41 national institutes and 2 national universities located in Tokyo had to move to the Tsukuba district – was almost accomplished in 1979. And in 1966, just before the liberalization of technology import, AIST of MITI launched a new institution of large-scale industrial technology development projects (MHD power generation, a new process of ethylene production, a desulferization process for exhausted gas from large-scale electric power plants, a super computer system, etc.). Furthermore, in 1967, the government provided a new tax reduction system for R&D expenditures, in order to strengthen R&D activities in industry.

Behind these policy directions it must be also recognized that the intensive analysis of "technological gaps" among member countries conducted by OECD (mainly gaps between the USA and other countries) during the latter half of the 1960s had a strong impact on the science and technology policy of member countries, particularly Japan. The results of OECD studies pointed out that the technological gaps were caused mainly by the following five factors: (i) R&D capabilities in the respective country; (ii) whether markets were of sufficient size

Table 5: Structural relation between the Income Doubling Plan and the first Recommendation of the CST

National Income Doubling Plan	The first Recommendation of the CST
(Socio-economic Goals) 1. to sustain economic stability 2. to achieve continous high growth rate (7.8% at annual average) 3. to improve living conditions 4. to realize full employment	(Goals) to achieve Income Doubling Plan: (Policy Priority) 1. to raise over-all S&T level 2. to recover technological gap between the advanced countries and Japan
(Role of the Government) to provide the base of private sector's activities: 1. to enhance social capital 2. to improve human ability and develop science and technology 3. to fulfill social security and welfare 4. to guide private industry – liberalization of trade and capital – promotion of trade and international economic cooperation – improvement of industrial structure and strengthening international copetitiveness	(Major Policy Approaches) 1. to improve quality of and increase scientists and engineers (during 10 years, shortage was estimated at 170,000) 2. to improve S&T information system 3. to promote the transfer of advanced technologies to small and medium-sized industries 4. to promote absorption of new knowledge from the advanced countries 5. to increase total national R&D expenditure equivalent to 2% of national income

Sources: Economic Council (1960); Council for Science and Technology (1960)

to justify the scale of the industry; (iii) the scale of demands and procurement of the respective government; (iv) the company's scale; and (v) whether or not the respective society had social and economic conditions receptive to the acceptance of such innovations.

From the viewpoint of the OECD analysis, it can be summarized that Japanese economic and S&T policy provided the industry with the appropriate environment for continuous innovations.

As regards the *large-scale sciences*, space and ocean development programs were initiated during the 1960s. Space R&D had been conducted by Tokyo University since the late 1950s, and the success of Sputnik in the USSR in 1957 resulted in the government's full initiative in the program. In order to coordinate overall space R&D activities and to formulate the national space development policy, the Space Activities Commission (SAC) was established in 1968 on the basis of several recommendations of the Space Development Council established in 1960. Then, in 1969, the National Space Development Agency (NASDA) was established, in order to conduct the practical purpose-oriented projects of the

National Space Development Program (e.g. meteorological satellites, broadcasting satellites, communication satellites, and launching systems). The Institute of Space and Astronautical Sciences of Tokyo University conducts scientific activities on space under the guidance of the SAC. This organizational setting has not changed until today. In addition, the government initiated the promotion of ocean science and technology from 1951 on and participated in international programs such as the International Survey Program of the Indian Ocean (1962) and the International Joint Survey of the *Kuroshio* in the framework of IQSY (1964). It also conducted a survey of abnormal cold masses occurring in Japanese waters. In 1968, the STA began to operate a deep-sea exploratory vessel, the *Shinkai 2000*. All in all, the basis of today's ocean development activities was built by the establishment of the Council for Ocean Development and Japan Marine Science & Technology Center (JAMSTEC) in 1971.

On the other hand, with respect to the pending question of overall coordination of science and technology policy also involving university research policy, significant debates arose among the Diet, the government and academia. As is described, the first CST recommendation in 1960 pointed out that the government should provide a new legal framework for science and technology, not only for its promotion but also to strengthen coordination of overall governmental S&T activities, including those from the university. Then, in 1962, the CST proposed the concept of the Basic Law for Science and Technology. At the same time, the *ad hoc* Committee for Science and Technology of the House of Representatives presented a draft of the Basic Law for Promotion of Science and Technology. And the JSC recommended to the government the framework of the Basic Law for Scientific Research. Taking into account these proposals and discussions, the CST recommended in 1965 the framework of the Basic Law for Science and Technology. Based on CST recommendation, the government submitted to the Diet the Draft of the Basic Law Science and Technology, in February 1968, which was rejected in December of that year. Briefly speaking, the main reason for this rejection came from the objection that universities' activities should fall outside science and technology policy. This problem has cast a shadow over the integrity of science and technology policy even until today; S&T policy priority determines basic research, but universities are still expected to play a major role in such basic research. Also, these discussions infer today's Japanese bureaucratic rivalry. Each ministry tends to conduct science and technology policy with attempts to expand its jurisdiction into the emerging areas of S&T.

4 Priority Shift of S&T Policy from Economic Growth to Social Welfare (1970-1980)

This period was a time when a series of significant changes occurred in domestic and international society. In industry as well, there emerged new key technologies – such as IC, LSI, computer, recombinant of DNA, etc. Naturally, there appeared new types of problems which required a priority shift of Japan's S&T policy. First, following the rapid growth of the Japanese economy in the 1960s, air and water pollution revealed themselves in the industrialized zones of Japan. These environmental pollution problems were perceived as a shadow on economic prosperity and provoked public antipathy to economic growth policy in the beginning of the 1970s. In response, the government provided an intensive legal framework, such as the Basic Law for the Prevention of Environmental Pollution in 1967, the Law for the Prevention of Air Pollution and the Law for Anti-Noise in 1968, the Law for the Prevention of Water Pollution in 1970 and the Establishment Law of the Environmental Agency in 1971. Second, with regard to international environment, it can be said that Japanese remarkable economic growth depended both on the fixed foreign exchange rate of the Yen (1U$=360 yen) and on a stable supply of petroleum. In the beginning of the 1970s, however, these two elements became unstable: namely, in 1971, the United State adopted the floating exchange rate (the yen-US$ rate changed to 1US$=308 yen); and in 1973, the first oil crisis occurred. Japan suffered from high inflation and a subsequent stagnation of economic activities, particularly in falling investments. These changes in domestic and international environments had a strong impact not only on economic and industrial policy, but also on science and technology policy.

In addition, during the past few decades, many industrial goods had widely diffused to the public – such as home-electrical and -electronic appliances and cars, etc. Reflecting on material sufficiency, the center of national consciousness or the public attitude towards science and technology were both gradually shifting toward those issues improving the quality of life (see Figures 4 and 5). There appeared a strong indication which suggested that the past role and future expectation of science and technology needed to be reexamined. Typical examples included anti-environmental pollution movement, the anti-nuclear power plant movement and monitoring food safety. A hot political issue was not only how environmental pollution can be overcome, but also how science and technology should be further developed in harmonization both with the individual and with society.

Reflecting upon these changes, the priority of national policy shifted from industry- and growth-oriented to social-oriented ones, as typified by the titles of the New Economic and Social Development Plan adopted in 1970 and the Basic Plan for Economy and Society adopted in 1973. Both plans proposed that the government should take steps towards the improvement of the general living

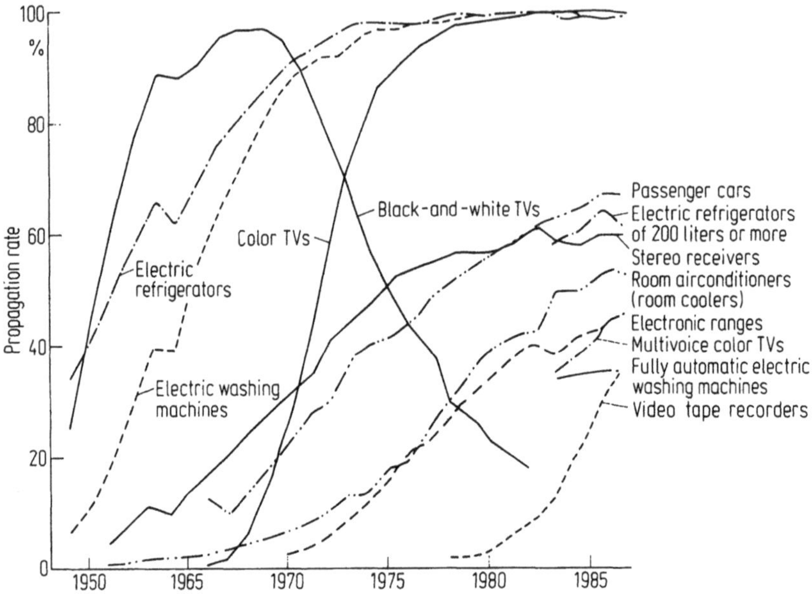

Figure 4: Growth of major home electrical appliances
Note: Growth before 1963 was for non-farming households in cities with populations over 50,000. The rate after 1964 was for all households.
Source: Economic Planning Agency (1985:125)

conditions of Japanese society, well-balanced with her economic potential, as well as the promotion of international cooperation. In other words, improvement of quality of life instead of economic growth alone, became the priority issue of the national policy. Particularly urgent policy objectives included environmental protection, energy security, and the development of social infrastructure to solve urbanization problems. In this context, the CST presented to the Prime Minister in 1971 the 5th Recommendation on the Overall Policy Framework for Science and Technology in the 1970s. Successively, the CST proposed the Fundamental Approaches of Overall Science and Technology Policy in 1977 (the CST 6th Recommendation).

Highlights of both recommendations are summarized as follows:

5th Recommendation:

Goals:

- to respond to social needs (quality of life, improvement of infrastructure, environment, international cooperation, etc.)

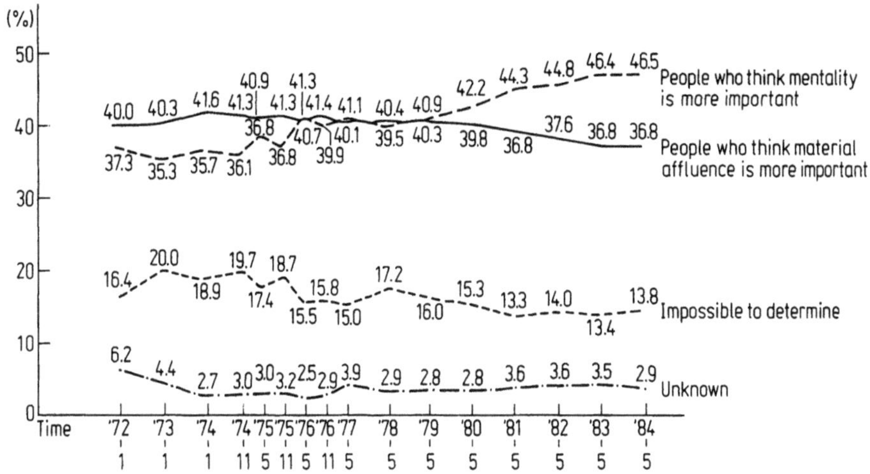

Figure 5: Changes in values among the Japanese population
Source: Commission on the History of Science and Technology Policy (1991:117)

Approaches:
- continuous efforts to pursue the lines of the 1st recommendation
- total R&D expenditures to exceed 2.5% of National Income

Priority R&D area:
- S&T for environment protection

Issues:
- technology assessment
- basic sciences for future new technology (life sciences)
- soft sciences

6th Recommendation:

Goals:
- to overcome socio-economic difficulties and to build up a strong national base (mainly, energy resources)

Approaches:
- follow-up efforts of the lines of the 5th recommendation
- total R&D expenditures to exceed 3% of National Income

Priority R&D area:
- S&T to solve resource problems

Issues:
- S&T for improvement of living conditions
- S&T for health
- advanced and basic S&T

– S&T for competition and international cooperation

The recommendations clearly focused on the socio-economic problems of that time. Highest priority in S&T policy was particularly given to the promotion of science and technology to prevent environmental pollution, to save energy and to develop alternate energy sources. In the environmental fields, public research organizations (including prefectural level ones) played an important role in monitoring pollution and in developing technologies to reduce pollutants from processing and factories. In industry – particularly iron and steel, metal refining, electric power generation, automobile and chemicals – high priority was given to the prevention of environmental pollution (see Figure 6). Taking an example from the automobile industry, the American anti-pollution regulation of 1970 urged firms to develop environment-friendly types of automobile engines.

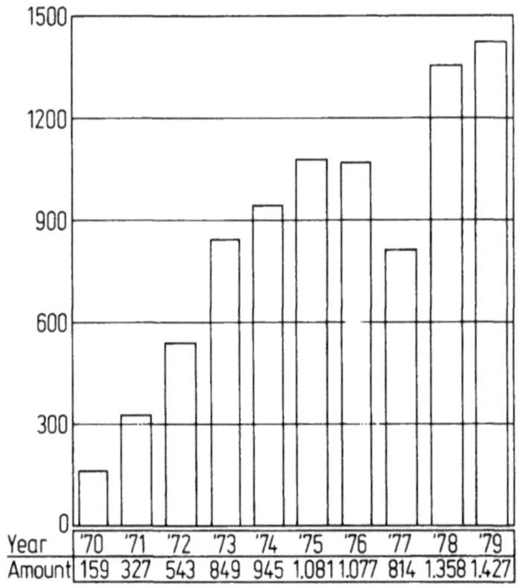

Figure 6: Trend of R&D expenditure for prevention of pollution (in 100 million yen)
Source: Commission on the History of Science and Technology Policy (1991:130)

In the fields of energy S&T, the government played an important role to overcome the energy crisis – in particular, by developing alternative energy sources on the basis of the CST recommendation of 1971. In 1974, for the promotion of nuclear power generation, the government provided new legislation with the aim of solving the location problems of nuclear power plants. Under the new legal framework, the government can give subsidies to the local governments (responsible for locating electric power generation plants) for any plans to improve infrastructure. Likewise, nuclear energy development was accelerated throughout the field by the government, with the improvement of nuclear safety in LWRs, development of fast breeder reactors and fuel cycle technology, management technology for radioactive waste, nuclear thermal fusion, etc. In other energy fields, the AIST of MITI started in 1974 Sunshine Projects within the framework of the Large-Scale Industrial Technology Development Programs, in order to develop alternative energy sources. Projects included gasification and liquidification process for coal, geothermal utilization, solar heat utilization and solar-electric cells, etc. In addition, the AIST in 1978 started Moonlight Projects within the same programs, in order to improve the development of energy-saving technologies in energy supply and transportation sectors, as well as in processing. In 1978, the STA also started projects concerning utilization of ocean and wind energy. With regard to automobiles, the MITI and the Ministry of Transportation initiated a project aiming at the reduction of fuel cost.

In addition, the CST suggested the following priority lists for science and technology:

- soft sciences (studies on relations between technology and society, policy research, etc.);
- life sciences (genetic engineering, molecular biology, etc.); and
- basic research targeting for new technologies (electronics, new material, etc.).

Considering the rapid development of electronic technology and the tendency for its diffusion in various sectors of industry, the MITI integrated two existing laws, i.e. the Law for the Promotion of the Electronics Industry and the Law for the Promotion of the Machine Industry, into a single law in 1971. The purpose of this was to fuse electronics technologies with the machinery area. Also, the AIST of the MITI supported a series of R&D programs for electronics; namely, a cooperative computer R&D program by six computer makers from 1972 to 1976, a super LSI development program from 1976 to 1979, and an R&D program of basic computer technology from 1979 to 1983 under the Act of Industrial Research Association. The STA promoted life sciences R&D projects in RIKEN from 1972 to 1981.

Although government policy greatly stimulated industry's R&D, it should be noted that any really intensive R&D efforts on the part of industry to overcome difficulties, resulted from the oil crisis of 1973 and 1979. In order to reduce production cost, industries made particularly enormous efforts to develop recycling technology for waste heat and energy-saving technology for raw material and resources, as well as the introduction of factory automation and robotics for saving labor. Based on those technological backlogs, Japanese industry attained an advanced level of industrial technology. Throughout the period, Japanese industry transformed its industrial structure from *energy-intensive* to *high-technology-oriented*. In other words, the *information-technology-oriented* structure took shape as early as the 1970s (Odagiri/Goto; for further information on this and the following period, see the *White Papers on Science and Technology*, published by the Science and Technology Agency in 1980, 1989, 1990 [full text in Japanese only]).

5 S&T Policy Built into Globalization (after 1981)

All in all, Japanese industry was able by the beginning of the 1980s to overcome the oil crisis, the rapid rise in the value of the yen and the subsequent worldwide economic recession. As a result, Japanese industry is assumed to have obtained a relatively dominant position in the world economy and presently expands its activities outside Japan, while most advanced countries suffered from long-term recession and 'stag-flation'. For example, Japan's GNP exceeded 10% of world GNP in 1980; and in 1988 it reached 14%. Also, Japan's trade balance recorded a surplus of 19.6 billion US$ in October 1977 and 101.4 billion dollars in 1986. This trade imbalance – and thus, the strong competitiveness of Japanese industry – resulted in the trade frictions between Japan and the other advanced countries as early as the late 1960s. These frictions brought about the following arrangement:

- the textile trade agreement between the USA and Japan in force since 1962;
- the export control of iron and steel to the USA since 1966 and to EC countries since 1972;
- the export control of TVs to the USA since 1968;
- the export control of machine tools to the USA since 1986 and the export price control to the EC since 1981;
- the export control of cars to the USA since 1981;
- the export control of VTRs to Australia since 1981 and to France since 1982;
- the semiconductor agreement between the USA and Japan, and between the EC and Japan, since 1986;
- MOSS (Market-Oriented Sector-Selective) began in 1985 between the USA and Japan (for telecommunications, pharmaceuticals, medical equipment, electronics, and forest products at the beginning; and in 1986, computers were added; and
- SII (Structural Impediments Initiative) between the USA and Japan since 1989.

Recently, trade frictions have appeared in the categories of high-tech products assumed to be strategic to both economic growth and security. In addition, both the expansion of world trade and the activities of multinational corporations tend to promote the globalization of Japan's economy. Inevitably, such friction problems become so complex that they involve the nation's structural and cultural issues, including Japanese market practices and general domestic social policy issues. In particular, since the beginning of the 1980s, Japan has faced structural adjustment issues. In response to these problems, *Maekawa's Report* (the report of a working group on structural adjustment, an unofficial advisory board to the Prime

Minister) proposed in 1986 a new policy direction for the expansion of Japan's imports and direct foreign investment. In 1987, the *Report of the Adhoc Committee of the Council of Economy* (the new Maekawa's Report) followed up along the same lines of the 1986 report and recommended that Japan promote foreign direct investments in the manufacturing industry, in order to expand Japan's import share of industrial products and also to contribute to the development of production capacities and to enhance employment opportunities in other countries. In 1989, negotiations on structural adjustment issues, including industrial restructuring, were conducted between Japan and the USA.

Amid the favorable performance of Japanese industry in the world market, the structural friction issue has become one of the hottest. Not surprisingly, there appeared to critics that Japanese industrial success was based on the good coupling of government intervention with industry's indigenous efforts. Japan is accused of being a "free-rider" of basic research which other advanced countries have contributed. In coping with newly-emerging international and domestic problems, S&T policy priority had to shift to the promotion of basic research, which is deemed to be of common interest to international society. For instance, in 1981, the STA established the following new systems to promote basic research:

- Special Coordination Funds allocated to basic research;
- programs of governmental laboratories under the CST's guidance; and
- Exploratory Research for Advanced Technology (ERATO) from the JRDC (a system to conduct research challenging open-ended themes under the leadership of an innovative and scientifically key individual whom the JRDC selects as leader).

The Ministry of Education expanded the amount of its grants to university R&D and to post-Doctoral fellowships in 1981. Likewise, the AIST of MITI started R&D Projects on Basic Technologies for Future Industries in 1981, which intends to support basic research aiming at the exploration of key technology needs of future industry.

In 1984, the CST proposed a new framework of science and technology policy focusing on (i) the promotion of basic research for creative science and technology, (ii) harmonization of S&T with people and society and (iii) internationalization of S&T activities in Japan (the 11th Recommendation of the CST). Through the recommendation, the traditional industry-oriented approach was cancelled. High priority was given to basic research in the fields of communications and electronics, material science and technology, life sciences, soft sciences, space science and technology, oceanography and earth sciences. It also recommended that government promote international cooperation as well as the internationalization of science and technology activities. Along these lines, the government took a series of actions, as follows:

- the budget for basic research increased in the ministries and agencies concerned;
- RIKEN established its Intentional Frontier Research System in 1986;
- the government provided in 1986 a new legal framework, the Act for Facilitating Research Exchange, in order to promote research cooperation among universities, governmental laboratories and industry – as well as international R&D cooperation of national laboratories;
- STA fellowships for foreign researchers started in 1988;
- the Human Frontier Science Program (international cooperation program) was initiated by STA and MITI in 1989; and
- Japan participated in the Space Station Program in 1984 (with the cooperation agreement settled in 1988).

All these new systems reflect the shift of priority in science and technology policy; and the budgets increase year by year. However, the total government budget for the promotion of science and technology has rather stagnated from 1980, since the government has kept the *no-increase budget policy* because of the large deficit (more then a hundred trillion yen). As a result, universities and the governmental laboratories suffered from a shortage of R&D-running funds. The R&D environment, including the quality of equipment and facilities, has not been reasonably improved (see Table 6).

Table 6: Trend of the budget for science and technology in Japan (in 100 million yen)

	Total Amount	Universities	(%)	National Labs.	(%)
FY1980	1,294,893	–	(–)	–	(–)
FY1985	1,532,869	534,251	(35)	221,391	(14)
FY1986	1,606,386	563,039	(35)	234,655	(15)
FY1987	1,662,336	594,243	(36)	245,929	(15)
FY1988	1,715,745	621,781	(36)	256,189	(15)
FY1989	1,815,196	657,517	(36)	275,452	(15)
FY1990	1,919,603	689,721	(36)	292,719	(15)

Source: Science and Technology Agency (1991:113-118)

The R&D expenditure share of the government decreases yearly. In contrast, Japanese industry has been expanding its overall R&D activities, including basic research, since the 1980s (see Table 1). As Figure 2 has shown with regard to the trend in the number of corporate laboratories established, the 1980s have been the second boom (after the 1960s). Most of the newly-established laboratories are targeted towards basic research. Roughly speaking, the major Japanese companies spend 10% of their total sales for R&D; and the expenditure for basic research amounts to 1% of their total R&D expenditures.

6 Conclusions

Through this historical review of Japanese S&T policy, the following can be viewed as tentative conclusions.

(1) Japan's S&T policy following World War II was successful in light of the country's quick restoration and subsequent industrial and socio-economic development. The S&T policy responded adequately to the industrial needs resulting from changing domestic and international situations. S&T policy-making was well incorporated into both industrial and economic policy.

(2) Through a process of success, and despite a gradual change in S&T policy goals, the institutional framework of S&T policy – particularly its administrative structure – has not changed since the establishment of the CST in 1959. The institutional continuity has several advantages that ensure incremental changes of S&T policy (not radical ones) and thus enabled Japan to keep long-term R&D programs stable, even if funds do not remarkably increase (e.g. energy R&D programs). On the other hand, it is claimed that some disadvantages appear with regard to the funding of newly-emerging S&T activities (e.g. a budget allocation system).

(3) At present, industry plays a major role in the branching-out of innovations. S&T policy should provide the appropriate environments for possible innovations. Taking into account the rule of market mechanisms, government intervention should be carefully conditioned. On the other hand, basic research is mainly conducted in universities and governmental R&D organizations, and the results are opened to the international academic community. Basic research also creates new scientific knowledge which stimulate innovation. Thus, the government is expected to promote basic research. Recently, Japan's S&T policy has given high priority to basic research. This target shift should be stressed in light of the declining R&D environment, at universities and national laboratories.

(4) Today, Japan is facing a phase of radical changes. The priority of S&T policy should shift to the international contribution; and R&D, oriented to scientific knowledge. From this point of view, it is a challenging question to ask how the existing institutional framework of Japan can remain effective regarding new S&T policy.

Acknowledgments

I would like to express my sincere appreciation to Dr. Y. Baba for his excellent suggestions and advice.

References

Commission on the History of Science and Technology (ed.) (1991) *Historical Review of Japanese Science and Technology Policy*, Tokyo: National Institute of Science and Technology Policy.

Council for Science and Technology (1960) *Juunengo wo Mokuhyou tosuru Kagakugijutu-Sinnkou no Songuteki Kiho-Housaku nitsuite* (The First Recommendation – Comprehensive Poliy Guideline for Promotion of Science and Technology 10 Years hence), Tokyo: Science and Technology Agency.

Economic Council (1960) *Kokuminshotoku Baizou-Keikaku* (National Income Doubling Plan), Tokyo: Economic Planning Agency.

Economic Planning Agency (1956) *Keizai-Hakusho* (White Paper on Economy, Fiscal Year 1956), Tokyo: Economic Planning Agency.

Economic Planning Agency (1985) *Kokumin-Seikatsu Hakusho* (White Paper on National Living), Tokyo: Economic Planning Agency.

Kodama, F. (1991) *Haiteku-Gijutu no Paradigms*, Tokyo: Chuuokoronsha.

Odagiri, H., A. Goto (1990) National Systems Supporting Technical Advance in Industry in Japan, unpublished manuscript, Tokyo.

Okita, S. (1960) *Shotoku-Baizoukeikaku no Kaisetsu* (Explanation Note of National Income Doubling Plan), Tokyo: Nippon Keizai Shinbunsha.

Science and Technology Administration Council (1951) *Kagakugi jutu Nenpo in 1951*, Tokyo: The Secretariat of the Council.

Science and Technology Agency (1991) *Indicators of Science and Technology*, Tokyo: Science and Technology Agency.

Statistics Bureau of the Prime Minister's Office (1953; 1954; 1955; 1956) *Kenkyukikan Kihontokei Chosa* (Statistics on Research and Organizations), Tokyo: Statistics Bureau of the Prime Minister's Office.

Statistics Bureau/Agency of Administrative Management (1961; 1976; 1984; 1990) *Kagakugijutu-Kenkyu* Chosa (Statistical Survey on R&D), Tokyo: Statistics Bureau/Agency of Administrative Management.

National Systems of Innovation and Technology Policy: The Case of Denmark

Bent Dalum

1 Introduction

During the accelerating internationalization in the 1980s, intensified in Europe by the establishment of the Single Market, technological change has been emphasized more than before as a major factor in economic growth. Technology policy has in many countries conquered a more prominent role on the policy agenda *vis-à-vis* standard macroeconomic policy prescriptions.

The spectacular success of Japanese economic performance was focused on in Europe and the United States and led to the proliferation of technology policy programmes in large as well as in small countries. "High-tech" programmes became the fashion. In many countries, these efforts were founded on fairly similar concepts; they were often 'imitated' in a cross-national sort of way. Programmes supporting information technology were given especially high priority at the political level, often in line with the ESPRIT programme when involving EC countries.

However, a more differentiated approach is apparently necessary. More funding for research in new "core technologies" – such as information technology and biotechnology – is not a sufficient, though probably a necessary condition for better long-term economic performance. When conceived oversimplistically as a magic formula, a blind "Science & Technology Push" strategy does not contain any guarantee for successful economic performance. Is so-called socially-oriented technology policy an alternative, and what may be the content of such a concept? The present chapter will try to sketch out a framework for that discussion.

It must be emphasized that recent developments in the orientation of Danish technology policy do not fit well into what one probably would expect to be the framework of a "socially-oriented" technology policy. The most salient feature has thus been drastic cuts in budget appropriations for industrial and technology policy measures in fiscal year 1990 and onwards.

To prevent misunderstandings, Danish industrial and technology policy has been fairly liberalistic in the postwar period. This is sometimes a surprise to foreign analysts, who would expect to find a more 'dirigiste' approach in a Scandinavian welfare state. However, as stressed a few years ago in the OECD review of Danish science and technology policy, the government initiated "a wide range of initiatives aimed at substantially strengthening science and technology policy" during the mid-1980s (OECD 1988:11). These new policy signals were to a certain extent brought to an end in December 1989, when cuts were approved by the government – published as a sort of barter for a decrease of 10 percentage points in corporate tax rates.[1]

The chapter takes its point of departure partly in discussions raised by that change and, more generally, in the concept of National Systems of Innovation (NSI).[2] Section 2 contains a brief summary of the most salient features of Danish technology policy in the 1980s. In Section 3, some important features of the international specialization patterns of small high-income OECD countries are discussed, followed by the fundamental question of whether it is at all relevant to focus on the national arena in Section 4. Finally, Section 5 contains the discussion of the concept of "socially-oriented" technology policy in a Danish context.

2 Recent Developments in Danish Technology Policy in Brief

From the beginning of the 1990s, Danish technology policy at first glance appears to have returned to its traditionally liberalistic flavour – although in a new disguise. However, the general policy approach is at the moment very much open to debate. Thus, future trends are difficult to predict.

Technology (and, more broadly, industrial) policy was not a central policy issue in Denmark during the 1960s and 1970s, although governments were dominated by coalitions headed by the Social Democrats. The policy was more an outcome of a lack of explicitly formulated goals rather than a result of conscious political decision-making. According to the OECD review (1988:23),

> Little specific legislation covers research and technology policy. To a large extent the policy is reflected in the size and relative weighting of the budget appropriations. The policy is also determined by the way in which these appropriations are actually allocated by the competent authorities, which thereby decide the main course to be followed in encouraging R&D and technology in both the public and the private sectors.

In practical terms, these authorities have acted under the guidance of committees, dominated by representatives from industry and trade associations. In the early to mid-1980s, this non-policy-like approach was somewhat changed.[3] The persistence of Danish balance of payments problems (permanent deficit in the quarter of a century from 1964 to 1989) did raise discussions of whether the cause could be exclusively formulated in terms of standard macroeconomic variables – such as (excessively high) increases in wage costs, government expenditure and private consumption. During the early to mid-1980s, more emphasis was put on "structural" variables, such as the composition of production and exports and their determinants, as a result of which R&D and technological innovation began to be taken more seriously.

Technology policy was changed by the introduction of several major programmes. The IT-oriented Technological Development Programme (TUP) of 1984 was by far the largest effort (for a brief summary, see OECD 1988:59-61). Other programmes were also initiated, of which a biotechnological research and development programme from 1987 was probably the most significant (see Arnfred et al. 1989). However, as pointed out by Christiansen (1989), the policy change was to a certain extent more programmatic than real. As far as the TUP programme was concerned, a significant part of the funds were integrated into already existing policy instruments. The very broad aim of the programme, i.e. more rapid diffusion of IT to all sectors of Danish industry and services, combined with administrative integration into existing schemes, resulted in continuity rather than any radical break.

In spite of that and several other, in our view, reasonably positive effects of the programme[4], it was closed before schedule at the end of 1988 – as a prelude to the "Great Cut" in the 1990 budget illustrated in Table 1. However, the latter has raised fairly extensive discussions of the content and aims of national technology policy strategies and efforts. The Dalum et al. (1991a) report, commissioned by the National Agency of Industry and Trade (NAIT) as a foundation for discussions of future approaches to industrial and technology policy in the new Industry and Trade Development Council (ITDC), has been part of this general debate. The approach of that report will be interpreted below as a contribution to the discussion of a framework for a "socially-oriented" technology policy.

3 International Specialization of Small OECD Countries

One of the most intriguing questions in much of the present international debate on technology policy is whether national policies have any future relevance at all in a "borderless world" (see e.g. Ohmae 1990), especially seen from a small-country point of view. As a background for that discussion, some salient features of international specialization will be presented.

In their effort to develop the causal chain from technical change to international trade and basing their reasoning on extensive empirical studies, Dosi et al. (1990) emphasize that international differences in technological capabilities are fundamental in explaining cross-national differences in levels and trends in exports, imports and income per capita (see e.g. the 'stylized facts' of their Chapter 3). They also demonstrate that the level and change of market shares among the major OECD countries apparently can be explained by country-specific absolute advantages or disadvantages – leading to favourable or unfavourable foreign income elasticities *vis-à-vis* a given country's exports.

Table 1: Research funds in Denmark, 1988-1993 (1988 = 100)

	1988	1989	1990	1991	1992	1993
Department of Industry	100	72	57	56	46	34
Department of Agriculture	100	113	117	119	101	103
Department of Energy	100	99	96	95	94	93
Department of Environment	100	109	110	101	103	82
Department of Education	100	103	102	99	99	98
Index	100	97	94	92	88	88

Source: Dalum et al. (1991a), adapted from Ministry of Education (1990) Regeringens *Stategiske Forsknings- og udviklingsprogram,* Copenhagen: Ministry of Education

In their approach, converging as well as diverging mechanisms coincide in shaping country and industry patterns. This dichotomy consists of development of the main technologies on the one hand, which vary across countries and across firms, and "strategies, context conditions and history" on the other (Dosi et al. 1990:89). The latter factors vary, however, not only among companies and countries but also across technologies. The technology factor, in very crude terms, nevertheless represents two dimensions: innovation, whether of product or process technology, leads to divergence; and diffusion leads to convergence among firms and countries. The national context has a role to play in the determination of long-term growth patterns, especially because of the inherent cumulative and path-dependent features of technology: "Once the cumulative and firm-specific nature of technology is recognized, its development over time ceases to be random, but is likely to be constrained to zones that are closely related technologically to existing activities. If those zones can be identified, measured and explained, it is possible in principle to predict likely future patterns of innovative activities in firms and countries" (Dosi et al. 1990:85).

Based on extensive empirical studies, using among others internationally comparable patent data[5], Dosi et. al (1990:102) summarize four characteristics of technological specialization:

- small countries are more specialized than large countries;
- there is a striking stability in each country's specialization pattern over time;
- specialization patterns are distinctly different from country to country; and
- relatively few large firms exert a significant influence on country patterns.

The first three of these features can also be found in the exports specialization patterns of the OECD countries.[6] Figures 1 and 2 show the Danish export specialization pattern 1961-1987 according to two different aggregations (shown in the Appendix). Specialization is measured by the Revealed Comparative Advantage index (RCA), showing the export structure for one country compared with the structure of total OECD exports. The RCA index shows, e.g. Danish exports of "resource-based products" as a share of total Danish exports divided by the same fraction for the OECD as a whole. If the index is above 1, the country is 'specialized' in that group of products, and vice versa. The weighted index is by definition 1.0 for the country total.

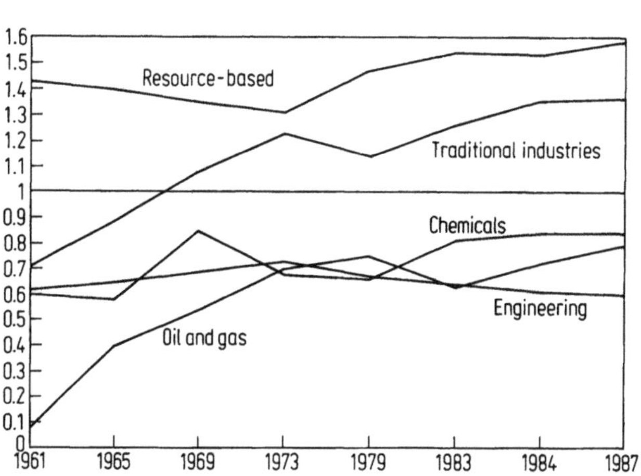

Figure 1: Export specialization in Denmark, 1961-1987
Source: IKE trade database

The importance of these patterns for national economic performance can be discussed in terms of the ideas behind the concept of National Systems of Innovation (Andersen 1992). Existing patterns of specialization of production and exports and linkages between domestic industries are conceived to be important for interactive learning processes. A significant part of technical innovation is, according to this approach, the result of incremental processes, highly determined by existing production patterns. The transmission mechanism between the export specialization pattern, on the one hand, and long-term economic performance on the other, is the foreign income elasticity of demand of the country's exports. If a given country at the outset is specialized in products with high present or future growth rates in world trade, it then has an advantage compared to other countries – and vice versa. Thus, specialization in high-growth products may be taken as an indirect proxy of foreign income elasticities of demand.

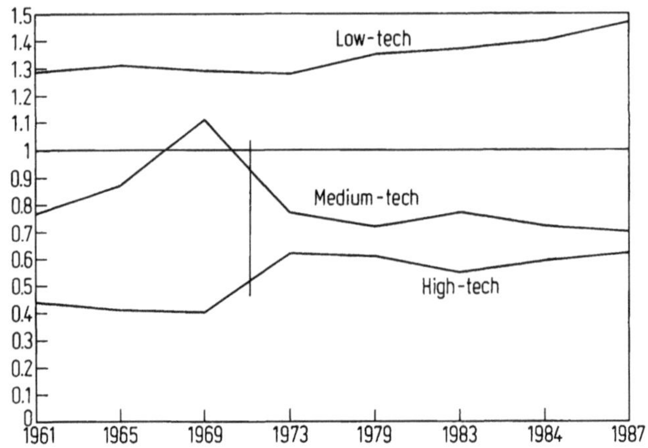

Figure 2: Export specialization in Denmark, 1961-1987
Source: IKE trade database

The general pattern of most small high-income OECD countries is similar to the Danish one, having been fairly stable over a time horizon of a quarter of a century.[7] The "standard" small country features are specialization in resource-based products and what have been called "traditional industries". On the other hand, the "standard" small country is not specialized in chemicals or engineering. In terms of the aggregation of technology intensity, the typical small-country pattern is characterized by specialization in "low-tech" products and the opposite concerning "high-" and "medium-tech" counterparts.[8] At the other end of the sample, specialization in high-tech products is a common feature of the three most powerful large OECD economies – USA, Japan and Germany. In terms of the sector aggregates in Figure 1, the only common feature is specialization in engineering.

Various country studies focusing on the importance of technical innovations for economic performance have isolated a group of countries – the United States, Japan, Germany and the two small countries Switzerland and Sweden. All these show significantly higher innovative performance than the other OECD countries when measured in relative terms (per capita, share of GDP, etc.) (see Patel/Pavitt 1987, 1989; Dosi et al. 1990; Fagerberg 1988b). In general, these countries have also been the most successful in terms of long-term economic performance, except in the case of the US balance of payments problems in the 1980s and the more recent Swedish economic problems. It is a common feature of these members of the "first division" of OECD countries that they have been specialized in engineering, at least since mid- to late 1960s. However, specialization in high-tech products is not a common denominator; Sweden has not been specialized in high-tech products during this quarter of a century. Export specialization in high-tech products is not in itself a sufficiently reliable proxy for a strong National System of Innovation, as indicated below by the Irish case. Specialization in engineering appears, on the other hand, to point in the same direction as the analyses of "high innovative performance" referred to above.[9]

The specialization pattern for most OECD countries has been surprisingly stable over the most recent quarter of a century (when neglecting the effects of oil and gas). The outstanding exceptions have been Japan and Ireland; both cases are unique and highly different. One lesson to be learned is that the cumulative and path-dependent character of technological change at the micro level is highly visible in the macro patterns. Fundamental restructuring of these patterns represents an enormous task for society at large and appears the be the exception rather than the rule.

History thus shows that there is no 'easy fix' for a small OECD country to become a "high-tech" economy – except in the case of Ireland as shown in Figure 3 and 4. However, specialization in exports of research intensive goods is by no means a 'safe' proxy for an indigenous national system of innovation. The lion's share of Irish high-tech exports has its origin in chemical plants and electronics assembly located by foreign multinational enterprises (MNEs), which function more like export enclaves than an integrated part of the Irish system of innovation.

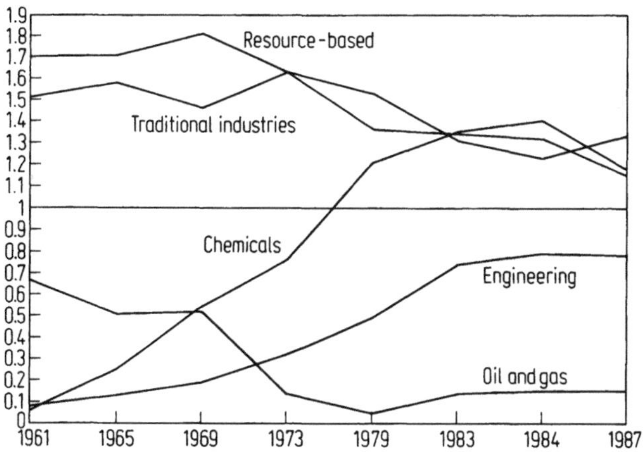

Figure 3: Export specialization in Ireland, 1961-1987
Source: IKE trade database

On the contrary, it appears necessary to take the point of departure in unique accumulated skills and experiences of the country – i.e. in the structure of production and exports already given and other important 'context variables', such as institutional structure and history in broad terms. At the present stage, it must be stressed

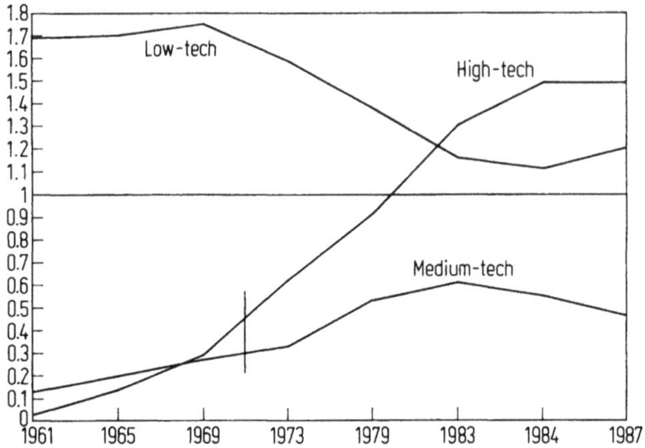

Figure 4: Export specialization in Ireland, 1961-1987
Source: IKE trade database

that technology policy programmes apparently must reflect these conditions, i.e. they must 'fit' well with the national context.

4 Nations and Globalization

You may argue that a major part of the discussion in the previous section only occasionally took into account the influence of MNEs on national patterns. Porter's (1990) analysis of "competitive advantage of nations" is based on a fairly large amount of industry studies, in which indigenous and foreign MNEs – as well as small and medium-sized firms – are present. These studies have been performed by research teams in ten countries under the guidance of a common framework – the so-called diamond shown in Figure 5. Porter stresses that there is no inconsistency between the globalization of industries and the significance of national (or regional) context variables.

The globalization of industries and the internationalization of companies leaves us with a paradox. It is tempting to conclude that the nation has lost its role in the international success of firms. Companies, at first glance, seem to have transcended countries. Yet what I have learned in this study contradicts this conclusion. ... Competitive advantage is created and sustained through a highly localized process. Differences in national economic structures, values, cultures, institutions and histories contribute confoundly to competitive success. The role of the home nation seems to be as strong or stronger than ever. While globalization of competition might appear to make the nation less important, instead it seems to make it more so. With fewer impediments to trade to shelter uncompetitive domestic firms and industries, the home nation takes on growing significance because it is the source of the skills and technology that underpin competitive advantage (Porter 1990:18-19).

His "diamond" framework consists of four determinants of competitive advantage. Factor endowments, including infrastructure in broad terms, is the realm of both the standard comparative advantage version of trade theory and the majority of analyses of international competitiveness of nations in terms of comparative costs. Strategy, structure and competition contains the dominant analytical approach in two of Porter's previous books (1980, 1985). It is important to emphasize that he extends the framework of the traditional industrial economics discipline to include company strategies as a major determinant.[10] Also added should be the emphasis on the effects of domestic rivalry on international competitive advantage.

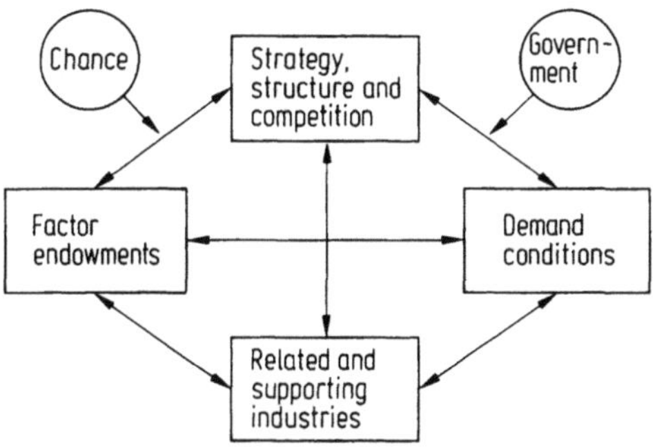

Figure 5: Porter's 'Diamond'
Source: Porter (1990:127)

The other two building blocks represent extensions of his previous work, both to a discernable degree inspired by Scandinavian research. The characteristics of domestic demand (size and, secondly but not least, qualitative features) exert a major influence on competitive advantage according, to the line of thought of Linder (1961); Vernon (1966); and Andersen et al. (1981a, 1981b). Finally, clusters of industries, containing both related and supporting industries, exert a major influence in creating external economies highly beneficial to international competitive advantage. In this case, Porter draws on Hirschman's (1958) linkage and complementarity approach, as well as on the Swedish "network" approach in international business represented by the Uppsala School (see e.g. Johanson/Mattsson 1989) and by one of their predecessors – Dahmen's (1970, 1988) concept of "development blocks". The importance of user-producer relations as the micro counterpart of (non-military) "industrial complexes", as elaborated by Lundvall (1985, 1988) and Andersen et al. (1981a, 1981b), is also a part of the foundation of Porter's eclectic framework.

The "diamond" is characterized as a mutually reinforcing system. Porter does not accept the concept of national competitiveness as expressed by a few quantitative indicators, such as wage rates and exchange rates. However, countries may be analyzed in terms of "clusters" of industries. "The basic unit of analysis for understanding national advantage is the industry. Nations succeed not in isolated industries, however, but in clusters of industries connected through vertical and horizontal relationships. A nation's economy contains a mix of clusters, whose makeup and sources of competitive advantage (or disadvantage) reflect the state of the economy's development" (Porter 1990:73).

A significant consequence of this line of thought could be sketched as follows: the more European integration is speeded up, the more national/regional context variables are important. This also applies to the competitive context for small firms: homogenous markets call for product differentiation. If a nation (or region) is not to any significant degree a 'home nation' for large MNEs, as is the case with Denmark, increasing globalization of industries and firms makes it even more important to exploit unique local knowledge assets. Whether this is done by Danish or foreign companies is a question of secondary importance. Further upgrading of such assets is the safest insurance premium to pay for future economic development, unless a country is endowed with 'manna from heaven' in terms of natural resource endowments – for instance, oil (which may degenerate industrial development in the long term).

As an illustration of the cluster approach, Table 2 shows Danish export specialization in an elaborated version of Porter's "Cluster Chart" (see Dalum et al. 1991a:38). The data, in which only product groups with an export specialization above 1 are represented, apparently confirms the importance of discernible clusters in the Danish export specialization pattern. Internationally competitive firms are not spread stochastically around the industrial landscape, but are rather – and to a significant extent – concentrated in clusters of industries which are part and parcel of a genuinely unique historical context. In many cases, though not all, the latter covers a unity of political, institutional, economic and geographic factors – i.e. nations.

5 'Socially-Oriented' Technology Policy in a Danish Context – Some Examples

In the previous sections much emphasis has been given to the discussion of why nations still matter. The cluster analysis has been seen as a useful analytical tool. It can also serve as a means of discussing policy.[11] The interplay between domestic demand – especially government expenditure – and private firms is an important policy issue added to the well-known emphasis, in all countries, on the quality of

Table 2: Shares of "clusters" in Danish Exports, 1961-1987, in %

	1961	1973	1987
1. Food "Cluster"			
Meat	30.9	18.8	11.4
Dairy products (2)*	14.3	5.3	3.9
Fish (2)	4.1	4.5	6.7
Cereals (4)	1.3	1.4	2.5
Animal Feed (5)	0.9	2.7	0.7
Other manufactures	4.5	4.9	5.0
Agricultural machines	1.1	1.4	1.1
Food-processing machines	0.4	0.4	0.5
Other related machines	0.4	0.7	1.2
Heating & cooling, pumps (24-25)	3.5	3.8	3.5
Total	61.4	43.9	36.5
Of which machinery =	5.4	6.3	6.3
2. Housing "Cluster"			
Wood and wood manufacturing (7)	1.2	1.2	1.6
Dyestuffs (17)	0.6	0.7	0.8
Domestic electric equipment (32)	0.3	1.5	1.1
Furniture (39)	1.4	2.4	3.9
Cement and cement manufacturing	0.3	0.3	0.4
Central heating equipment (32)	0.1	0.4	0.2
Total	3.9	6.5	8.0
3. Health "Cluster"			
Pharmaceuticals (18)	1.5	1.7	2.3
Electromedical equipment	0.1	0.1	0.6
Other medical equipment	1.0	1.2	1.3
Total	2.6	3.0	4.2
4. Transport "Cluster"			
Ships	3.4	5.5	2.0
Sum of "Clusters"	71.3	58.9	50.7

* Numbers in brackets refer to the United Nation's Standard International Trade Classification, Revision 1 (SITC, Rev. 1).

Source: OECD trade data base, Institute of Production, Aalborg University (IKE trade database)

science and technology infrastructure. Existing and/or potential new clusters (at present at an embryonic stage) may form the starting point for focusing technology policy efforts.

"Social orientation" of technology policy requires, in my view, that efforts are directed towards satisfaction of the needs of the general public by means of new product or process technology – without explicit protectionistic purposes. The latter has often been the case in e.g. the telecommunications and weapons industries. History has seen many (non-military) cases where such fulfilment of certain 'social' needs has led to international excellency among private firms who solved the problems in question. The development of the strong competitive advantage of Danish hearing-aid firms is an outstanding example (Jorgensen 1986; see also Dalum et al. 1988; and Lotz 1991). However, such cases have most often been studied *ex post*. What is proposed here is to learn from these examples in *ex ante* policy development.

In a Danish context a "socially-oriented" technology policy could thus be programmes focused on development of new (and better) food products and food-processing technologies, which may have less detrimental environmental effects.[12] Many analysts argue that Danish agriculture and food-processing in general is following a 'dead end' – i.e. a trajectory of steady productivity increases in agriculture by means of fertilizers, exploitation of economies of scale by merging small farms to large entities, etc. The same logic appears, at the moment, to be the main driving force in recent outstanding structural change in the Danish food-processing industry in terms of significant increases in merger and acquisition activity. Another area could be urban renewal and energy conservation technology for buildings, taking its point of departure in the housing cluster.

The approach presented can also be used to guide discussions of the content and relevance of the many and proliferating high-tech programmes, mentioned in the introduction. The high-tech industries usually cover electronics, fine chemicals, pharmaceuticals and aerospace. Policy issues and instruments may very well differ between these sectors. The electronics industry will serve as a useful example. Full 'coverage' has not been intended.

Export specialization of the total Danish electronics industry had been fairly low between 1961 and 1987 (0.4 – 0.7). In terms of export structure, around 40% of Danish electronics exports are measuring and controlling instruments and electromedical equipment in 1987 – a significantly higher share than in most other OECD countries, as can be seen in Figure 6. Telecommunications is the third subsector where Denmark has been specialized, but only during the 1970s. That sector contains wired telephone systems, mobile communications and marine communications. The competitive advantage of Danish firms has been fairly strong in the two 'wireless fields', partly concealed by the statistical aggregate telecommunications in which wired equipment by far has the highest weight.

The high degree of specialization in instruments and medical electronics has been created by small and medium-sized niche firms, of which several are among

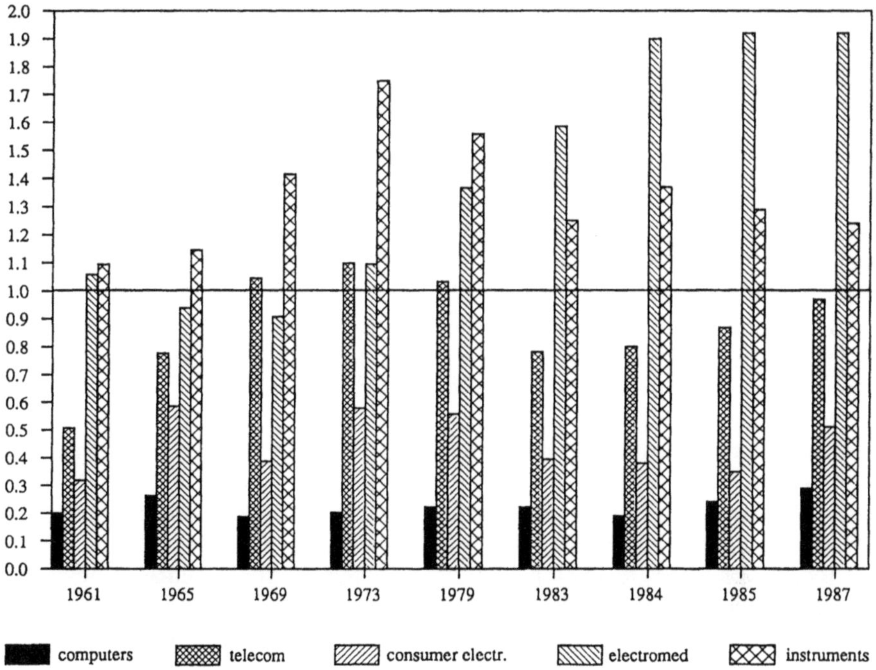

Figure 6: Export specialization in Denmark – Electronics, 1961-1987
Source: IKE trade database

the world leaders in their narrow fields. These firms are mainly owned by their founders or near relatives; and the background of their international success is often a combination of the domestic research system, the quality of domestic demand in the initial phase of the product cycles, and outstanding entrepreneurship.

The competitive advantage of many Danish instrument firms has to a high degree been based on a combination of the R&D system and the quality of the public-sector part of domestic demand – e.g. from hospitals and/or the more general health care system. In the narrow sence, it appears intuitively plausible to talk about a 'socially oriented' technology policy context here. But the policy proposals should not be confined to just more research in these niche fields. Of equal importance may be the quality of e.g. the health care system at large, in its continous function as a prototype user/developer *vis-à-vis* firms. However, such a definition may be too narrow – or even 'romantic' in its more or less explicit purpose of reaping some productive side effects from welfare state governmental consumption.

The sector of marine communications has the same industry characteristics as the instrument sector; while the field of mobile communications previously was dominated by two 'old', originally indigenous firms, both acquired by foreign multinationals in the mid-1970s (Storno by General Electric, and subsequently by Motorola from 1986; and A.P. Radiotelefon by Philips). The most important factors behind these companies in the 1980s have been a combination of regulation by the Nordic PTTs – who launched the pathbreaking Nordic Mobile Telephone system (NMT) in 1981 – and local research excellence. The Danish mobile communications industry also contains two small Danish newcomers (founded in the 1980s), of which one has been aquired by a German company. The NMT system was a major 'social innovation', even brought about by cooperation between four telecommunications monopolies (the PTTs) in the Nordic welfare states.

"Social orientation" may thus be conceived as fulfilment of demands expressed by the general public and transmitted into the political decision-making process at some stage. Besides the health sector and telecommunications[13], such demands are represented by e.g. less pollution of soil, air and water; exploitation of alternative energy sources and derived change in energy distribution; expansion of collective transport systems; etc. Small welfare states, such as Denmark, may be in a favourable position to develop solutions for such problems, because they inherently require a combination of social and technological innovations.

"Socially-oriented" technology policies may thus be conceived as policies designed to solve technological bottlenecks which may deter social innovations. The selection criteria for investing scarce governmental resources in technology policy programmes could be existing competitive strength in discernible clusters of industries and/or commitment to participate in spurring the creation of new ones. The recent Danish Enviromental Technology Development Programme, from spring 1991, indicates that such an approach has (re)captured a more prominent role in governmental policy.

Notes

[1] It was also of importance, that the decision coincided with appointment of a new minister of industry, who defended the policy profile as an offensive strategy. The rhetorics surrounding the new approach was to some extent conceived as a neo-liberal (read Thatcherite) move by the general public.

[2] For an introductory discussion of that concept, see Andersen/Lundvall (1988). For a more extensive treatment, see Lundvall (1992).

[3] For a discussion, see Christiansen (1989, 1988) and Sidenius (1989).

[4] At least that is how Dalum et al. (1991a) have interpreted the evaluation reports of the programme.

[5] The indicator of technological specialization is calculated by the same expression as used and explained below for the analysis of export specialization.

[6] The empirical material used in this section is presented in more detail in Dalum (1992). The trade analysis is based on the OECD trade database at the Institute of

Production, Aalborg University (the IKE trade database) – developed by the author and Jan Fagerberg. See also Fagerberg (1988a).

[7] The volatility of oil and gas disturbs the general pattern of stability, independent of country size.

[8] This aggregation focusses on the research content of the products, and not on process technology – i.e. a partial analysis of the export specialization pattern. In interpretation of the results, it is important to have this reservation in mind. In spite of these, the aggregation of technology intensity (based on research intensity) is considered to be useful by the present author. There is thus a high correlation between high-tech and high-growth products (for detailed documentation, see Fagerberg 1988a). Notice also that the definition of high- versus medium-tech is different in 1961-1969 as opposed to 1973-1987 – indicated by a vertical line in the figures. For a conceptual discussion and empirical analysis of Sweden along the same line of thought, see Edquist/McKelvey (1991).

[9] The lack of Swedish specialization in high-tech goods has, however, recently been interpreted as a problem for both present and future Swedish long-term productivity growth and economic performance in general. See Edquist/Lundvall (1993) and Edquist/McKelvey (1991).

[10] In Porter (1980), strategies are applied as "structural" variables in the analysis of the internal substructure of industries – especially in defining "strategic groups" as an analytical tool for applied industry analysis (see e.g. chapter 7).

[11] The following discussion is based on Dalum et al. (1991a). It should be stressed that the present author finds Porter's eclectic framework useful as device for structural analysis and as a point of departure for policy discussions. However, I have several critical reservations on the actual analysis of and proposals for government policies in his chapters 12-13; for a further elaboration see Dalum et al. (1991b).

[12] A food-processing technology programme has actually been started in Denmark recently, despite cuts in appropriations in December 1989.

[13] This would be true assuming that a modern telecommunications network is a *sine qua non* for a high income welfare state – although the most visible feature of mobile communications initially may have appeared to be the service of 'yuppie' demand.

APPENDIX – Table A.1

Product group	SITC
Products based on natural resources	
1. Animals, meat and meat preparations	00, 01, 091.3, 411.3
2. Dairy products and eggs	02
3. Fish and fish preparations	03, 411.1
4. Cereals and cereal preparations	04
5. Animal Feed	08
6. Skins and leather manufactures	21,61
7. Wood and wood manufactures	24, 63
8. Pulp and paper	25, 64
9. Textiles	26, 65
10. Iron ore	281
11. Iron, steel and iron alloys	67
12. Aluminium	684
13. Other products based on natural resources	Rest 0-4 (excl. 33, 34), 62, 66 and rest 68
Oil and gas	
14. Oil and gas	33, 34
Chemicals	
15. Organic chemicals	512
16. Inorganic chemicals	513, 514
17. Dyestuffs, colouring materials	53
18. Pharmaceuticals	54
19. Fertilizers	56
20. Plastic materials	581.1, 581.2
21. Other chemicals	Rest 5
Engineering, electronics and transport equipment	
22. Power-generating machinery	711
23. Machinery for special industries or processes	712, 715, 717, 718, 719.3, 719.5, 719.8
24. Heating and cooling equipment	719.1
25. Pumps and centrifuges	719.2
26. Typewriters and office machines	714.1, 714.9
27. Computers and peripherals	714.2, 714, 3
28. Semiconductors	729.3
29. Telecommunications equipment	724.9
30. Machinery for production and distribution of electricity	722, 723, 729.9
31. Consumer electronics	724.1, 724.2, 891.1
32. Domestic electrical equipment	725
33. Scientific instruments, photographic supplies, watches and clocks	726, 729.5, 729.7, 861, 862, 864
34. Road motor vehicles	732
35. Aircraft	734
36. Ships, boats and oil rigs	735
37. Other engineering products	Rest 7 (excl. 719.4)
Other industrial products	
38. Manufactures of metal	69, 719.4, 812.1, 812.3
39. Furniture	82
40. Clothing	84
41. Industrial products, not elsewhere specified	Rest 8-9 (excl. 861, 862, 864, 891.1)

Note: The products are classified according to United Nation's Standard International Trade Classification, Revision 1 (SITC, Rev. 1). The products shown are mixed across 2-, 3- and 4-digit groups in SITC nomenclature.
Excl. = the mentioned groups excluded
Rest = products not already included

APPENDIX – Table A.2

Product	SITC
High-technology products	
16. Inorganic chemicals	513, 514
18. Pharmaceuticals	54
20. Plastic materials	581.1, 581.2
22. Power-generating machinery	711
27. Computers and peripherals	714.2, 714, 3
28. Semiconductors	729.3
29. Telecommunications equipment	724.9
30. Machinery for production and distribution of electricity	722, 723, 729.9
33. Scientific instruments, photographic supplies, watches and clocks	726, 729.5, 729.7, 861, 862, 864
35. Aircraft	734
Medium-technology products	
15. Organic chemicals	512
17. Dyestuffs, colouring materials	53
19. Fertilizers	56
21. Other chemicals	Rest 5
23. Machinery for special industries or processes	712, 715, 717, 718, 719.3, 719.5, 719.8
24. Heating and cooling equipment	719.1
25. Pumps and centrifuges	719.2
26. Typewriters and office machines	714.1, 714.9
31. Consumer electronics	724.1, 724.2, 891.1
32. Domestic electrical equipment	725
34. Road motor vehicles	732
36. Ships, boats and oil rigs	735
37. Other engineering products	Rest 7 (excl. 719.4)
38. Manufactures of metal	69, 719.4, 812.1, 812.3
Low-technology products	
1. Animals, meat and meat preparations	00, 01, 091.3, 411.3
2. Dairy products and eggs	02
3. Fish and fish preparations	03, 411.1
4. Cereals and cereal preparations	04
5. Animal Feed	08
6. Skins and leather manufactures	21,61
7. Wood and wood manufactures	24, 63
8. Pulp and paper	25, 64
9. Textiles	26, 65
10. Iron ore	281
11. Iron, steel and iron alloys	67
12. Aluminium	684
13. Other products based on natural resources	Rest 0-4 (excl. 33, 34), 62, 66 and rest 68
14. Oil and gas	33, 34
39. Furniture	82
40. Clothing	84
41. Industrial products, not elsewhere specified	Rest 8-9 (excl. 861, 862, 864, 891.1)

Note: The products are classified according to United Nation's Standard International Trade Classification, Revision 1 (SITC, Rev. 1). The products shown are mixed across 2-, 3- and 4-digit groups in SITC nomenclature.
Numbers 26, 31 & 34 are high-tech products, 1961-1969. In 1973-1987 they are put in the medium-tech group.
Excl. = the mentioned groups excluded
Rest = products not already included

References

Andersen, E.S. (1992) Approaching National Systems of Innovation from the Production and Linkage Structure, in: Lundvall (1992), 68-92.

Andersen, E.S., B. Dalum, G. Villumsen (1981a) *International Specialization and the Home Market*, Aalborg: Aalborg University Press.

Andersen, E.S., B. Dalum, G. Villumsen (1981b) The Importance of the Home Market for the Technological Development and the Export Specialization of Manufacturing Industry, in: C. Freeman (ed.) *Technological Innovation and National Economic Performance*, Aalborg: Aalborg University Press, 49-102.

Andersen, E.S., B.-A. Lundvall (1988) Small National Systems of Innovation Facing Technological Revolutions: An Analytical Framework, in: C. Freeman, B.-A. Lundvall (1988), 9-36.

Arnfred, N., A. Hansen, J.L. Pedersen (1989) Bio-technology and Politics, Danish Experiences, in: Pedersen (1989), 159-184.

Christiansen, P.M. (1988) *Teknologi mellem stat og marked*, Aarhus: Politica.

Christiansen, P.M. (1989) Danish Technology Policy – From Market Direction to Structural Change, in: Pedersen (1989), 31-60.

Dahmen, E. (1970) *Entrepreneurial Activity and the Development of Swedish Industry 1919-39*, Homewood: American Economic Association Translation Series.

Dahmen, E. (1988) 'Development Blocks' in Industrial Economics, in: *Scandinavian Economic History Review* 1: 3-14.

Dalum, B. (1992) Export Specialisation, Structural Competitiveness and National Systems of Innovation, in: Lundvall (1992), 191-225.

Dalum, B., J. Fagerberg, U. Jorgensen (1988) Small Open Economies in the World Market for Electronics: The Case of the Nordic Countries, in: C. Freeman, B.-A. Lundvall (1988), 113-138.

Dalum, B., K. Moller, U. Jorgensen, F. Valentin (1991a) *Internationalisering og erhvervsudvikling – Et debatoplaeg om fremtidig dansk erhvervspolitik*, Copenhagen: Industri- og Handelsstyrelsen.

Dalum, B., K. Moller, U. Jorgensen, F. Valentin (1991b) Porter og den erhvervspolitiske debat, in: *Samfundsokonomen* 7: 33-41.

Dosi, G., C. Freeman, R. Nelson, G. Silverberg, L. Soete (1988) *Technical Change and Economic Theory*, London: Pinter.

Dosi, G., K. Pavitt, L. Soete (1990) *The Economics of Technical Change and International Trade*, Hemel Hempstead: Harvester Wheatsheaf.

Edquist, C., B.-A. Lundvall (1993) Comparing the Danish and Swedish Systems of Innovation, in: R. Nelson (ed.) *National Innovations Systems – A Comparative Study*, Oxford: Oxford University Press, 265-298.

Edquist, C., M. McKelvey (1991) Diffusion of New Product Technologies and Productivity Growth in Swedish Industry, Working Paper No. 91-15, Berkeley: Center for Research in Management, University of California.

Fagerberg, J. (1988a) Technology, Growth and Trade – Schumpeterian Perspectives, Ph.D thesis, Sussex: Science Policy Research Unit, University of Sussex.

Fagerberg, J. (1988b) Why Growth Rates Differ, in: Dosi et al. (1988), 432-457.

Freeman, C., B.-A. Lundvall (eds.) (1988) *Small Countries Facing the Technological Revolutions*, London: Pinter.

Hirschman, A.O. (1958) *The Strategy of Economic Development*, New Haven: Yale University Press.

Johanson, J., L.-G. Mattsson (1989) Internationalisation in Industrial Systems – A Network Approach, in: N. Hood, J.E. Vahlne (eds.) *Strategies in Global Competition*, London: Routledge, 287-314.

Jorgensen, U. (1986) Elektronikbranchens etablering og strukturelle wdvikling, Research Report No. 11, Copenhagen: Institut of Social Science, Technical University of Denmark.

Linder, S.B. (1961) *An Essay on Trade and Transformation*, Stockholm, New York: Almquist & Wiksel and Wiley.

Lotz, P. (1991) Demand-Side Effects on Innovation: Reflections on the Development of Danish Medical Devices, mimeo, Copenhagen: Institute of Industrial Research, Copenhagen Business School and Sloan School of Managenement, MIT.

Lundvall, B.-A. (1985) *Product Innovation and User-Producer Interaction*, Aalborg: Aalborg University Press.

Lundvall, B.-A. (1988) Innovation as an Interactive Process: From User-Producer Interaction to the National System of Innovation, in: Dosi et al. (1988), 349-369.

Lundvall, B.-A. (ed.) (1992) *National Systems of Innovation -Towards a Theory of Innovation and Interactive Learning*, London: Pinter.

OECD (1988) *Reviews of National Science and Technology Policy – Denmark*, Paris: OECD.

Ohmae, K. (1990) *The Borderless World*, London: Collins.

Patel, P., K. Pavitt (1987) Is Western Europe Loosing the Technological Race?, in: *Research Policy* 2-4 (16): 59-85.

Pavitt, K., P. Patel (1989) The International Distribution of Technological Activity, in: *Oxford Review of Economic Policy* 4 (4): 35-58.

Pedersen, J.L. (ed.) (1989) *Technology Policy in Denmark*, Copenhagen: New Social Science Monographs.

Porter, M.E. (1980) *Competitive Strategy – Techniques for Analyzing Industries and Competitors*, New York: The Free Press.

Porter, M.E. (1985) *Competitive Advantage – Creating and Sustainaing Superior Performance*, New York: The Free Press.

Porter, M.E. (1990) *The Competitive Advantage of Nations*, London, Basingstoke: Macmillan The Free Press.

Sidenius, N.C. (1989) *Dansk Industripolitik – Nye losninger pa gamle problemer*, Aarhus: Hovedland.

Vernon, R. (1966) International Investment and International Trade in the Product Cycle, in: *Quarterly Journal of Economics* May (80): 190-207.

Towards a Social Orientation in Finnish Technology Policy

Torsti Loikkanen and Esko-Olavi Seppälä

1 Introduction

The first part of this paper examines the evolution of the Finnish R&D system, together with its national S&T policy. If Finland is compared to other countries, rapid progress has occurred on both of these fronts; moreover, the latest policy documents lay down that this development is to continue.

Towards the end of our paper, we shall assess the orientation of national technology policy in light of the Sundqvist Report (OECD 1988b). Our study will also avail itself of that report's conclusions on technology policy and social orientation.

National technology policy would appear to involve two distinct tendencies. We shall refer to the first as the "strong development line" and to the second, correspondingly, as the "weak line". The first of the two is tied to the promotion of technology, whereas its counterpart is connected to assessment and impact analysis. We shall analyze these from the dual perspective of national S&T policy documents, on the one hand, and implemented policy measures on the other. It is our opinion that the leeway available for integrating these policy tendencies will depend on how recent proposals for institutionalizing futures studies, innovation research and S&T policy analyses are advanced in Finland.

We shall also discuss a special instance of the integration of environmental and technology policies. It illustrates a few national characteristics which have a bearing on the development of Finnish S&T policy.

2 Policy for R&D

Generally speaking, science and technology did not become a major focus of interest in Finland until the 1960s, when enterprises and other organizations increased their investment in research and product development (for the following, cf. Lemola/Seppälä 1987). When the children born during the exceptional 'baby boom' of the 1940s reached the age when they began to enroll as students at higher-education institutes, Finland passed a special act on the expansion and development of her university system. The act remained in force from 1967 until 1986, a period which witnessed a dramatic rise in university staff, students and facilities. A further act, which reinforced this trend, entered into force in 1987.

The country's universities expanded on a nation-wide scale. There were only two university centres in Finland in the 1950s: Helsinki and Turku, with a combined total of less than 20,000 students. Today there are more than 110,000 students in twenty institutions of higher learning in ten university towns. Similarly, the number of university teachers and researchers has soared from roughly 1,500 to almost 8,000.

The late 1960s were also notable for vigorous scientific research development. The end of the decade saw the various research councils become united under a single organization which was named the Academy of Finland. The academy was given legal guarantees that it would be the recipient of significantly higher funds until the mid-1970s. This was how the bodies behind Finland's higher education and research – and related development policies – were put on a modern footing.

Nevertheless, the age of technology did not really transpire until the 1970s. The state began to allocate support for firms' research and product development in the late 1960s, when the Ministry of Trade and Industry was empowered to distribute greater funding. Although the first oil crisis interrupted the growth both of the universities and of the Academy of Finland, enterprises and the government continued to increase their backing of technical-industrial R&D. A specialized body known as the Technology Development Centre (TEKES) under the auspices of the Ministry of Trade and Industry, was established in 1983 to serve as an expert organization in R&D matters. This allowed the public organization of applied technical research and development to reach the level on which higher education had already stood for fifteen to twenty years.

A governmental advisory body, the Science Policy Council of Finland, was established in 1963; it was a ministerial body supplemented by scientific experts. Earlier on, during the debate about whether the council should be created, mention was made of two shortcomings which the body might help rectify (Seppälä 1990:169). The first was that the advice available to Cabinet members on scientific issues would be inadequate unless the research community had direct, institutionalized contacts with ministers. The second was that, without such a body being established, the government would have no permanent system for reviewing plans of the promotion of research – an obvious impediment to purposeful science policy. Because of this, a government report on technology policy was submitted to Parliament in 1985. In accordance with the report's recommendations, the Science Policy Council was also envisaged as a cooperative organ for technology policy (Council of State 1985). In formal terms, this involved the abolishment of the Council and the establishment of a Science and Technology Policy Council in its stead. These steps were taken in 1987. The council is chaired by the Prime Minister – as its predecessor had been from the very outset.

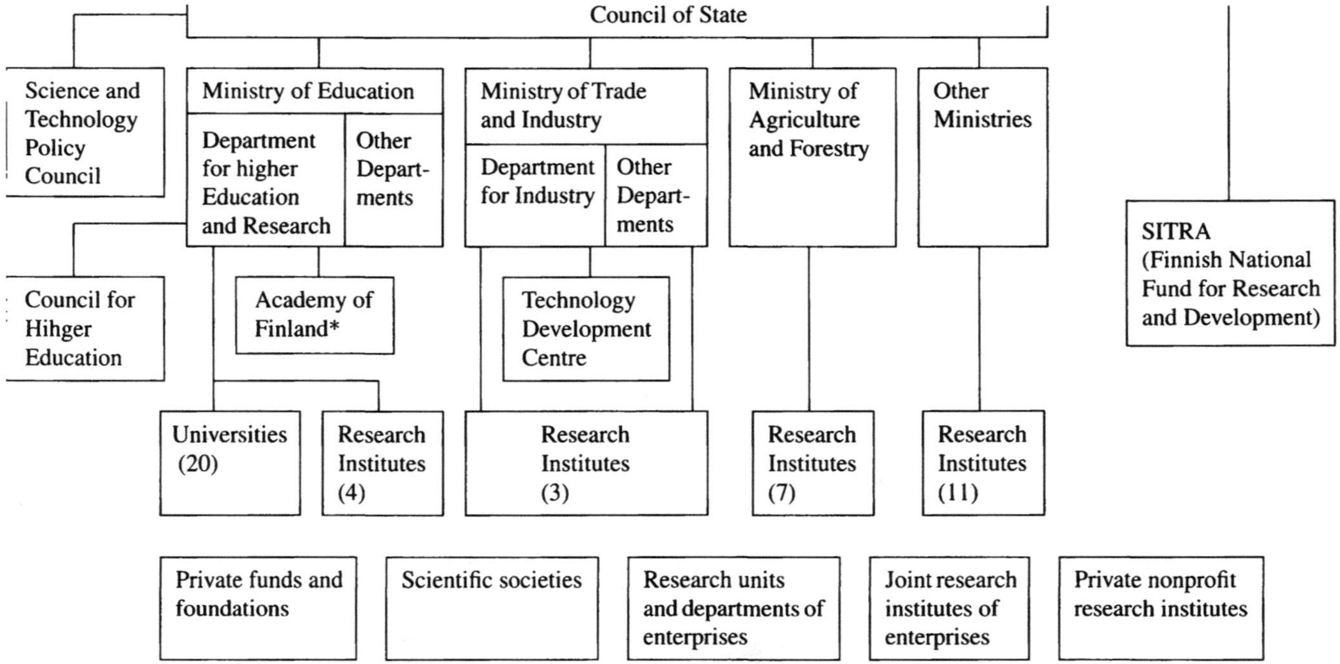

* The Academy of Finland is a central body for science administration, consisting of the Central Board of Research Councils and seven Research Councils.

Figure 1: Organization of R&D in Finland

The Finnish Parliament has no specific organ for dealing with science and technology issues, and neither does the Council of State. As we pointed out earlier, the Science and Technology Policy Council is an advisory body – not part of the Cabinet's apparatus. The government allocates a key R&D role to two planning and financing organizations, the Academy of Finland and the Technology Development Centre. Both bodies operate entirely under the state budget. A further organ which deserves mention is the Finnish National Fund for Research and Development (SITRA), a public fund with resources emanating from basic capital and from returns on successful projects. It mainly finances the development of enterprises' production activities but is also entering the venture capital market.

Finland's university system is regionally decentralized, and individual universities are also regarded as regional development know-how centres. The universities have a dual role – teaching and scientific research. As a result, they are responsible for most of Finland's basic research. The University of Helsinki is the country's largest, having some 28,000 students and a teaching staff of approximately 1,800.

All in all, twenty-five state research institutes operate under various ministries. The most important is the Technical Research Centre of Finland (VTT). It has an annual budget of FIM 1 billion, and about one third of its operating expenses are covered directly by the state budget. The remaining two thirds are met from research projects commissioned by enterprises, as well as from research funds made available by the state sector in general and TEKES in particular. There are three other especially prominent state research institutes – the Geological Survey of Finland, the Agricultural Research Centre and the Finnish Forest Research Institute. The four major institutes command over 80% of all the resources allocated to state research institutes, leaving less than 20% for the other 21.

Today, in-company R&D accounts for more than 60% of Finland's overall R&D activity (see Table 1). Finnish businesses have formed few joint research bodies. The largest, the Finnish Pulp and Paper Research Institute, jointly owned by the wood-processing industry, employs more than 300 people and has an annual budget of approximately FIM 90 million. Each separate enterprise covers most of its own R&D costs, and the self-financing ratio has remained constant at around 90%. The remaining 10% is met by the state, other companies and parties overseas.

If we remember that the volume of the Finnish GDP increased by 80% between 1971 and 1989, it is apparent that R&D volumes almost quadrupled within twenty years in terms of funding.

The 1980s were notable for a rapid increase in R&D in the business enterprise sector. Nevertheless, the volume of R&D – and its importance relative to Finland's GDP – has grown in all sectors, as illustrated in Table 2.

Table 3 focuses on the financial flows between the sectors, showing that they equalize differences in the funding growth figures.

The flow with the steepest growth rate has been the outside funding of university R&D. In 1990, more than 55% of university R&D staffyears were paid for by

Table 1: The proportion of research expenditure accounted for by different sectors, 1971-1989 (in %)

	1971	1977	1983	1989
Business enterprise sector	54	52	57	61
Public sector and private non-profit sector (not including universities)	25	27	21	19
Universities	21	21	22	20
Total	100	100	100	100
Million Finmarks	436	1,299	3,696	9,018
General expenditure on R&D as % of GDP	0.89	1.07	1.34	1.82

Source: Central Statistical Office

Table 2: R&D in different sectors, as % of GDP in 1983 and 1989

	1983	1989
Business enterprise sector	0.75	1.10
Public sector and private non-profit sector (not including universities)	0.29	0.35
Universities	0.30	0.37
Total	1.34	1.82

Source: Central Statistical Office

external sources. The most important endowers were the Academy of Finland's various research councils and a range of ministries. Yet although the contract research financed by enterprises is increasing, it still accounts for less than 5% of universities' overall R&D funding.

Table 4 shows trends in the business enterprise sector. It portrays the R&D expenditure figures of certain product groups and also illustrates what proportions of production's value they accounted for. If we combine all product groups, we see that R&D expenditure has increased by more than two percentage points within ten years. Moreover, the figure for 1989 is 4.3% of the value added, illustrating further growth of R&D intensity.

The R&D activity of the Finnish business sector displays a high degree of concentration. The number of companies engaging in research and development has increased constantly and now stands at more than 1,200. Notwithstanding, more than half of the sector's overall R&D expenditure is due to five enterprises only.

Table 3: The average annual real growth of R&D in Finland between 1893 and 1987 (in %)

Sector	As source of funds	As performer of R&D
Business enterprise sector	12	11
Public sector and private non-profit sector (not including universities)	9	< 9
Universities	7	> 9
Total	10	10

Source: Central Statistical Office

Table 4: R&D expenditures of enterprises in some product groups and the proportional share of the value added in 1977 and 1987 (millions of Finmark [MFIM] and %)

	1977		1987	
Product group	MFIM	%	MFIM	%
Food	27	0.6	178	1.6
Pulp and paper	69	1.9	247	2.0
Drugs	39	11.7	209	29.3
Chemical products	53	3.0	397	7.2
Fabricated metal products	10	0.5	96	2.0
Machinery	119	3.1	639	7.3
Computers/	77	7.3	207	16.6
Electrical machinery			218	7.2
Electronic equipment and components	52	8.2	540	12.1
Instruments	31	16.2	149	16.7
All product groups	675	1.6	4,002	3.7

Source: Central Statistical Office

The rapid growth of research and development in Finland has brought problems of recruitment and intellectual resources to the fore. The distribution of the best-qualified personnel is uneven. Over 30% of the research staff working in universities have postgraduate degrees; the corresponding figures for the public and business sectors are, respectively, 10% and a mere 5%.

Enhancing the postgraduate education has been a special priority in those fields which have grown especially rapidly. Information technology is one case in point –

measures to increase Ph.D numbers were taken as far back as the early 1980s. Research, however, indicates that there is a considerable need for new intellectual resources in all disciplines (see e.g. Pimiä 1989). This is indeed the main factor behind university development policy today: systematic researcher education and internationalization are its chief objectives. The recruitment and training of other R&D personnel has also become more important.

Summarizing R&D resource trends, it is notable that the R&D expenditure of business enterprises has grown especially rapidly since 1977 – by 10% a year in real terms. The sole sector of state R&D financing to have grown with equal speed, is the funding of research related to technology and industrial development. It is evident, then, that Finland has seen intense national investment in technology over the past 12 to 15 years; and this is a trend which will continue, as is confirmed by the new review of the Science and Technology Policy Council (Science and Technology Policy Council 1991). The review lays down guidelines for national S&T policy in the 1990s. These policy trends and their predecessors will be discussed in more detail below.

3 Finnish (Science and) Technology Policy

The genesis of Finland's national science and technology policy is the subject of the next two sections of our paper. We shall focus on the most important events and national programmes. First of all, let us look at a list of the central milestones:

- Technology Committee (1980);
- government decision-in-principle on technology policy (1982);
- founding of Technology Development Centre (TEKES), first national technology programmes (1983);
- governmental report to Parliament on Technology Policy (1985);
- founding of the Science and Technology Policy Council (1987);
- evaluation of technology programmes (1988-1990), the second programme generation; and
- guidelines for Science and Technology Policy in the 1990s by the Science and Technology Policy Council (1991).

We shall now introduce national policy programmes and documents. We shall then move on to analyze what role they have played in practical policy measures, together with the overall Finnish developments.

3.1 Prior to the 1980s

As we noted earlier, Finland's science policy apparatus antecedes her organization of technology policy (for this, cf. Lemola/Lovio 1989). The latter was only put on its present modern footing in 1983, when TEKES was established. Finland, in other words, was a latecomer in the field of technology policy (see Section 4 for details).

The key instrument in the formation of the principles for a new policy was the Technology Committee, which was appointed in 1979 in response to the need to (i) form a more systematically goal-oriented technology policy, (ii) assess the rate of technological development and study its societal repercussions, and (iii) channel public debate about automation. The chief reason for the establishment of the committee was the disquietude fostered by the potentially negative effects of automation – the threat of structural unemployment had been aired with particular emphasis – and the necessity of anticipating the direction and rate of technological change. The labour movement had followed international trade union trends and launched a thorough and ongoing discussion of the repercussions of automation. Scant attention had previously been paid to the societal implications and impact of technological advance.

Representatives of the major labour organizations were appointed to the committee, together with spokesmen for science and technology, industry and administration. A political support group was established by the governmental parties and the largest opposition party, the conservative Coalition party; and this gave the committee's voice added force.

While the original motivation behind its creation was anxiety over the injurious consequences of automation, the committee finally resulted in a strong, general belief in the blessings of technological progress and a conviction that Finland too ought to avail herself of the new vistas opened up by modern information technology. It was thought that this would entail strengthening policies which aimed at expanding R&D, together with promoting goal-orientation and planning in technology.

Perhaps the Technology Committee's greatest import was not the concrete proposals it made, but rather its role as a cooperative organ which efficiently legitimized a technology policy geared to speeding up the application of automation and information technology. In essence, this policy aimed at promoting development by renewing enterprises and raising their level of technology.

3.2 From the Technology Committee to the Government's Report to Parliament in 1985

Turning to the promotion of technology and the acceleration of technological development, the follow-up to the Technology Committee's proposals soon led to the establishment of TEKES (the Technology Development Centre). Efforts to promote a more comprehensive technology policy thinking remained on a general level. In 1982, the government had taken a decision-in-principle on technology policy which reflected a wider, more societal view, although it gave priority to speeding up the application of automation and information technology (Council of State 1982). The overall tone of this 'decision-in-principle' was very similar to the strategy of positive adjustment recommended by the OECD at that time.

Towards the end of 1983, the Prime Minister's office set up an expert group to monitor the implementation of the "decision-in-principle". In addition, the group was assigned the task of preparing a technology policy report to be submitted by the government to Parliament. What was especially interesting here, was that the leading role in coordinating technology policy was now being played by the Prime Minister's office. Moreover, the group's members tended to take a social-policy view of technology. A broad view of technology policy therefore characterized the preparations for drawing up the Government Report.

The report itself, submitted to Parliament in the autumn of 1985, departed from the assumption that maintaining a high and steady level of employment – considered essential – entails annual economic growth of at least 3%. A further postulate was that such growth was only possible if Finnish production fared well on the international market. The report went on to advance arguments for accelerating technological development and measures for promoting technology transfer, international cooperation and the introduction of new technologies. Education was seen as an essential factor.

The success of measures taken to further the renewal and diversity of production, to prevent marginalization in working life and to encourage redeployment and mobility in employment were all, in the report's view, closely tied to education. Similar considerations applied to encouraging participation in social and cultural life.

The report then addressed the social impact of the new technologies. It referred to the effect of technological progress on employment, to cuts in the average working hours, to the prospect of more diverse leisure, and to the promotion of national culture. Yet, "It is still possible that actual choice will remain limited; here the possible turning point is the common desire to influence future developments. Measures should be taken to ensure that the influx of cultural impulses into Finland is not narrowed down or diminished. Problems arising from the application of new technology which influences our environment and working conditions should be investigated and forestalled well in advance" (Council of State 1985:II,2).

As had been the case earlier, there turned out to be one fundamental difficulty in implementing the report's recommendations: whereas technology development proposals could be built on existing research and other activities, there was no comparable ready-made basis for impact analyses.

3.3 The Science and Technology Policy Council

The Government Report put an end to the coordinating role of the Prime Minister's office in technology policy. The government stated, however, that a comprehensive line of thinking should be preserved in technology policy coordination. This stance led to a reform of the Science Policy Council, as described in Section 2. The organ became the Science and Technology Policy Council under a Decree promulgated in late 1986. Membership of the council was extended to also include representatives of the social partners.

The new council published its first triennial review in 1987. It noted that the rapid development of technology, as well as an understanding of its significance, had resulted in the implementation of a deliberate technology policy in most developed industrialized countries. The review declared that making full use of the new opportunities would require the public authorities to support the development and application of technology. They were further exhorted to take measures targeted towards the general framework under which business functioned, towards better maintenance and the further development of the technological infrastructure, and elimination of the harmful effects of technological change.

The council examined the R&D activities of the business enterprise sector, of applied technological research, of international technology cooperation and of impact analysis. The position taken on technology policy preparation represents a new approach: "The preparation of technology policy in Finland has primarily been based on decisions reached by broad-based committees and working groups. Finland should follow the practice of the other Nordic countries and develop research which would serve technology policy-oriented decision-making on different levels. Consideration should also be given to organizing this so-called innovation research on a more permanent footing than its current basis" (Science and Technology Policy Council 1987:111).

The firmest proposals in the 1987 Review concerned increases in R&D financing. The major part of increases were still to be allocated to research promoting technology and to product development. It was also recommended that public financing should be increased to cover 10% of the overall R&D expenditure in the business enterprise sector by 1992. The 1987 OECD median in this regard was 12.75%; in 1989, the Finnish figure was still lower than 5%.

The most recent S&T policy document, the Science and Technology Policy Council's 1990 review, strongly emphasizes the importance of developing the national system of innovation. Knowledge and know-how are viewed as the key factors for social and economic development in the 1990s. New investments in education, training and R&D are considered crucial for promoting social and technological innovation. Improving the quality of R&D, increasing the use of selective methods in financing, and paying more attention to utilization of new knowledge are all central objectives. The review also focuses on internationalization processes in economy, technology and business activities.

Evaluation and assessment are considered inseparable from other activities. Research on the development of the national system of innovation needs to be strengthened. The review holds that developments within innovation research, future studies, and science and technology policy research are central in this regard (Science and Technology Policy Council 1991).

4 Towards a Social Orientation?

4.1 The Social Orientation

The OECD Report on New Technologies in the 1990s (OECD 1988b) introduces a socio-economic strategy which highlights two key areas. It first stresses the importance of technological development and the smooth introduction of technology. Research into social innovation in technological development plays an important role for instance, in solving the 'productivity puzzle'. The report secondly underlines the need for extensive studies of technology and its effects. Technological innovation and social change must be seen as an integrated process and managed as such.

As a direct consequence of these observations, it follows that policies for technological change need not only R&D but also need to give prominence to the labour market, the education system, enterprises and various other social aspects. Policies in the area of S&T will and should not be restricted to questions of the economic integration of technological change. Economic, social and ethical factors will increasingly determine which technical innovations will be adopted.

Information technology is a central factor in structural change and productivity growth as well. It is vital, however, to realize that the potential inherent in technology does not automatically translate into increased productivity and thereby into higher rates of output, earnings and employment. The two key requirements for a long-term socio-economic strategy are an advanced R&D capacity, together with social and institutional innovation to facilitate rapid diffusion.

Measures taken at the enterprise level are obviously pivotal. The effective implementation of technological change in a company requires a strong commitment by management to deepen the involvement of those affected by new products, processes and systems. On the macro-level, this emphasizes the concept of the 'active society'. The active society is a new pattern where policies are designed to help as many people as possible play a positive role – individuals should perceive themselves as making a useful contribution to economic and social life and be viewed in a similar light by others.

One of the important characteristics of the active society is a responsive education system. This is important both to society in general, which requires education and training throughout life, and to efforts to bring about technological change.

This section will discuss integrating a social and environmental orientation within Finnish technology policy. Theoretically, two alternative strategies of integration can be distinguished. First, there is an anticipative and preventive strategy which takes social and environmental consequences and potential negative or positive impacts into account in advance. The reverse also applies: threats, possibilities and other aspects of technological change are included in sectoral social and environmental policy measures beforehand.

The second strategy, a reactive and curing strategy, refers to social and environmental impact analysis which is conducted retrospectively. The general consensus of the last twenty to thirty years is that anticipatory strategies are beneficial to individuals, industries and also society as a whole. This is why anticipatory procedures and methods have been developed and applied in many countries.

We shall now discuss whether Finnish policy may be said to have a social and environmental orientation and assess the role of the two above strategies.

4.2 The Role of Technology Policy for Finnish Economic and Social Development

Finland has a small open economy with a one-sided industrial structure based on forest industries. The country has tended to be very dependent on forest industry exports. Its industrialization proceeded after World War II slower than in most of the Western countries, and extensive growth continued for a long period. Finnish industry continued, technologically speaking, to lag behind its competitors until the late 1970s. Since then, the country's industry has undergone rapid structural change. New industrial technologies have been introduced and applied. Social investments, on their part, did not soar to the same levels as their Swedish and Dutch counterparts. It is also important to note that the mass emigration from rural districts to industrial centres and towns, one associated with industrialization, did not occur until the 1960s – again later than in most other developed OECD countries. Until the mid-1970s, postwar Finland's GDP increased slower than the figures for other Western countries. Yet after the major recession of 1975-1977 came to an end, the growth of the Finnish GDP remained until the 1990s higher than the OECD average (see e.g. Hjerppe 1988:49).

4.2.1 Developments before the 1980s

The postwar period of stable economic growth in the developed economies lasted until the early 1970s. One of the hallmarks of that period was a growth in the importance of technology, induced by increasing R&D expenditure and by techno-economic competition between the East and the West. This stable growth generated strong confidence in the accuracy of forecasting. Forecasting methods for medium and long-term developments were elaborated and applied in order to support planning and decision-making.

Finland followed these international trends, though with a lag of several years. The factors which brought technological development to the forefront in the late 1960s were a strengthening of internationalism and a liberalization of trade. Finland's production structure and technological capacity were seen to be low compared with her competitors, and this brought those sectors under pressure. R&D was not yet an inherent part of Finnish production nor of its development in the early 1970s. Consequently, the government then began to support industrial R&D and to provide higher-education institutions and research institutes with more resources. In addition, the state began to organize goal-oriented technological research.

In most of the industrialized countries, the impact of technology generated widespread anxiety in the 1960s, due to exacerbated societal and environmental problems. There was also worldwide concern – the Club of Rome's views, for instance, received extensive publicity. By the early 1970s, progress towards socially- and environmentally-oriented thinking had occurred. Numerous countries, for example, established environmental administrative organs. Technology assessment was institutionalized as well; the USA was the first country to make such a move and established in Congress the Office of Technology Assessment (OTA) in 1972. Concern over injurious social and other impacts of technology was also prevalent in Finland; but environmental protection was the sole field to receive new resources.

A period of uncertainty, labelled by slackening economic growth and other disturbances, followed the oil crisis. In most countries, this resulted in economic and technological policies being reassessed. A general reappraisal was also fostered by the rigidity of economic structures – and difficulties faced in responding to challenges arising from demand swings – fluctuation in the size and composition of the labour force, new technologies and stringent pollution control. In the OECD, these challenges culminated in the formulation of what were termed *Positive Adjustment Policies* (OECD 1983).

This climate of uncertainty weakened belief in conventional forecasting. It was recognized that forecasts, besides economic and technological trends, should also include social and ecological aspects. In addition, "soft" values, alternative movements and calls for the public's wider democratic participation in planning and decision-making all won advocates. The final outcome was the institutionalization of technology assessments in many countries (see Smits/Leyten 1988).

When the depression passed, Finland's main concerns were the competitiveness of her industry and the effects of new technologies on employment. The country's emerging technology policy was faced with the problem of how to modernize traditional sectors, how to accelerate the creation of new high-tech production, and how to manage technological development to ensure that it would have a positive effect on employment. In such cases, however, prevailing technology policy paradigms laid strong emphasis on applications and tended to sustain the existing production structure rather than seek new innovative fields.

The first clear understanding of the importance of technology assessment in Finland was voiced in 1973. The Science Policy Council requested a report on how the impact of technology was assessed in other countries and of its potential implications for Finland. The report (Kohi 1976) did not, however, lead to concrete steps in the institutionalization of technology assessment. The work of the Advisory Board for Information Technology also underlined the social ramifications of technology. Moreover, the establishment of the National Board of Labour Protection in 1973 was a reflection of government-level concern about the impact of technology in work environments. Notwithstanding, Finland in the late 1970s was characterized by a techno-economic approach to science and technology.

4.2.2 Towards Social Orientation in the 1980s

Beginning in the late 1970s, many countries' attitudes towards economical, technological and environmental change moved technology policy in a direction which, besides looking at the promotion of technology, also focused on social and environmental aspects. New developments in production technologies – a gradual transformation from traditional Fordist mass production to new flexible systems – played an important part. In the early 1980s, debate in Finland centred on new technologies' effects on employment, on the one hand, and on promoting companies' techno-economic competitiveness on the other. These two approaches labelled the formulation of technology policy. Policy measures were chiefly directed towards ensuring competitiveness.

When the 1980s came, Finnish companies began to target their resources to fields where they were strong and possessed high, specialized expertise. A strategy of flexible specialization gradually made itself felt; some industries adopted new technological methods with great rapidity – the JOT principle in engineering companies is a good case in point (see Ollus et al. 1990). R&D and other intangible investments increased rapidly (see Table 4). The 1980s were also a decade for the internationalization of Finnish companies. Finnish firms were interested in both European research and technological cooperation, and their production networks enlarged rapidly. The postwar shift from the forest sector to the metal industry accelerated; and this was especially apparent in export figures. The industrial labour force decreased, and the number of people employed in the service sector showed a marked rise. Employment rates improved; and in the late 1980s, labour shortages made themselves felt from time to time. This resulted in competition for qualified employees. As a corollary, more attention was paid to education and to the development of work organizations (Ollus et al. 1990:183-184).

New developments in technology and production brought about new requirements for how the labour force should be organized. There were, for instance, growing calls urging a flexible use of manpower, as opposed to vertically- and horizontally-fragmented labour utilization. By the same token, it was suggested that skills needed to be more intensively utilized. Alasoini (1990) argues that the Finnish rationalization model has three distinguishing characteristics. First, Finnish companies tend to be small in size; and this forces them to be market-flexible. The removal of international trade barriers and the expansion of markets may well result in increased pressure to concentrate still more on small quality series and to develop functional flexibility in labour use. Secondly, over the last 20 to 30 years, Finland's industrial labour force has become used to frequent changes caused by foreign trade fluctuations and the rapid modernization of the country's industry. It appears that most of the work force has experienced rationalization measures positively; and this is probably one of the reasons why the trade union movement has firmly committed itself to a national policy of industrial modernization. Thirdly, the Finnish industrial labour force has a comparatively high average skill level, and the country's labour market is less segmented than in many other parts of the industrial world. The prospects for functional flexibility in labour utilization are consequently good.

The manner in which Finnish industries developed during the 1980s, had much in common with the OECD's Technology/Economy Programme (TEP) and its analysis of the new techno-economic system. How, then, has the Finnish technology policy apparatus responded to the challenges posed by the new techno-economic system and technology's social and other impacts?

Finland's most notable response was to establish the TEKES (Technology Development Centre; see Section 2). The Centre's main duty is to finance and administer technological R&D. The first generation of national technology programmes was run in the mid-1980s, and that was TEKES' first real test. The programmes increasingly focussed TEKES' attention on technology promotion, the 'strong line' of technology policy. Programme areas had earlier been identified by a group of experts who based their proposals on initiatives posed by research institutes, institutions of higher education and industry. Nevertheless, the first technology programmes lacked a comprehensive analysis or view of the condition of Finnish industry and of prospects for society as a whole.

The founding of the Science and Technology Policy Council (see Section 2) was a second milestone. The body brought about a more explicit division of labour between science and technology policies than before, besides its being evidence of official recognition of the need to coordinate the targets and programmes of scientific and technological administrations in an integrated manner. The appraisal of the first national technology programmes provides one example of how the new coordination mechanism functions. The programmes were assessed between 1988-1990 on the instigation of the Science and Technology Policy Council. The task of appraising the completed programmes was entrusted to foreign experts, who reported that the first programmes had given Finland a new cooperation climate and a new interaction network (see e.g. Tomner 1990).

Now that we have reached the early 1990s, measures are being taken to launch a second generation of technology programmes. These programmes were prepared by a special working group between 1989 and 1990 (Ministry of Trade and Industry 1990). The working group identified 11 new areas for programmes and laid down that special attention should be paid to the maintenance and development of competitiveness. Environmental issues were also emphasized – a fact which may reflect a new way of thinking, strengthening the 'weak line' in technology policy. The programmes underline the integral role of environmental considerations in the development of all technologies. Turning to the social dimension of technological development, on the other hand, the group merely observed that the most important social sectors – health care, transport, telecommunications and education, for example – should be scrutinized separately. These areas were seen to require a strong technology input.

In the 1980s, Finnish technology policy chiefly focussed on technology promotion. Nevertheless, there were also indications that social and environmental orientations were being recognized. In 1987, an Institute of Labour Research was established at the University of Tampere. The expert committee on the organization of futures studies also included technology assessment in its work. Futures studies and science studies are promoted, and the latter are also being institutionalized by the Academy of Finland. The problems of innovation research, moreover, have been recognized and new activities are under way (see Lemola et al. 1990). These trends reflect a serious attempt to guarantee better continuity in research, vital for strengthening technology policy's 'weak line'.

4.2.3 Technology Assessment Efforts

While there are no systematically-organized forms of technology assessment (TA) procedures and no legislation which would have a similar effect, several large TA projects have been carried out in Finland. They have generally been organized by committees of expert and special working groups. The Energy Committee, appointed by the government in 1987, for instance, carried out TA research on the social, health and environmental impacts of different energy options. The central organizations for TA are VTT and the Finnish National Fund for Research and Development (SITRA). VTT develops and conducts TA both as part of its technical research activities and as part of its own research programmes. VTT's TA work spans numerous technological fields and is usually ordered by various private and public organizations.

One of Finland's most ambitious TA efforts was the country's contribution to the Technology-Economy-Society (TES) programme of the International Institute for Applied Systems Analysis (IIASA). The Finnish national project was steered by SITRA and carried out by VTT, the Research Institute of the Finnish Economy (ETLA) and other institutes (see Ollus et al. 1990). In addition to technical aspects, the study also looked at economic and regional impacts. While there are but few assessments of technology's social and cultural aspects, one study deserves to be singled out: the one on information technology and working environment, carried out by the Work Environment Fund and VTT.

During recent years, there has been lively discussion about attaching TA activities to the work of Finland's Parliament. The idea has been fostered by the Finnish Society of Scientists and MP's, TUTKAS, along the lines of other European countries.

4.2.4 Intermediate Summary

Before the 1980s, the main emphasis of technology-related policies was on competitiveness and technology promotion. In addition, a few proposals had been made for socially- and environmentally-oriented technology policy. In most policy fields, however, development was relatively autonomous. Social and environmental orientations have been emphasized since the early 1980s, but the number of concrete measures taken has remained low. With this in mind, we shall now turn our attention to the factors which hamper and promote integrative technology policy trajectories in Finland.

A few special characteristics facilitate the adoption of new technologies, their consequences for employment and other social ramifications. First, some favourable factors support the modernization of Finnish industry (see Alasoini 1990). Moulding the work force's attitudes and adapting them to fit in with changing labour processes, without flattening the skill hierarchy, have tended to be the most important ways of reacting to technological advance and changing product markets. The reorganization of work is accepted as a 'technological necessity' by most workers and local trade unions. Secondly, Finland's labour market bargaining model may have been a partial contributory factor in facilitating the adoption of new technology. New mechanisms for income policy and centralized agreements were implemented in the late 1960s; since then, problems relating not only to wage policies but also to working conditions, education etc. have been solved using that apparatus. Consequently, strikes and other working conflicts have tended to revolve around wage policies. Cuts in the labour force brought about by new technologies have rarely caused strikes, and there has been no real protest against the new technology (see e.g. Kohtanen/Kauppinen 1988). These two factors thus seem to have lessened the need for other preventive procedures for controlling the social impacts of new technologies.

The international competitiveness of Finnish industry has been one of the pivots of economic, industrial and technology policies since the mid-1970s. Strong support for the promotion of technology – and, by association, the 'strong line' of technology policy – may have impeded socially-oriented policies. We should, however, remember that technology policies which have competitive targets do not necessarily conflict with social orientations.

Problems related to technology, social, ecological and other factors imply interdisciplinary scientific cooperation. This approach was not much in evidence in Finland before the 1980s. There was also a dearth of research on technology, innovation policies and associated theoretical and empirical issues. All these research lacks have hampered Finland's evolving towards a socially-oriented technology policy.

4.3 A Case: Integration of Environmental and Technology Policies

Finland is a country whose special characteristics include her geographical position and natural surroundings – forests, lakes and rivers, a Nordic climate and sparse habitation. With the exception of her forests, however, Finland has few natural resources. These national hallmarks have implications for human-nature interaction. For this reason, the country has long appreciated the threats which industrial production and technology posed for nature, as well as the necessity of protecting the environment. Milestones in the development of environmental protection include the promulgation of the Nature Conservation Act in 1923 and the Water Act in 1962 (see e.g. OECD 1988a:112-115). The Finnish economy has chiefly been based on forest industries whose production processes and technologies became relatively advanced at an early stage. In non-renewable, raw material sectors, such as the basic metal industry, non-waste technologies were mainly developed because of scarce resources (e.g. worldwide-diffused fuel-saving flash-smelting technologies since the 1940s).

Trends in the integration of environmental and technology policies can be assessed either from the point of view of environmental or from the standpoint of technology policies. Two phases may be distinguished in Finland. The first phase lasted from the early 1960s until the late 1970s and tended to take technological aspects into account in the formulation of environmental policies, rather than vice versa. Subsequently – especially since the early 1980s – environmental considerations were increasingly integrated within technology policies.

4.3.1 Developments Prior to the 1980s

In most Western countries, the rapid economic growth which characterizes the postwar period resulted in pressures on the environment. Finland began to be concerned about the impact of technology on the environment in the late 1960s. In the early 1970s, economic development was considered auspicious enough to allow resources to be allocated to improving environmental protection and pollution control. Finland's National Board of Waters was established in 1970, and the Department of Environmental Protection of the Ministry of the Interior was founded in 1973. Effective measures were implemented, especially in the water pollution sector. A long-term water pollution control programme was compiled for 1974-1985.

To some extent, technological and social aspects were already taken into account in environmental protection at that time. Water pollution control procedures included flexible scope for introducing preventive technology in industry. Discharge licences, for example, are issued on a case by case basis; the obligations to which licence holders are liable, may be postponed for a few years to enable preventive clean technology solutions to be implemented in conjunction with upcoming process renovation. The licensing procedure also includes social aspects such as "opening meetings", "enquiry meetings" and "oral hearings". Here, people who have rights or interests which are likely to be affected by the project in question, may put their views forward.

The oil crisis and the slump of the mid-1970s brought about a different set of circumstances. Economic policy directed resources to recovery, frequently shying away from 'non-profitable investments' – environmental protection, for instance – in favour of competitive investments. The subsequent rapid growth of the Finnish GDP resulted in strong environmental pressure. The country's industrial structure also changed rapidly, especially in the pulp and paper industries. When more integrated processes were developed in this sector, the principles of low-waste technology were increasingly applied – with obvious beneficial consequences for the environment. Energy-saving technologies were increasingly developed and applied – stimulated by higher oil prices, new technology's potential and technology policy measures as well. All these developments had positive environmental implications, in addition to the effects of direct pollution control measures.

4.3.2 Towards an Integrative Approach

The late 1970s and early 1980s were an active period in environmental policy. The Waste Management Act and the Air Pollution Control Act were introduced, for example. Environmental research was planned and promoted, and environmental technology acquired a more important role. The Research Council for the Environmental Sciences was established in 1983 in connection with the Academy of Finland. Resource-saving technology was viewed as an urgent priority in expert studies. More crucial steps, however, were taken in renewing environmental administration. The key features here were the establishment of the Ministry of the Environment in 1983 and the reorganization of water administration in 1986. The scope of the new ministry covered environmental protection and nature conservation, physical planning and construction – as well as housing. The National Board of Waters was subordinated to the Ministry of the Environment and renamed the National Board of Waters and the Environment.

As we have seen, a distinctive and purposeful technology policy took shape in Finland in the early 1980s. Subsequent technology policy trends have emphasized the integrative and preventive strategy of environmental considerations and the positive role of environmental technology. Permanent cooperation between technological and environmental authorities also began in the early 1980s.

These developments did not, however, result in immediate and automatic change in integrating technological and environmental aspects in practice. Instead, the question at stake was securing wide recognition of the need for a new integrative approach among the officials and other parties who had influence on the formulation of technology and environmental policy. The 1980s subsequently saw greater emphasis being laid on the integrative approach. We shall now discuss a few examples.

In common with the industrialized world as a whole, concern for the environment has become a high political priority in Finland. Environmental issues are prominent in the political platforms of most of the country's established parties, and the Greens have 10 of Finland's 200 parliamentary seats. Environmental considerations have also attracted more and more trade union activity. In addition, a large number of small voluntary organizations – showing interest in the links between development, ecology, technology and 'soft alternatives' – have come into being. Yet despite the large body of evidence which indicates that the public is convinced of the importance of environmental issues, environmental debate has remained scant.

From the perspective of the environment, one of the fundamental questions of democratic society is how public opinion is taken into account when the surroundings in which people have to live, are planned and organized. Sophisticated environmental impact assessment (EIA) should act as an acknowledged indicator showing how well these considerations have been recognized. Finland had earlier included some EIA aspects in her water pollution control procedures. A few companies also had experience in applying environmental impact assessment. The large energy concern, Imatran Voima, for instance, had conducted EIA when planning and constructing power plants and electricity power lines. By and large, the necessity of making EIA an essential part of all environment-related planning and decision-making, was recognized in the early 1980s.

Despite the fact that some sophisticated EIA elements may be observed in water management, it remains true that Finland still has no comprehensive formal legislative or planning requirements for environmental impact assessment. That is why the OECD, in its review of Finland's environmental policies, recommended that formal, obligatory, and simple procedural requirements for an EIA process be created – and, most importantly, that they be applied at an early stage in the overall decision-making process (OECD 1988a:210-214, 222). Now, in the beginning of 1990s, there are preparations going on concerning e.g. making EIA compulsory in connection with government propositions to the Parliament.

Environmental issues began to be studied in the 1960s, especially in the water pollution control sector. Research linking environmental and technological aspects in all environmental sectors, increased markedly in the late 1970s and the 1980s. Studies were promoted by the 1984 Research Cooperation Agreement between the Ministry of the Environment and VTT. Progress was also achieved in the promotion of low- and non-waste technology (see e.g. OECD 1988a). Other technology-related policies took an increasing number of environmental issues into account. The energy sector – the recent research programmes of the Ministry of Trade and Industry, for instance – has paid more and more attention to the environment. TEKES has launched an environmental technology programme and has also decided to include environmental considerations and to integrate them in to all the R&D projects it finances. Similarly, VTT now views environmental technology as one of the central future technologies.

There are also examples of economic and social research which has analyzed the links between environmental and technological considerations. Technology will have an important role in a research programme of sustainable development launched by the Academy of Finland. In general, however, studies linking the environment, technology and other relevant factors are still in their infancy. Joint analysis of these elements interacting with the context of sustainable development, still needs to be improved.

Education is of central importance in creating the prerequisites for anticipatory policy. Some universities – e.g. the Technical University of Helsinki – have courses on environmental technology. The situation is worse in other sectors, which has prompted OECD experts to point out the need for developing comprehensive environmental education (OECD 1988a:228-229). The quality of the environmental education given in secondary schools and universities should be improved, for instance.

4.3.3 Intermediate Summary

Before the early 1980s, environmental policies, technology and other relevant sectors were inconsistently linked. At that time, environmental strategy was more reactive than anticipative. Efforts were subsequently made to move towards preventive strategy, i.e. towards a systematic integration of policy sectors. Despite numerous positive programmes, endeavours and examples, however, evolution has remained sluggish in practice.

A range of factors continue to impede the integration of environmental and technological policies. One of them is that the obvious need to integrate environmental considerations within technology policy, was recognized rather late in Finland when compared to many other industrialized countries. This might be due to the fact that pollution problems did not make themselves felt in Finland as soon as they did elsewhere. The relatively late establishment of the Ministry of the Environment may have constituted a further impediment. One of the ministry's most important tasks is – or should be – effectively integrating environmental considerations within other policy fields. Integration would also provide for close cooperation between public authorities – in this case, between the environmental and other branches of administration. Finland's moves in this direction have been rather ineffective as yet, due perhaps to the traditionally narrow sectoral views of different administrative branches. The traditional role of the country's political parties may also have hampered effective policies. The Centre party, for example, is traditionally the mouthpiece of farmers and landowners who have sometimes felt threatened by proposed environmental legislation.

On the other hand, some trends have promoted the integration of environmental and technological policies. Finland has actually been in the vanguard in some fields of integrating environmental and technological considerations. The country has a wealth of experience in planning and implementing water resource management. The people immediately affected by the projects in question have been encouraged to participate – an essential element of sophisticated EIA. Finland's water pollution control procedures also include elements which help advance the application of preventive technologies. A number of technologically-, economically- and environmentally-progressive industries developed and applied clean technologies, which facilitated the creation and application of clean technologies as part of an integrative and anticipatory strategy. Despite their drawbacks, furthermore, Finland's consensus politics have proved able to generate discussions and solve problems between different interest groups by creating a range of pollution control programmes.

4.4 Assessment of Finnish Development

During the 1960s and 1970s, technology-related policies chiefly focussed on promoting competitive technology; no more than a few proposals and measures were advanced to integrate social and environmental considerations within technology policy. Until the late 1970s, technological aspects were taken into account in environmental policies – rather than vice versa. In other technology-related policy fields, development was comparatively autonomous; and there were no more than a few signs of a social orientation. Since the early 1980s, however, social and environmental points of view have been strongly emphasized. Societal and environmental issues have been stressed in technology policy documents and, in particular, environmental considerations have been increasingly integrated within technology policy. Nevertheless, no more than a limited number of concrete measures have been recorded.

In some fields, Finland appreciated comparatively early the necessity of integrating societal and environmental issues within technological development. We gave some examples in our conclusions to sections 4.2 and 4.3. The environmental sector, for instance, has some examples of EIA-type procedures in water management; and open-minded industries have developed clean technology. Turning to the social sector, the labour market bargaining system may be regarded as a factor facilitating the adoption of new technology. Moreover, Finland's consensus politics have brought about discussions and solved problems between different interest groups. In fields where social and environmental considerations have already been integrated, the practical measures should be further strengthened to bring added benefits in the future. Knowledge of EIA in water management, for example, is transferable to other environmental sectors.

Nevertheless, early positive developments in the integration of the social and environmental aspects of technological development are rather the exception than the rule. Generally speaking, in most areas the impacts of technology will necessitate an integrative approach in the future.

5 Conclusions

In Section 2, we pointed out that R&D investments have grown very rapidly in Finland. While the greatest rise has occurred in the volume of applied technical research and development, the whole R&D system has enjoyed the fruits of increased resources. One may justifiably say that Finnish science and technology experienced balanced development in the 1980s.

A similar balance can be found in national technology policy programmes and documents, a matter we talked about in Section 3. They represent – more or less – a holistic view, a comprehensive approach to technology in society. The policy line has not been restricted to just promoting technological development; social and environmental impacts, and how they should be assessed, have also been emphasized. Recent moves have striven to enlarge technology policy's scope to include the entire national system of innovation, knowledge and know-how as its driving forces.

The practical role of official policy statements, however, mainly lies in the sphere of promoting activities. Pronouncements have exerted demonstrable influence on Finland's national development, as Section 4 makes clear. Investing in enterprises' and society's ability to create new technology, and to then apply it to products and production, has been the chief outcome of the national effort. In our opinion, this part of the policy – the integration of technology with economy, or the strong development line – has been quite successful until the early 1990s at the very least.

Much less has been achieved in socially-oriented technology policy – that is, the integration of social, environmental and technological policy aspects. This weak line of technology policy has suffered from a lack of what might be termed "institutionalized interests". Development proposals have not become rooted in practice, not even research practice, thereby causing a situation where reiterating proposals has often been the only way to proceed.

Could there be other opportunities to strengthen social and environmental orientations, or – in the longer run – to develop a new, comprehensive technology trajectory which would integrate those various orientations which still lie apart? In addition to internationalization processes, much will depend on Finland's overall economic development. It is obvious that there will be progress in the direction of the environment. The social orientation, on the other hand, needs new supporters – despite the fact that it has the backing of the trade union movement. Irrespective of whether this support comes from political parties or other organizations, it is clear that the "active society" will remain a vision without institutionalized demand for a social orientation.

References

Alasoini, T. (1990) *Tuotannolliset rationalisoinnit ja teollisuuden työvoiman käyttötapojen muutos* (Productional Rationalizations and Changes in Industrial Labour Use), Helsinki: Ministry of Labour.

Central Statistical Office of Finland (1974-1990) *Official Statistics of Finland. Education and Research*, Helsinki: Government Printing Centre.

Council of State (1982) Valtioneuvoston periaatepäätös teknologiapolitiikasta (Decision-in-Principle on Technology Policy), mimeo (in Finnish).

Council of State (1985) *Report to Parliament on Finland's Science and Technology Policy*, Helsinki: Government Printing Centre.

Hjerppe, R. (1988) *Finnish Economy 1860-1985. Growth and Structural Change*, Helsinki: The Bank of Finland.

Kohi, P. (1976) *Tekniikan vaikutusten arviointi – Technology assessment*, Helsinki: SITRA.

Kohtanen, J., T. Kauppinen (1988) *Työtaistelut Suomessa vuosina 1971-1984* (Labour Conflicts in Finland in 1971-1984), Helsinki: Delegation for Labour Relations.

Lemola, T., T. Loikkanen, R. Lovio, R. Miettinen, P. Vuorinen (1990) *Teknologia tutkimuksen kehittäminen Suomessa* (The Development of Technology Research in Finland), Helsinki: TEKES.

Lemola, T., R. Lovio (1987) Transformation of the Finnish Industry in the 1980's, in: V. Stolte-Heiskanen (ed.) *Science Policy Studies from a Small Country Perspective*, Helsinki: The Academy of Finland, 42-59.

Lemola, T., R. Lovio (1989) The Relations between Science and Production in Finnish Industry, in: *Science Studies* 1 (2): 41-50.

Lemola, T. E.-O. Seppälä (1987) Technology Assessment – Finland, in: *Technology Assessment, an Opportunity for Europe*, The Hague: Dutch Ministry of Education and Science, 37-63.

Ministry of Trade and Industry (1990) *Teknologiachjelmatoiminnan linjat 1990-luvulle* (Lines of Technology Programme Activities for the 1990s), Helsinki: Government Printing Centre.

OECD (1983) *Positive Adjustment Policies. Managing Structural Change*, Paris: OECD.

OECD (1988a) *Environmental Policies in Finland*, Paris: OECD.

OECD (1988b) *New Technologies in the 1990s. A Socio-economic Strategy*, Paris: OECD.

Office of the Prime Minister (1982) *The Report of the Finnish Technology Committee*, Helsinki: Government Printing Centre.

Ollus, M., R. Lovio, J. Mieskonen, P. Vuorinen, J. Karko, S. Vuori, P. Ylä-Anttila (1990) *Joustava tuotanto ja verkostotalous* (Flexible Production and the Network Economy), Helsinki: SITRA.

Pimiä, M. (1989) *Forecast of the Need of Researchers up to 2000. Publications of the Academy of Finland 6/89*, Helsinki: Government Printing Centre (English summary.)

Science and Technology Policy Council of Finland (1987) *Science and Technology Policy Review 1987*, Helsinki: Government Printing Centre.

Science and Technology Policy Council of Finland (1991) *Review 1990. Guidelines for Science and Technology Policy in the 1990's*, Helsinki: Government Printing Centre.

Seppälä, E.-O. (1990) Science and Technology Policy Advice in Finland, in: W.T. Golden (ed.) *Worldwide Science and Technology Advice to the Highest Levels of Governments*, New York: Pergamon Press, 169-176.

Smits, R., J. Leyten (1988) *Key Issues in the Institutionalization of Technology Assessment*, Futures 1 (20): 19-36.

Tomner, S. (1990) *Rapport över utvärdering av TEKES teknologiprogram* (Report of the Evaluation of TEKES Technology Programmes), Helsinki: Ministry of Trade and Industry.

Technology Policy in Spain: Issues, Concerns and Problems

Luis Sanz-Menendez and Emilio Muñoz

1 Introduction

The relation between science and technology is becoming narrower every day and their development has implications for economy and consequently, society, influencing the quality of life and working conditions of citizens.

The levels of economic and scientific/technological development appear to be strongly associated, at least over the last thirty years. Several empirical studies (Pavitt 1979, 1980; Soete 1981; Fagerberg 1987; Quintanilla 1992) have identified the indicators of economic or scientific and technological development which have a tendency to vary most significantly, according to countries and periods.

The dominant conception of scientific and technological relations with economy has been characterized, in recent decades, by a linear model (e.g. Salter 1960). Some works, influenced by organization theory, began to develop alternative approaches, such as the evolutionary models of technological change (Nelson et al. 1976; Nelson/Winter 1977; Nelson/Winter 1982; Dosi 1988, 1991). The elimination of automatism from the linear model has helped to concentrate attention on a single serious issue, the global functioning of the science-technology-industry system.

More recent research (e.g. Mowery/Rosenberg 1989; Rosenberg 1991) has emphasized the importance of considering the efficiency of the scientific/technological system as a whole, and not in its particular elements, such as the scientific, technological or commercial aspects. The system is a mechanism which incorporates inputs and outputs, it is predisposed to its own transformation, a machine whose internal configuration and organization can assure a variety of functions. The contribution of science and technology to the development and competitiveness of economies appears to be marked by institutional and organizational factors which ought not to be ignored.

Certain studies of an historical character have contributed to the identification of the decisive processes which helped to develop and configurate the institutions in different countries which can be included in the so-called science-technology-industry system. Most of these works, however, refer to the United States (e.g. Noble 1977).

In the case of the development of R&D activities and of specific organizations, from the commercial perspective, the progression of organizational transformations in corporations appears to have played a major part in the development of *organized capitalism*. In fact, the evolution of industrial organization and the appearance of planning or programming units contributed to the establishment of a space for the institutionalization of innovative activities in businesses (Noble 1977; Chandler 1962, 1977). On the other hand, as it has been suggested (Schumpeter 1939), the seed of many businesses in new sectors of production can be found in the early processes of technological innovation.

Public intervention and concern of an explicit nature, seem to be derived from the programs for scientific/military development of the inter-war years (Ronayne 1984). The history of the scientific/technological effort in conjunction with the first steps taken in developing its results are associated with the development of the scientific/technological potential related to war.

In science and technology, as in economic development, every country is a product of its history, of its past, of its degree of economic development, and of its state policies, often closely associated with external military concerns. In this way, the situation of the science-technology-industry system appears to be associated, in a decisive way, with the attained levels of economic development.

However, to explain the development of public policies, analysts have fallen into two major currents: on the one hand, those who based their views on *prerequisites*, and argued that the causes of the development of scientific and technological policies are generally associated with the degree of economic and social modernization; and on the other hand, those who maintain the *diffusion* thesis, who concentrate on the study of the imitation of scientific policy programs between countries, or of their diffusion via organisms of a multinational character (like UNESCO or OECD).

The situation in Spain in the mid-1970s, during the democratic transition, where scientific and technological issues are concerned, was far removed from the rest of Europe. Undoubtedly, the scientific/technological vacuum in Spain was much greater than the economic one (Sanz/García 1990). Spain spent a mere 0.3% of its GDP in the mid-1970s and early 1980s on the development of R&D activities, one of the lowest percentages amongst the developed countries (OECD 1971).

The early years of the transition helped to freeze the situation, within an international context of a general stagnation of R&D (Gerelli 1983). National priorities were concentrated on political reform and the development of social compensations aimed towards the organization of the welfare state. The absence of science and technology in the Pactos de la Moncloa is very significant.

It was not until the electoral victory of the Socialist party, which renewed the political leadership of the country, that the importance of science and technology grew, together with a corresponding public interest.

The diagnosis which was made, after the establishment of the first socialist Cabinet, was very clear (Muñoz/Ornia 1986). The Spanish scientific/technological system was characterized by: (i) a small size and very little research; (ii) the large size of the public sector (universities and public research centers), to the detriment of business; (iii) a heavy imbalance of human resources against engineering and technology; (iv) a lack of cooperation between the directors and agents of R&D programs; (v) serious regional imbalances in scientific/technological material resources; and (vi) an absence of public policy coordination.

After a number of years of efforts of a budgetary, organizational and incentive-producing nature, the situation of the science-technology-industry system in Spain appears to have evolved favorably, although in 1992, Spain still spends less than 1% of its GDP on scientific and technological projects. A recent study (Quintanilla 1992) has indicated that Spain has, amongst all other EC countries, experienced the greatest and most sustained growth in R&D projects in the final years of the decade of the 1980s. However, in spite of this it is still well below the community average and the leading country, Germany. In the field of public policy in science and technology there have also been some developments; among others, the growth of public spending, the creation of instruments of intervention like the National Research and Development Plan, and the fixing of priority research areas.

Nevertheless, historical inertia in the working of the science-technology-industry system, the effects of policies, and the process of accelerated change have brought about situations of concern within this improved context. There is still a general deficit of engineers and technologists, financial investment is very low, the proportional contribution of personnel by universities for R&D continues to be very high, and the covering of technological imports is very low.

Social and ecological concerns in Spain are underdeveloped, against that which could be expected, given the magnitude of these problems. There is a high degree of unemployment, growing desertification, and evident problems with respect to biodiversity, but there is a lack of awareness in public opinion and in the key role of government determination of political agendas on these matters. The Spanish political scene has a low incidence of interest group pressure. Thus, the priorities of Spanish scientific and technological policy have been characterized by this underdevelopment, by the need to prioritize the growth of the scientific/technological system, and by the creation of hitherto unselective policies.

In the present analysis, the aim is to examine history and policy, and to explore the way in which political processes, in the context of the Spanish democratic transition and consolidation, have helped to create science and technology programs and policies. The introduction of science and technology initiatives on the political agenda is often the result of the action of organized social and economic interest. But the sensibility of public opinion is also an element of great importance in explaining the hierarchy of priorities in public action. The state and governments are also organizations and sources for the development of autonomous initiatives. It is therefore the case that their institutional structures can affect the political process decisively, and contribute to the emergence or consolidation of a certain type of public policy, in this case, science policy. On the other hand, science and technology policy transforms social and political reality.

We aim to describe the basic characteristics of the Spanish science-technology-industry system and to explore the most significant science and technology policy proposals which have been considered in the last ten years. This study therefore operates in two streams of analysis: the reality of the science-technology-industry system in Spain and the actions of public policy in relation to this system.

The conclusion one might draw from the present study is that, in this context, science policy initiatives respond to the processes of the diffusion of policies and knowledge from international sources, and not from specific interests or bureaucratic action. This is the case in spite of the fact that the establishment of science and technology policies after the socialist election victory in 1982 was characterized by a strong political and administrative leadership.

The fixing of ecological and social objectives is rather meagre, and in any case, very recent. International pressure (from the EC) on government action appears to have been decisive, given that the awareness in Spanish public opinion on these questions is still very low.

2 Antecedents of the Spanish Situation in Science and Technology: 1940-1975

The present situation of science and technology in Spain is to a large degree a result of its most recent past. The country's economic configuration after the Civil War (1936-1939), the political alignment of the Francoist dictatorship with the fascist powers during World War II and the exclusion of Spain in the reconstruction of Europe (the Marshall Plan), were to have decisive consequences for science and technology and the type of policy, or absence of policy, adopted by the Spanish state.

2.1 From Self-sufficiency to the Stabilization Plan: 1939-1959

The economic effects of the Civil War, but in particular the situation of political isolation the Francoist dictatorship found itself in after the end of the Second World War, resulted in a reconstruction policy based on autarquía, or self-sufficiency (see Ros-Hombravella 1973; Carreras 1987; García-Delgado 1987). This economic policy strategy (interventionism and isolation) and its accompanying circumstances did not ensure pre-war production levels until well into the 1950s; the most recent historiography talks of the "costs" of this period for the Spanish economy (Carreras 1987).

Concretely, the self-sufficient model required the application of many of the working methods of economic intervention by the state used during the war: administrative freezing of prices, narrow control over private investment, the creation of national interest industries, direct state investment via the National Institute for Industry (INI), the centralization of the provision of raw materials and of importation permits, and so on. This was all in addition to labor and commercial relations determined by a rigid, authoritarian and paternalistic model. This self-sufficient and interventionist strategy was also a response to the ideological and doctrinal position of the winning forces of the Civil War. At the same time, the protectionist attitude was also a continuation of policies which have characterized Spain since the end of the nineteenth century (Espina 1992).

The two first decades of the Franco regime were characterized by the move towards the industrialization of the country, a process which was considered to have failed and which was awaiting success. The documents of the time testify to the main objective: to increase production.[1] State fiscal policy was put at the service of industrialization.

There was a general conviction that the country was backward, as a result of an unfortunate economic administration; it was supposed that the state was the only agent capable of breaking away from the preceding situation, and for this reason the INI was created on September 25th, 1941. Four essential postulates ought to be noted (Martín-Aceña/Comín 1992:429-430): (i) the need to establish an industrialization effort which would ensure the economic take-off of the country and the launching of a powerful economy; (ii) the recognition of the inability of private initiatives to take on investment responsibility and to make up for the endemic lack of enterprise; (iii) the association between industrialization and national defence; and (iv) the acceptance that there were no appropriate credit systems to finance programs which would promote these objectives.

In the hope of attaining its objectives, the INI designed a program of action which covered a varied group of industrial sectors: solid and liquid fuels, iron, steel and aluminium, fertilizers, cellulose and artificial fibers, engines, industrial vehicles and automobiles, machines and tools, electrical goods, naval construction, and the production of electrical energy. Within this ambitious program, the reality of the results would expose the specialization of the Spanish economy around the primary sector and raw materials, and an orientation, in order to promote industrialization and the substitution of imported goods, towards difficult and unattractive sectors for private enterprise (Myro 1981).

All of these circumstances coincided with the European post-war years and with the attempts brought about by President Roosevelt in the United States (Bush 1945) to engineer the first 'non-military' actions in science and technology policy, and which would eventually be picked up and channeled by different states and international organisms such as UNESCO and OECD.

In Spain, at that time, there was no explicit science and technology policy but, as Nieto (1990) points out, the key to scientific action, from 1939 onwards, was held by the Higher Council for Scientific Research (CSIC). At the same time as the industrialization process adopted by the regime, the government of the day was able to see the need for research in a modern society, and created the CSIC by legislation on November 24th, 1939[2], later modified by another law on July 22nd, 1942. It states that the fundamental organ of support for scientific research is the state, and consequently, the coordination of all activities and institutions engaged in science is its responsibility. Thus the purpose of the CSIC was to encourage, orientate and coordinate scientific research in Spain. The idea was very ambitious, given that in the organization's statutes it was required to coordinate all of Spain's research centers, the relation between research and the economy and the possibility of territorial expansion.

It is the case that for certain reasons (Muñoz/Ornia 1986:38, 49; Nieto 1990:31-33; Muñoz 1990:18-19) this institution, a potential instrument of scientific policy, which was also attempting to connect with industrial and technological development, did not become functionally operative. This failure lies in the conflicting views of the CSIC's creators and in the passive attitude of those which opted for this model.

The universities, for their part, had suffered severe losses as a result of the Civil War, and in the context of a general absence of a research infrastructure, they were hardly in a position to promote scientific activity. On the other hand, they have experienced a complex relationship with the CSIC which has not provided very satisfactory results (Lora Tamayo 1990).

2.2 The Period of "Developmentism" in Spain: 1960-1975

At the end of the decade of the 1950s, there was a change in Spanish economic policy. After the end of political isolation (Spanish-American Pacts, 1953), the Franco regime changed its strategy – in 1959, it presented the Stabilization Plan to the International Monetary Fund. There was an opening up to the exterior, tourism was promoted and industrial development was sought by means of applying foreign technology for domestic use, attempting to make the best use of the country's cheap but underqualified labor force. The protectionist measures which characterize Spain's recent history remained unaltered.

This period in Spanish history, which covers the beginning of the 1960s until the oil crisis, yielded positive results, increasing production and placing Spain amongst the industrialized countries.

In spite of the notable advances in Spanish economic development, the scientific and technological situation remained far behind. In 1963, the R&D expenditure did not reach 0.2% of the GDP. The situation, in the years 1967-1975, did not vary substantially, given that Spanish R&D hovered around 0.3% of the GDP. This was one of the lowest percentages of the developed countries, be they free market economies or planned economies (OECD 1971). In this context, it can be said that science policy did not improve at all and continued in this way. Spain lacked a science policy (Nieto 1982; Muñoz 1982), although there clearly was a will to increase state support for scientific and technological research.

In this period, development plans were launched as a means to guide and generate economic inertia. These plans included R&D in their guidelines, as a response, among other things, to the OECD reports. In one of these, published in 1971, it was recommended by the international organism that "Spain should spend 0.5% of its GDP on R&D with no further delay" (OECD 1971:67), underlining that this was feasible with little effort. The development plans had considered substantial increases of resources for R&D, to try to reach 1% of the GDP by the 3rd Plan (1972-1975). However, the difference between the resources requested for scientific research and those eventually conceded was significant.[3] The fundamental problem was that Spain lacked a modern domestic fiscal policy. This is reflected in the fact that the first fiscal law was only introduced during the democratic transition.

In any case, it is worth reminding ourselves that the 1st Development Plan (1964-1967) had as its fundamental objective, within the field of R&D, to develop the scientific infrastructure. The National Fund for the Development of Scientific Research was created, run by the Commission for the Assessment of Scientific and Technological Research (CAICYT), founded in 1958 and boosted in 1963 with the aim of taking up the tasks the CSIC had been unable to put into practice, and to thus compensate for the deficiencies in science policy. The 2nd Development Plan (1968-1971) tried to improve the personnel and installation infrastructure, as well as supporting research in industry via the concerted research plans – a mechanism in which the state participated financially in the R&D projects of businesses paying up to half its costs – and the funding of official research via a grant system for projects. The 3rd Development Plan changed the course of action by aiming towards technological development and concentrating its focus on given projects in various sectors.

However, these plans of action aiming to lay the foundations for a science and technology policy and to increase the funding of R&D in Spain were suspended in 1976. The 4th Development Plan, prepared for the period 1976-1979, went into crisis and eventually collapsed. The provisions which were laid out in the 4th Plan for investment in R&D were not specifically included in the annual budgetary laws. This situation put Spanish scientific and technological research on the brink of chaos, shortly after having achieved the minimal conditions for growth.

This crisis in Spanish science policy coincided with the reduction in R&D in the international context of the developed countries, after the oil crisis. This kind of delay is a feature which seems to appear regularly in the history of R&D funding in Spain, on a comparative level with the advanced nations, to the point of becoming a constant. On the other hand, science and technology in Spain are invariably victims rather than motors of economic policy. This is a condition which prevents prioritization to ensure a smooth growth rather than progress which is subject to unpredictable decision-making.

In any case, in those years, most of the work carried out as a result of science policies was taken up by research centers (among which, as well as the CSIC, the Nuclear Energy Group [JEN] was very important), although during the 1960-1975 period the Spanish administration put various mechanisms into practice to stimulate the development of research in industry. One of these formed a part of the concerted plans, mentioned above, which was a system used in various European countries, aiming to promote research in business. It also hoped, at the same time, to be of general benefit and to create a research infrastructure in private industry. In the course of this action, it was also hoped that some contact between business and public research centers would be stimulated, generating a possible collaboration between both elements of the science/technology system. The formula undoubtedly contributed to the promotion of R&D initiatives in businesses, but the collaboration between businesses and research centers did not yield very good results, given that it tended to favor contacts based on subcontracting services. On the other had, businesses complained of a certain slowness in the bureaucratic procedure, which first examined applications, then made a selection, then proceeded with funding. In short, the difficulties of accommodating administrative functioning to business activity became evident.

Another measure, analogous with the experience of other countries, as in Germany with the Fraunhofer Society, was the creation of the research associations in 1961. The main aim of these associations was the development of programs of common interest for the businesses of a particular sector. The results of these initiatives could benefit any of the businesses involved which contributed to the development of so-called cooperative research. From 1962 until 1980, up to 29 of these were created, the majority (16-18) of which ran reasonably well. The formula was improved in 1970 to ensure the best possible use of human resources, scientific and technical personnel available.

It is significant that these initiatives, managed via CAICYT, were ultimately administered by the Ministry of Education and Science. The Ministry of Industry had stood aside from technological promotion, perhaps excepting the development of nuclear energy via the JEN, the importance of which we remarked above. In any case, the precarious position of science and technology in Spain was well reflected in the figures for the technological deficit, together with the meagre support given to R&D by the business sector.[4]

3 The Position of Science and Technology during the Democratic Transition

It is evident from the above that in the mid-1970s, at the beginning of the democratic transition[5], the legacy of the dictatorship was a very negative one. The following factors determined the parameters of science and technology in Spain during this period:

1. The historical narrowness of an industrial system designed for self-sufficiency during the 1940s and 1950s, which was too closed and not at all competitive on the international markets. This provided a very weak structure for Spanish businesses, whose competitive capacity on the domestic market was determined by tax protection. Foreign markets were only sold agricultural produce or products that had been transformed via the cheap labor available. This inferior competitive capacity, combined with the specialization in Spanish industry in branches of intensive capital (Segura 1983), using more equipment than producing more technology (as Pavitt 1984 points out), has been a decisive factor in the tiny investment in R&D by Spanish enterprise.

2. The development model adopted after the Stabilization Plan was based on the investment of foreign capital in Spain to compete on the domestic market by creating production centers in Spain. From the technological point of view, it ought to be noted that at this stage the massive importation of technological goods began (Sánchez 1984; Buesa/Molero 1989), but without the construction of a national scientific/technological base. This only contributed to the increase of the technological balance of payments deficit.

3. The scientific/technological system was thus very deficient, and research was predominantly concentrated in public centers. The situation of the universities, in the middle of an expansion process but without resources, was marginal, and only some public research centers played a significant role. The devastating effect of the war on the academic system and the conversion of the universities into centers of learning rather than original work, contributed to a major brain drain. It was not until 1960 that the best scientists began to study abroad, but R&D was still seriously marginalized and fragmented. This was the result of the dispersion of those rivalries created by the Franco regime.

4. From these years onwards, the only priority in science and technology policy was to increase resources for R&D, in such a way that the actions which were carried out were not very selective in their aims. The only objective was to increase the miniscule funding of R&D in Spain.

5. The few initiatives which were taken to support, rationalize and integrate public action in R&D came from the Ministry of Education and Science, via CAICYT. The Ministry of Industry was practically absent, and in any case tended to approach problems from the sectorial point of view, as we have pointed out.

This was the historical position at the end of the dictatorship. Added to the problems of instability during the democratic transition, are the developments of the world economic crisis, as well as the recession in the evolution of R&D expenditure which began in the mid-1960s (Gerelli 1982).

The official development plans which had been made at the end of the 1960s were suspended given the social pressure to give priority to the task of consolidating the democratic process. The complete absence of a reference to science and technology in the Pactos de la Moncloa is significant. This was an important agreement between the Spanish political parties which served as a basis for the democratic transition.

On the other hand, to the absence of political priorities for science and technology, the grave problems of the universities and businesses must be added. The shock to the system brought about a major industrial crisis which, in conjunction with an inability of businesses to be innovative, underwent a reconversion, and both production and employment dropped. The task of industrial reconstruction lasted all the way up to the first socialist legislature. Moreover, since the end of the 1060s, university expansion and its hyperpoliticization led to a complete paralysis in academic research and activity.

It can be said, with complete certainty, that the combined effect of the economic crisis of the mid-1970s and the Spanish democratic transition left little room for the development of policies and actions for science and technology (García-Delgado 1990), and this contributed to an even greater depression in the Spanish R&D system.

At this time, only the creation of the Center for the Technological Development of Industry (CDTI), with the support of the World Bank, in 1977, appears to be of any significance. It aimed to compensate for the lack of government action in the technological development of businesses. Furthermore, in 1980, the Ministry of Industry and Energy continued to advance towards the development of technology, establishing an organism for Industrial Innovation and Technology, to replace the one for Industrial Promotion.

The situation which the first socialist government of the transition was faced with, at the end of 1982, left much to be desired:

- the underdevelopment of the science-technology-industry system, characterized by: a very low investment in R&D in relation to the GDP, both in absolute and in relative terms; and a major imbalance in the execution of R&D in favor of the public sector;
- a lack of cooperation in R&D activities among businesses, universities and public research centers;
- a very high technological dependance on other countries, confirmed by a chronic deficit in the relation between payments and income for technology[6];
- an inability of policies to coordinate scientific research and technological development;

- very little selectiveness in objectives and priorities for research and funding;
- the predominance of a 'developmentist' focus, and a complete absence of ecological or social criteria in R&D activities.

4 Public Policy Actions in Science and Technology

The change in political responsibilities which occurred in Spain in 1982, propelled by a debate on the necessary modernization of the country, determined the establishment of a process of reform of the science/technology system and the reorganization of public involvement. The question became a priority on the political agenda of the government as a result of the initiatives of people who took up posts in the Ministries of Education and Science and of Industry and Energy. At last it became possible to attempt the application of the recommendations of international bodies, such as the OECD and UNESCO.

Public priorities are normally determined by a process of rationalizing reality, where organized interests, public opinion, or the attitudes of politicians or bureaucrats might play relevant roles.

In the case of Spanish scientific and technological policy, those who began to exercise a rationalized intervention were politicians and senior civil servants of the administration, influenced decisively by international bodies. Their aim was also to deal with major challenges for modernization. Their activities consisted in increasing resources for the science/technology system, to promote efforts in this field from all sides, and to improve public coordination. It was hoped that this would encourage research in universities and businesses, for which the state prepared a program of financial incentives, increasing funding for research in universities and businesses, for which the state prepared a program of financial incentives, increasing funding for research in universities and increasing grants for businesses to promote their R&D. The creation of instruments of public policy was also a feature of this process, as we shall later see.

4.1 The Reform of the Science/Technology System in Spain in the 1980s

The reform process began as an initiative of the Ministry of Education and Science, and developed after an evaluation of the features of the science/technology system. The conclusions (Muñoz/Ornia 1986) pointed out that Spain lacked a coherent model of science policy, or more precisely, an adequate coordination between science policy and technology policy with the subsequent economic and social development. All of the attempts carried out until then, to promote scientific and technical research, and then technological development and innovation, had been built around a model of spontaneity, inspired by the British and North Americans, and lacking any sense in a country like Spain, which could not then support such activities financially, and which lacked the adequate human and institutional infrastructure for scientific and technological activities.

The difficulties which determined the character of these policies in Spain had the following roots: one was related to the deficient administrative organization of the science/technology system, which required an immediate structural and organizational reform. Others, of a more general character, were derived from the little attention received by scientific research and technological innovation, as well the sectorial distribution of public funding of R&D in Spain, by the country's economy and enterprise.

4.2 The Instruments of Reform

After the process of analysis, reform was channelled two ways: the first was normative, and was embodied in the law of General Support and Coordination of Technological and Scientific Research; the other, of an operative nature, was the National Plan for Scientific Research and Technological Development (National R&D Plan).

These means aimed to encourage and promote all activities related to scientific research and technological development, from the creation and expansion of the research infrastructure, to the general promotion of knowledge in all disciplines and their concrete application to goods and services. At the same time, they attempted to establish coordination mechanisms between agents involved in the scientific/technological process.

The law, passed in 1986, which takes its name from a specific article of the Spanish Constitution (149.1.15), covered the coordination and support of all R&D activities. The general coordinating function was guaranteed, in the absence of a specific ministerial department, by an Interministerial Commission of Science and Technology (CICYT), integrated by representatives of ministries with direct responsibilities in scientific research and technological development, but whose importance corresponded, in this order, to the Ministry of Education and Science and to the Ministry of Industry of Energy. In fact, the law was the result of an agreement between both ministries.

The support provided by the National Plan for Scientific Research and Technological Development (National R&D Plan) which was aimed towards a generalization of the total financial effort of the country in matters of scientific research and technological development, both in terms of current expenditure and in terms of capital, covering the entire group of programs which would be carried out by the public sector as well as by the private sector.

Universities and businesses, together with public research organisms, were the executors of the R&D programs of the National Plan. This was explicitly pointed out by the law in Article 5: "to support scientific research and the technological development of businesses". This reference acquires particular significance in relation to the connection between research and innovation, so necessary to attain an efficient technological policy.

In Spain, as we have noted above, different means of support were developed for industrial growth and innovation (summarized in Figure 1), but their effectiveness was fairly limited.

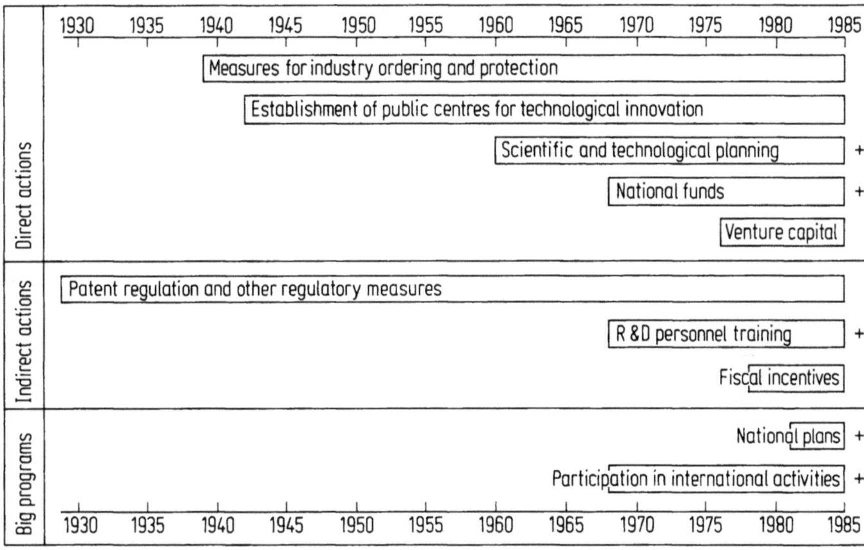

Figure 1: Timetable of steps for promoting technological innovation in Spain before the Science Act 1986
+ Initiatives influenced by the Science Act
[] Periods during which actions have been operative

The National R&D Plan was established, therefore, as a means to integrate actions of scientific policy and of technological policy, with the collaboration of the most pertinent authorities, the Ministries of Education and Science and of Industry and Energy. However, it must be recognized that there was little selectiveness of the research objectives outlined in this plan. It was conceived as an aggregate of eighteen priority programs, covering almost all the objectives of the European Community guidelines for research programs.

Since it covered all the interests and needs of the science/technology system, within the National Plan, the CDTI acquired particular relevance in its capacity to evaluate the technological and economic interest of projects, as well as promoting the industrial use of technologies, developed by public laboratories, universities and research centers, and procuring the technological returns wherever there was Spanish participation in international programs.

4.3 Technological Policy and the National R&D Plan

In 1981, the R&D expenditure of public and private enterprise reached a level of 35,000 million pesetas (25% of capital and 75% of current expenditure). In 1983, business expenditure reached 43,000 million pesetas – 0.19% of the GDP – split evenly by the public and the private firms. This activity was characterized by an imbalance in sectorial distribution, given that more than half of the expenditure on R&D was concentrated in three sectors, chemical industries 21.9%, electrical goods and electronic products 18.5%, and transport equipment 15.7% (Martín/Rodríguez 1985; Saez 1991). Consequently, there was still the effect of the policies of large scale importation of technology, characteristic of the era of "developmentism", as we have noted. This was further aggrieved by the limited capacity for adaptation of imported technology.

For example, in the years 1978-1980, Spanish businesses spent, on average, 0.25% of their added value on R&D, while the expenditure on technology was about 0.34% of the same value (see Table 1).

The technology policy included in the scheme of the National R&D Plan was a generalist policy, of a reactive character and, in many ways, prone to imitation, given that Spanish businesses appeared to act in accordance with a model which clearly depended on the exterior.

The National Plan for Scientific Research and Technological Development, in the period 1988-1991, recognized two fundamental objectives of technology policy, both of which were rather broad: (i) to increase the added value of industrial activities via the incorporation of new technological elements in industrial processes and products ("diffusion"); and (ii) to develop new technological fields in Spain so that the country could offer an adequate scientific and technological potential ("innovation").

This policy admitted the scale of the effort required at that time, while accepting the intrinsic differences between the two objectives, their general nature, and up to a certain point, their spontaneity in that they would require an effort of will. The National Plan, in its early stages, depended for its success upon its becoming an instrument of economic and industrial policy aimed towards modernization and flourishing in Spain. In order for this to be possible, it hoped to accommodate the adaptive capacities of industry to Spain's real potential with the aim of eliminating the bottlenecks which are an obstacle to technological development and to the satisfaction of social demands. As we have pointed out, the National Plan gave the CTDI ample powers to carry out these tasks.

Table 1: Interindustry comparison between Technological payments (TP) and Business Enterprise Expenditure on R&D (BERD) in Spain (average from 1978 to 1980), in %

Industries	TP	TP/GDP	BERD	BERD/GDP
Agriculture	0.80	0.03	1.45	0.04
Mining and Petrochemical	1.44	0.28	4.95	0.71
Power, Gas and Water	6.63	0.94	2.82	0.03
Metal Industries	3.49	0.77	6.99	1.15
Industry of Non-Metallic Minerals	0.94	0.20	3.71	0.59
Chemical Industry	15.92	2.43	22.54	2.54
Manufacture Metal Goods and Machinery	17.56		6.55	
Manufacture Electrical and Electronic Goods	11.73	1.88	18.18	1.17
Manufacture Motor Vehicles and Transport Equipment	28.41	2.80	11.70	0.85
Food, Drinks and Tobacco	2.26	0.27	2.43	0.21
Textile Industry	2.90	0.25	1.12	0.07
Wood and Printing	0.95	0.20	0.92	0.14
Other Manufacturing Industries	2.78	0.29	8.21	0.63
Building	2.66	0.11	0.38	0.01
Service Industries	1.53	0.01	8.05	0.04
Total	100.00	0.34	100.00	0.25

Source: Banco Bilbao: GDP, Instituto Nacional de Estadistica: BERD, Ministerio de Industria y Energía: TP, as cited in Muñoz/Ornia (1986:100)

4.4 Technology Policy and the Ministry of Industry and Energy

The elaboration of science and technology policy in the first socialist term had some interesting precedents, with plans of a sectorial character which were launched by the Ministry of Industry and Energy at the end of the 1970s and the beginning of the 1980s.

The most significant precedent, an expression of the interest raised by new technologies, was the National Electronic and Computing Plan (PEIN). In 1981, a committee of experts was created for the elaboration of the foundations of a

National Plan for Electronics Industry, taking its conclusions to the government in 1983.

Moreover, at the same time as the elaboration of the National R&D Plan, the Ministry of Industry and Energy defined its own framework of industrial policy, published in 1987, under the title España en Europe: *Un futuro industrial (La política industrial en el horizonte de 1992)* (Spain in Europe: an Industrial Future [Industrial Policy Towards 1992]). This work analyzed the situation in Spanish industry, compared it against its European counterparts, and suggested policies and programs designed to attain the aims and objectives in industry in relation to the EC, by 1992. As a result of this action, some of the original programs were continued, such as PEIN, the Plan for Advanced Automatization (PAUTA), as well as the establishment of others, like the Plan for Supporting Research in the Pharmaceutical Industry (FARMA).

Special mention ought to be made, in relation to the social and ecological concerns of R&D, that in 1989 the Environmental Program for Industry and Technology was approved, to be applied in the period 1990-1994. Its essential objective involves the development of industrial procedures and new technologies which reduce the generation of pollutants. This program runs parallel to the Plan of Technological and Industrial Action (PATI) from 1991 to 1993, which groups together the majority of technology programs of the Ministry of Industry and Energy (PEIN, PAUTA, FARMA, etc.). This plan was initially established as complementary to the National Plan, operating within the private sector. However, early data are beginning to reveal that the National R&D Plan and PATI are not complementary but competitive. A major distorting element has been introduced with the creation of two units of political and administrative control, attempting the same task without coordination.

5 Large Policies, Little Ecological and Social Awareness

Only recently has there been an evaluation of the last ten years of Spanish science and technology policies. Some are simply triumphalist in their appraisal of the recent events (Rojo 1991; Lafuente/Oro 1991; CICYT 1991). Others, in spite of their recognition of the drastic improvement of the situation, still detect grave structural deficiencies or worrying tendencies (Quintanilla 1992; Sanz/Pfretzschner 1992).

These data (see Table 2) show that Spain has been the EC country with the highest rate of growth in recent years. The creation of science and technology policies has contributed to the growth in Spanish participation in European R&D, representing 4% of the European total in 1990, compared with the 2.4% of 1984. Much of the merit is due to the financial effort of the state and in the increased contribution of businesses. In fact, the participation of private enterprise has reached 57% of the national total, and at the same time, the size of the public financial contribution to private R&D has risen from 2% to 16% in 1990.

The establishment of these science and technology policies has not helped to defend Spain from some of the difficulties which are approaching with the Single European Market, in particular for industries which will require a high technological level (Martín 1990).

So, as we have pointed out, the existing difference in science and technology between Spain and the rest of the European Community is very significant. It can be expressed in the figure that Spain spends something under 0.9% of its GDP on R&D, while the EC average is over 2%.

But the difference is not only quantitative, it is also qualitative. When the rest of the EC the main concerns centred around new social and ecological approaches, which aimed to incorporate the quality of life into R&D, Spain was still in the middle of the struggle to increase its R&D effort.

The introduction of concerns of a social and ecological character for the development and practice of R&D are hardly existent, secondary, and, in any case, very recent, especially when compared to the basic objective when increasing the size and effort of research in Spain. Amongst the primary objectives of R&D, only recently, around 1988-1989, has the environment begun to occupy a relevant place.

We must remember that the first public plan for scientific research and technological development, the National R&D Plan, began to operate in 1988. It contained some priority clauses related to aspects of the quality of life and to environmental problems, such as desertification (Crespo et al. 1991:245-246), and some social aspects, such as the labor market, in relation to R&D, but with hardly the appropriate resources to carry them out.

In the plans of the Ministry of Industry and Energy, as we have said, environmental industries have become the targets of technology policy since 1990 (De Andrés 1991).

An analysis of the official statistics for R&D reveals that during the decade of the 1980s, social and environmental concerns were marginalized within the whole of the R&D investment in Spain, both by businesses and by public research centers (see Table 3).

Since 1988, there has been a gradual tendency towards an increase in priorities of this kind in Spanish R&D. There has undoubtedly been a modification of the system of public priorities, probably determined by the need to deal with the EC environmental norms urgently.

However, this recent progress cannot be used to ignore that the expenditure per capita which is budgeted for these purposes in Spain is clearly inferior to the same average spending in the EC. Although, in relative terms, in so far as Spain is concerned, the main imbalance is not as notable in ecological and social objectives but in the expenditure for technological/industrial objectives.

Table 2: Gross Expenditure on Research and Development (GERD), European Community

Million current PPP (US$)*

	1984	1985	1986	1987	1988	1989	1990
Belgium	1,679.7	1,788.7	1,851.1	1,956.9	2,060.0	2,167.8	2,262.0
Denmark	715.2	785.3	879.0	971.5	1,075.9	1,191.6	1,297.4
France	13,627.7	14,571.1	15,146.6	16,237.7	17,511.6	19,176.4	21,129.4
Germany	18,589.5	19,983.6	21,224.9	23,103.7	24,578.3	26,867.6	28,688.3
Greece	176.3	201.7	205.6	222.0	251.7	275.1	300.7
Ireland	177.8	202.9	223.6	252.0	269.0	299.6	332.5
Italy	5,963.4	7,014.5	7,432.7	8,277.8	9,164.2	10,360.0	11,077.0
Netherlands	3,059.0	3,437.3	3,808.1	4,148.3	4,259.1	4,626.5	5,052.6
Portugal	202.4	226.0	254.4	285.2	319.8	358.5	401.9
Spain	1,398.9	1,629.3	1,909.4	2,164.6	2,640.6	3,174.1	3,760.2
UK	13,417.2	14,443.5	15,548.3	16,230.8	17,042.4	18,092.5	19,207.2
Total EC	59,007.1	64,284.9	68,483.7	73,850.5	79,173.4	86,589.7	93,482.3

Share of total EC GERD, in millions current PPP (US$)*

	1984	1985	1986	1987	1988	1989	1990
Belgium	2.9	2.8	2.7	2.7	2.6	2.5	2.4
Denmark	1.2	1.2	1.3	1.3	1.4	1.4	1.4
France	23.1	22.7	22.1	22.0	22.1	22.2	22.6
Germany	31.5	31.1	31.0	31.3	31.0	31.0	30.7
Greece	0.3	0.3	0.3	0.3	0.3	0.3	0.3
Ireland	0.3	0.3	0.3	0.3	0.3	0.4	0.4
Italy	10.1	10.9	10.9	11.2	11.6	12.0	11.9
Netherlands	5.2	5.4	5.6	5.6	5.4	5.3	5.4
Portugal	0.3	0.4	0.4	0.4	0.4	0.4	0.4
Spain	2.4	2.5	2.8	2.9	3.3	3.7	4.0
UK	22.7	22.5	22.7	22.0	21.5	20.9	20.6
Total EC	100.0	100.0	100.0	100.0	100.0	100.0	100.0

Per Capita Population, in current PPP (US$)*

	1984	1985	1986	1987	1988	1989
Belgium	170.4	181.6	187.7	198.3	208.5	218.1
Denmark	139.9	153.6	171.6	189.5	209.7	232.2
France	248.0	264.1	273.4	291.9	313.4	341.5
Germany	303.9	327.5	347.6	378.3	400.0	433.4
Greece	17.8	20.3	20.6	22.2	25.1	26.6
Ireland	50.4	57.3	63.1	71.1	76.3	85.2
Italy	104.7	122.8	129.9	144.4	159.5	180.1
Netherlands	212.1	237.2	261.3	282.9	288.6	311.6
Portugal	20.0	22.3	24.9	27.8	31.0	34.7
Spain	36.5	42.3	49.4	55.7	67.7	81.0
UK	237.5	255.1	273.9	285.1	298.6	316.3
Total EC	183.9	199.9	212.4	228.5	244.0	265.5

* PPP = Purchasing Power Parity

Source: OECD (1991:2); INE (1991) (Spain)

Table 3: Gross domestic expenditure on R&D by chapters of NABS in Spain (million of current pesetas)

Government Sector

	1982	1983	1984	1985	1986	1987	1988	1989
Total	27,618.7	30,749.5	32,756.4	37,558.9	49,217.1	58,187.6	66,685.4	77,137.0
Agriculture	5,031.5	5,814.6	7,452.3	8,374.9	9,722.7	11,410.2	14,877.0	17,778.8
Industry	4,348.8	4,460.8	4,267.7	174.5	1,159.9	607.4	11,091.7	12,572.8
Defense	2,948.7	3,511.5	3,387.6	3,937.6	2,847.8	4,253.7	7,591.6	4,135.4
Environment	186.4	178.6	0.0	0.0	60.7	133.6	1,831.0	4,608.3
Health	356.7	973.7	1,302.8	1,296.3	1,809.1	5,121.7	8,402.8	10,398.3
Social	0.0		69.3	31.7	242.1	271.7	124.7	1,911.1
Others	2,877.8	1,698.5	1,255.0	6,578.5	13,189.3	13,372.2	21,742.6	25,732.3
Non classified	11,868.8	14,111.8	15,021.8	17,165.5	20,185.4	23,017.1	1,024.1	0.0

Business Enterprise Sector

	1982	1983	1984	1985	1986	1987	1988	1989
Total	46,861.8	52,142.9	65,411.0	85,795.2	110,338.2	126,706.6	163,370.1	191,153.2
Agriculture	6,999.9	841.6	1,477.5	1,875.8	2,749.0	4,721.0	5,917.2	5,651.8
Industry	29,828.6	33,460.9	33,045.8	43,116.4	60,667.3	68,344.1	89,285.9	97,207.8
Defense	1,199.3	945.6	6,609.3	10,925.5	12,865.3	14,001.8	15,998.3	16,956.5
Environment	258.0	4.8	87.3	151,4	175.6	336.2	136.5	235.2
Health	5,075.6	4,482.7	5,285.8	6,218.5	8,686.7	10,641.1	13,288.9	16,273.2
Social	13.9	13.7	23.4	67.0	368.2	252.1	231.5	192.9
Others	3,486.4	12,393.6	18,882.0	23,440.6	24,826.1	28,410.3	38,511.9	54,635.8
Non classified	0.0	0.0	0.0	0.0	0.0	0.0	0.0	0.0

Agriculture: Agricultural production and technology
Industry: Industrial production and technology
Defense: Defense
Environment: Control of environmental pollution
Health: Protection and improvement of human health
Social: Social Structure and relationships
Others: Other civil research
Data for Higher Education Sector not available

Source: Instituto Nacional de Estadística (INE), Spain

A comparative analysis of public budgetary credits which in 1989 were worth more than 60% of the total spending on R&D in Spain, can give a good idea of the degree of priority and importance of the social and ecological concerns (see Table 4).

Thus, the R&D boom in Spain occurred around 1988-1989, coinciding with a greater orientation towards ecological and social objectives in the same country.

Both in actual spending, as well as in budgetary measures, the Spanish situation seems to have improved, but there is still a major gap in the state's priorities. Within the small expenditure, the presence of ecological and social questions is very small, in spite of the comparative growth of these priorities. In our opinion,

Table 4: R&D financing by chapters of NABS*, 1989, in ECU per head at 1989 values and exchange rates

			Large countries				Medium and small countries					
	ESP	EC	FRG	FRA	UK	ITA	NL	BEL	DK	GRE	POR	IRE
Total expenditures	46.1	129.9	183.9	217.2	126.3	100.6	127.1	80.7	147.7	15.8	14.2	34.3
3. Environment	1.0	2.6	6.3	1.5	1.6	2.1	4.8	1.0	4.5	0.5	0.4	0.3
4. Health	3.0	5.1	6.3	7.2	6.4	4.7	3.3	2.1	2.6	1.3	0.6	1.4
6. Agriculture	2.6	4.8	3.8	8.6	5.6	2.7	5.7	4.7	11.4	2.6	2.5	6.7
7. Industry	7.8	17.7	23.5	28.2	10.7	17.8	28.7	15.4	23.0	1.9	2.5	11.4
8. Social	0.6	2.5	4.5	1.0	2.7	3.6	2.6	0.5	7.0	0.7	0.2	2.9
10. GUF	7.5	30.3	60.3	24.1	22.9	33.0	44.3	18.6	52.3	6.1	5.2	8.9
11. Non-oriented	4.8	15.8	25.6	32.9	6.0	10.7	14.2	18.9	32.9	0.6	0.3	0.7
13. Defense	9.7	31.5	23.5	83.6	57.1	6.8	4.1	0.5	0.6	0.3	0.0	0.0

* NABS Objectives:

1. Exploration and exploitation of the earth
2. Infrastructure and general planning of land-use
3. Control of environmental pollution
4. Protection and improvement of human health
5. Production, distribution and rational use of energy
6. Agricultural production and technology
7. Industrial production and technology
8. Social structure and relationships
9. Exploration and exploitation of space
10. Research financed from general university funds (GUF)
11. Non-oriented research
12. Other civil research
13. Defense

Source: Eurostat

this change is due to the growing mid- and long-term influences of the European Community.

The late inclusion of ecological and social objectives in R&D activities can be explained by a poor social awareness and the absence of these questions on the political agenda. On the other hand, it is significant that Spanish environmental groups are not represented in Parliament or in local government.

A determining factor in the formation of public policies on science and technology, and in particular for the importance of social and ecological concerns within these policies, is the degree of urgency and importance given to these concerns by Spanish opinion.

Public opinion polls have helped to establish differences in values between countries fo the European Community. When the so-called welfare indicators[7] of EC countries are examined, it can be seen that, while in Spain only one out of four citizens considered that the state of the environment was an indicator of well-being, the EC average was one-in-three. Awareness of environmental issues was, indeed, very high in Luxembourg, Denmark, Italy, the Netherlands and Germany. Below the EC average, at around one-in-five, were France and the United Kingdom; and further behind still, Portugal and Greece (CEC 1986).

It is for these reasons, namely that Spaniards have a low perception of environmental problems and that there is a near absence of any organization of social interests, that it has been maintained that the Environmental Program for Industry and Technology (PITMA) is better designed for the "diffusion" of environmental policies than private initiatives. The former system has arisen as a result of EC norms, and the latter tend to arise in countries with a higher level of development, especially those with highly organized pressure or interest groups.

Another significant point about the political importance or environmental problems is that until 1991, the administrative responsibility for these matters in central government was held by a General Secretariat for the Environment (at the third level of government, after ministers and secretaries of state). It was not until the ministerial modifications of 1991 that political responsibility was taken to a higher level, with the establishment of the Secretary of State for Environmental Policy and Water.

In any case, public R&D initiatives with social and ecological approaches are few and evidently uncoordinated, and they are at present dispersed among various ministerial departments.

6 Conclusion. Convergence with Europe: How Far are We?

A surface reading of the indicators of science and technology in use in Spain, together with their recent evolution, could give us a optimistic view of the Spanish situation. However, a comparative analysis of the variation between the percentage of the GDP invested in R&D and the per capita income, expressed logarithmically, show that Spain is at the top of the group of Europe's least developed countries, with Ireland, Greece and Portugal (Quintanilla 1992). Logarithmic representation shows, in addition, that there is an area – between 1 and 1.5% of the GDP spent on R&D – beneath which any increase in expenditure on science and technology generates results of virtually no significance.

The influence of R&D in relation to economic development can be described in a metaphor involving a passenger on a conveyor belt. If the passenger advances in the same direction as the conveyor belt, his speed gains substantially. If he does not move, his speed will remain identical to that of the conveyor belt. The image of the passenger and the conveyor belt suggests the relation between investment in R&D and economic development: overcoming the area of difficulty (expressed in the logarithm) is indispensable if R&D is to acquire any social significance.

In these circumstances, it can hardly be surprising that technology policy and its partner, science policy, in Spain, have been of little historical relevance, suffering from a lack of strategy and an appropriate model. Some actions and mechanisms put forward in recent years have managed to generate resources and initiatives, but at the same time, they have exposed basic structural deficiencies and have been useful in the elaboration of an early evaluation of that which has been achieved and of the problems still to be confronted. Among these, for obvious reasons of time and lack of experience, it is evident that the biggest problems are the social construction of technological development and the social compatibility of technical change.

To conclude and summarize, the improvement in social and ecological terms of science and technology policies in Spain is characterized by certain facts and circumstances which would explain the situation of Spanish underdevelopment in relation to the rest of the European Community. The perception and concern of public opinion for environmental problems in Spain is very low. There is an important gap, in terms of the level of economic and scientific/technological development, between Spain and the EC average.

This low perception of social and ecological problems, in conjunction with the institutional weakness and the absence of strong interest groups in democratic civil society, have ensured that environmental issues find their way to the political agenda via two streams of access: international diffusion of approaches, and the personal leadership of particular politicians.

In any case, we must remind ourselves that the introduction of items or issues on the Spanish political agenda, in the fields of science and technology in recent years, have been aimed towards the mobilization and increase of the practice and resources of R&D, and not by discriminating or well-chosen concerns.

Notes

[1] The text entitled *The Industrial Policy of the New Spain*, published in 1942 by the Ministry of Industry, quoted the head of state saying that there was a need to "increase production". The industrial policy was founded on this idea.

[2] It must be pointed out that the creation of the CSIC occurred at the same time as that of the *Centre National de la Recherche Scientifique* (CNRS) in France. Both institutions appeared in 1939, although in circumstances which were radically opposed in historical and sociological terms. The French organism was the reflection of a popular government, upon which a rapid conflagration was looming, while the CSIC appeared at the end of a civil struggle, and with an entirely different ideological framework to its counterpart, as well as a different orientation in its aims and objectives. The CNRS was preparing to develop aspects of technology and experimentation, and the CSIC was to begin a kind of crusade (it was said that among the essential functions of the institution, that it must "put scientific production at the service of the spiritual and material interest of the nation").

[3] The OECD (1971) points out that the 2nd Development Plan required more than 41,000 million pesetas investments, although few more than 6,000 million were conceded.

[4] Several studies of the years 1967-1974 pointed to this situation. This was the actual period in which the development plans operated. For example, the CSIC (1977) carried out one of the few studies in existence on the matter from which the following conclusions can be drawn:

In 1974 there was, finally, a substantial increase in spending on R&D by businesses (145% in fixed pesetas), which indicates a positive response of the concerted plans.

Most spending was very concentrated in sectorial terms (the chemical industry, the pharmaceutical industry, transport materials, and electrical and electronic goods). The proportion of R&D expenditure with respect to the added value of each sector or branch was very varied, from a minimum of 0.0077% for agriculture, forestry, sheep, pig and cattle farming, and fishing, to a maximum of 1.98% for the chemical industry, pharmaceuticals, and oil and petroleum products).

Spending came almost entirely from the businesses themselves, with no significant funding from loans or grants from the mentioned concerted plans which the public administration had been supposedly conceding.

Capital spending was relatively low, which indicates a structural weakness in businesses in relation to R&D.

The distribution of spending was split thus: 2.4% in basic research, 30.7% in applied research, and 66.9% in technological development.

Businesses of a critical size (say those with more than 250 employees) constituted 60% of those businesses involved in R&D and their spending constituted 91% of the sector's total expenditure.

In any case, the size of R&D units was small; 82% of companies, in spite of their activity in the promotion and management of R&D, employed less than ten scientists and engineers to carry out these tasks.

[5] We must remember that General Franco died in 1975. The first democratic General Elections took place on the 15th of June of 1977. The new Constitution was approved in December 1978. In 1979 there were new elections. In October of 1982, the Socialist party won its first victory.

[6] For example, the relation between payments for technology bought abroad and domestic expenditure on R&D has always been significant: the United States invested 200 times more on technological development than it bought, France and Germany five times, and Italy 1.3 times, while in Spain the business sector invested half of what it spent abroad on technology in R&D.

[7] The system consists in indicating the percentage of interviewees in each one of the countries which pointed out that the question was an indicator of well-being.

References

Buesa, M., J. Molero (1989) *Innovación industrial y dependencia tecnológica en España*, Madrid: EUDEMA.

Bush, V. (1945) *Science, The Endless Frontier: A Report to the President on a Program for Postwar Scientific Research*, Washington, DC: Government Printing Office.

Carreras, A. (1987) La industria: atraso y modernización, in: J. Nadal, A. Carreras, C. Sudria (eds.) *La economía española en el siglo XX. Una perspectiva histórica*, Barcelona: Ariel, 280-312.

CEC (Commission of the European Community) (1986) *Eurobarometer, 1986: Los europeos y el medio ambiente*, Brussels: CEC.

Chandler, A.D. (1962) *Strategy and Structure*, Cambridge, MA, London: MIT.

Chandler, A.D. (1977) *The Visible Hand. The Managerial Revolution in American Business*, Cambridge, MA, London: The Belknap Press of Harvard University Press.

CICYT (1988) *Plan Nacional de Investigación Científica y Desarrollo Tecnológico*, Madrid: Ministerio de Educación y Ciencia.

CICYT (1991) *Memoria de desarrollo del Plan Nacional de I+D in el período 1988-1990*, Madrid: Comisión Interministerial de Ciencia y Tecnología.

Crespo, A., M. Costa, M. Millán (1991) Política de I+D en materia medioambiental: perspectiva científica, in: R. Dorado, J. Rojo, E. Triana, F. Martinez (eds.) *Ciencia, tecnología e industria en España*, Madrid: Fundesco, 239-260.

CSIC (1977) *La investigación científica y el desarrollo tecnológico in España, 1967-1974. Evolución y comparación con algunos países europeos*, Madrid: Gabinete Técnico del CSIC.

De Andrés, C. (1991) Tecnologías industriales medioambientales, in: R. Dorado, J. Rojo, E. Triana, F. Martinez (eds.) *Ciencia, tecnología e industria en España*, Madrid: Fundesco, 261-277.

Dosi, G.(1988) Sources, Procedures and Microeconomics Effects of Innovation, in: *Journal of Economic Literature* September (XXVI): 1120-1171.

Dosi, G. (1991) Perspectives on Evolutionary Theory, in: *Science and Public Policy* 6 (18): 353-361.

Espina, A. (1992) *Recursos humanos y política industrial*, Madrid: Fundesco.

Eurostat (1991) *Estadística de financiación de la I + D*, Luxemburg: Eurostat.

Fagerberg, J. (1987) A Technology Gap Approach to Why Growth Rates Differ, in: *Research Policy* (16): 87-99.

García-Delgado, J.L. (1987) La industrialización y el desarrollo económico de España durante el franquismo, in: J. Nadal, A. Carreras, C. Sudria (eds.) *La economía española in el siglo XX. Una perspectiva histórica*, Barcelona: Ariel, 164-189.

García-Delgado, J.L. (ed.) (1990) *La Economía Española en la Transición y la Democracia (1973-1986)*, Madrid: Centro de Investigaciones Sociológicas.

Gerelli, E. (ed.) (1982) *La politica per l'innovazione industriale: problemi e proposte*, Milan: Angeli.

Gerelli, E. (1983) *Per una politica dell'innovazione industriale*, Milan: Angeli.

INE (various years) *Estatística sobre las actividades en investigación cientifica y desarollo tecnológico*, Madrid: INE.

Lafuente, A., L. Oro (1991) Evolución del sistema de ciencia y tecnologia en España. El Plan Nacional de I+D, in: R. Dorado, J. Rojo, E. Triana, F. Martinez (eds.) *Ciencia, tecnología e industria en España*, Madrid: Fundesco, 33-123.

Lora Tamayo, M. (1990) Recuerdos del Consejo Superior de Investigaciones Científicas en su 50 Aniversario, in: *Arbor* 529: 99-115.

Martín, C. (1990) Espagne, in: P. Buigues, F. Ilkovitz, J.F. Lebrun (eds.) *L'impact sectoriel du marché intérieur sur l'industrie: les enjeux pour les Etats membres*, Brussels: CEC, 221-233

Martín, C., L. Rodriguez (1985) Política de promoción del Cambio Técnico y Reindustrialización, mimeo D.T.8502, Madrid: Fundación Empresa Pública.

Martín-Aceña, P., F. Comín (1992) El Estado en la industrialización española de posguerra: el Instituto Nacional de Industria, in: L. Prados de la Escosura, V. Zamagni (eds.) *El desarrollo económico en la Europa del Sur: España e Italia in Perspectiva histórica*, Madrid: Alianza Universidad, 421-461

Ministerio de Industria y Energía (1990) *España en Europa un futuro industrial (La politica industrial en el horizonte 1992)*, Madrid: Ministerio de Industria y Energia.

Mowery, D., N. Rosenberg (1989) *Technology and the Pursuit of Economic Growth*, Cambridge, GB: Cambridge University Press.

Muñoz, E. (1982) Política científica, innovación y empleo, in: A. Nieto, J.M. Gómez Fatou, E. Muñoz, J.A. Muñoz-Delgado, T. Mendizabal, C. Llaguno, F. Catalina (eds.) *Apuntes para una política científica. Dos años de investigación in el CSIC: 1980-1982*, Madrid: Consejo Superior de Investigaciones Científicas, 219-226.

Muñoz, E. (1990) El CSIC, una síntesis de tradición y futuro. Media centuria in la balanza de la ciencia española, in: *Arbor* 529: 13-28.

Muñoz, E., F. Ornia (1986) *Ciencia y Tecnología una oportunidad para España*, Madrid: Ministerio de Educación y Ciencia-Aguilar.

Myro, R. (1981) La estrategia de la Empresa Pública concurrencial en España: un análisis de la actuación del Instituto Nacional de Industria, in: *Investigaciones Económicas* 15: 23-50.

Nelson, R., S. Winter (1977) In Search of a Useful Theory of Innovation, in: *Research Policy* 1 (6): 36-76.

Nelson, R., S. Winter (1982) *An Evolutionary Theory of Economic Change*, Cambridge, MA, London: The Belknap Press of Harvard University Press.

Nelson, R., S. Winter, H.L. Schütte (1976) Technical Change in an Evolutionary Model, in: *Quarterly Journal of Economics* 1 (XC): 90-118.

Nieto, A. (1982) La investigación científica en España, in: A. Nieto, J.M. Gómez Fatou, E. Muñoz, J.A. Muñoz-Delgado, T. Mendizabal, C. Llaguno, F. Catalina (eds.) *Apuntes para una política científica. Dos años de investigación en el CSIC: 1980-1982*, Madrid: Consejo Superior de Investigaciones Científicas, 203-205.

Nieto, A. (1990) El CSIC durante el periodo de la consolidación democrática in: *Arbor* 529: 29-48.
Noble, D.F. (1977) *America by Design*, Oxford, New York: Oxford University Press.
OECD (various years) *Science and Technology Indicators*, Paris: OECD.
OECD (1971) *Políticas nacionales de la Ciencia: España*, Madrid: Ministerio de Educación y Ciencia.
Pavitt, K. (1979) Technical Innovation and Industrial Development. 1. The New Causality, in: *Futures* 6 (11): 458-470.
Pavitt, K. (1980) Technical Innovation and Industrial Development. 2. The Dangers of Divergence, in: *Futures* 1 (12): 35-44.
Pavitt, K. (1984) Sectoral Patterns of Technical Change: Towards a Taxonomy and Theory, in: *Research Policy* (13): 343-373
Quintanilla, M.A. (1992) El Sistema español de ciencia y tecnología. (Proyecto EPOC), in: *Arbor* 554-555: 9-251.
Rojo, J.M. (1991) El sistema de I+D: fortalezas y debilidades, in: R. Dorado, J. Rojo, E. Triana, F. Martinez (eds.) *Ciencia, tecnología e industria en España*, Madrid: Fundesco, 15-32.
Ronayne, J. (1984) *Science in Government*, Caulfield East, Baltimore: Edward Arnold.
Ros-Hombravella, J. (1973) *Capitalismo español: De la Autarquía a la estabilización (1939-1959)*, Madrid: Cuadernos para el Diálogo.
Rosenberg, N. (1991) Critical Issues in Science Policy Research, in: *Science and Public Policy* 6 (18): 335-346.
Saez, F. (1991) *Tecnología y empleo en España*, Madrid: Instituto de Estudios y Análisis Económicos.
Salter, W.E.G. (1960) *Productivity and Technical Change*, Cambridge, GB: Cambridge University Press.
Sanchez, P. (1984) *La dependencia tecnológica española*, Madrid: Ministerio de Economía y Hacienda.
Sanz, L., C.E. García (1990) Presente y perspectivas de la brecha tecnológica en la Comunidad Europea, in: M. Gamella, M. Hernandez (eds.) *Nuevas Tecnologías y Orden Económico Internacional*, Madrid: Fundesco, 17-48.
Sanz, L., J. Pfretzschner (1992) Política científica y gestión de la investigación: El CSIC (1986-1990) en el sistema español de ciencia y tecnología, in: *Arbor* 557: 9-51.
Schumpeter, J.E. (1939) *Business Cycles*, New York, London: Mac Graw-Hill.
Segura, J. (1983) Crisis y estrategias alternativas: el caso española, in: *Pensamiento Iberiamericano* 3: 93-112.
Soete, L.L.G. (1981) A General Test of Technological Gap Trade Theory, in: *Weltwirtschaftliches Archiv* 4 (117): 638-660.

Technology Policy under Conditions of Social Partnership: Development and Problems of an Integrated Strategy in Austria

Georg Aichholzer, Renate Martinsen and Josef Melchior

1 Introduction

Austria is among the advanced industrialized nations that have only recently introduced an active technology policy. For this reason, it is all the more interesting to analyze which more or less successful strategies the country has pursued in the course of its relatively young history of technology policy, to assert its position in the international technology race. The paper starts with a brief outline of the decisive structural elements, i.e. the chances and restrictions offered and/or imposed both by the status of a small state with a specific economic profile and a particular political culture, and by a number of current challenges. It subsequently describes the specific varieties of research policy and the introduction of an active technology policy, understood as innovation strategy affecting all parts of society. This leads to a characterization of the current policy mix, as well as of the progress and problems arising during its implementation, in particular regarding the mandate of an anticipatory consideration of social and ecological goals. The conclusion primarily concentrates on the contribution made by the policy of social partnership, one often deemed unique; and takes it up once more in the context of the current discussion on technology policy.

2 Structural Conditions of Technology Policy in Austria

2.1 The Problem of Small States

" 'Small' and 'large', not only interpreted in their economic but also in their political and cultural context, ostensibly represent relevant distinctive features influencing the particular shape of technological and economic developments" (van Tulder 1989:1). According to Katzenstein (1985:191-211), a number of convincing arguments permit us to speak of a specific group in connection with small states[1]: from an economic point of view, items worthy of particular mention here are the openness of national economies, the adaptation pressure exercised by economic 'giants', and the production for small segments of the world market. From a political point of view, there seems to be an affinity towards corporatist decision structures, along with compensatory interventions by welfare-state measures to absorb the effects of the high exposure on the world market. Moreover, the small domestic market and scarce research resources favor structural conservatism, i.e. the narrow industrial basis usually permits only limited internal competition and makes a sanctioning policy more difficult *vis-à-vis* established sectors.

As the above-mentioned problem aspects imply, the 'smallness' of a nation apparently constitutes a handicap for its technological scope of action. According to this theory, the increased economic adaptation pressure exerted on small states constitutes a disadvantage for competitiveness, which is largely dependent on technological innovation potential. However, according to Katzenstein's theory (1985), the adaptation pressure – permanently exerted on small states in any case – increases their ability to "actively adapt", thus giving them the "plus" of being particularly flexible. It is doubtful whether this optimistic assessment can also be upheld in times of an amazingly dynamic international development of technology. Yet the (implicitly contained) statement is certainly correct: the political and institutional ability to productively assimilate international developments is decisive for the economic and welfare state balance of a small state. In this sense, and despite a common constituent framework of small countries, there are definitely differences regarding the success of the respective specific national adaptation strategy.

2.2 Economic Structure

In a global perspective, the performance of the Austrian economy is quite satisfactory, even when compared with other European OECD countries (see, for example, Fesselhofer 1992). The Geneva World Economic Forum puts Austria in the seventh place of its 'hit list' on competitiveness in the 22 countries of the OECD (*Der Standard*, August 20, 1992:1). However, this distinction must be put into perspective by adding some supplementary aspects. The industrial sector especially remains a principal headache. The lacking dynamic of this focal sector leads to a complex of problems showing that Austria's position among the "top ranks worldwide" is both precarious and endangered. Even though large sectors of the Austrian industry are currently being re-structured, considerable 'structural deficiencies' still remain:

- *Outdated production structure:* Austrian industry is characterized by being 'raw-material-heavy', whereas the technology-intensive branches of the production structure are underrepresented. Austria therefore also has the worst relation among all the European industrial nations, between exports and imports of high technology (Glatz 1991:122-123; Tichy 1987:67-68; Kuntze 1990:65; OECD 1992a:132).

- *Low degree of active internationalization:* Despite of recent improvements the internationalization of industry shows an unfavorable balance with an excess of passive internationalization. The ratio between foreign direct investments in Austria and Austrian direct investments abroad is estimated to be 2.5:1 (Glatz 1991:123-124; Tichy 1987:69; Kuntze 1990:65).
- *Too little specialization:* Austria's export returns per weight standard are substantially lower than those of its competitors in the highly-developed industrial nations. We may thus conclude that only very simple, scarcely-specialized standard products are exported. The situation even deteriorated during the 1980s and is aggravated by a lack of market leadership (Tichy 1987:69-70).
- *Lack of large domestic corporations:* Austria's private industry consists predominantly of small and medium-sized firms. There is a lack of firms operating on a worldwide basis, functioning as "flag-ships" that could breach and thus open up foreign markets. With its rather small-structured industry, Austria will find it increasingly difficult to attain sufficient economies of scale in research, organization, production, and marketing (Glatz 1991:122; Kuntze 1990:65).
- *Large "protected" sector:* The strong state protection of parts of the domestic market, in particular in service and trade, leads to low competition and increased price levels, as well as impeding structural changes (Glatz 1991:125; Kuntze 1990:65).
- *Low R&D share:* In an international comparison, the expenditures for R&D, though decisive for the technological competitiveness of a nation, are modest in Austria (Glatz 1991:125-126; Tichy 1987:72-76; Kuntze 1990:64-65).
- *Financial weakness:* The essential weaknesses of the Austrian capital market are a dominance of financing by outside capital (in particular through banks and insurance companies) and a low degree of financing through internal and external equity capital (Glatz 1991:126; Kuntze 1990:64).

Finally, a recent attempt to at least partially redress these shortcomings is worth mentioning: pushed by the primacy of budget consolidation, the growing deficits of the nationalized economy during the 1980s led to a completely new structure and partial privatization of this industry. Despite initial successes in the attempt to establish a powerful Austrian multinational concern, while concurrently launching an important instrument of industry and technology policy, these efforts eventually failed and once more entailed restructurization attempts currently still under way. The turn in economic policy from an Austro-Keynesian to a more supply-oriented strategy constitutes the background of this development.

2.3 The Policy Pattern of Social Partnership

In an international comparison, Austria is regarded as a model of a very pronounced "neo-corporatism" and a "cooperatively-concerted policy" (Lehmbruch 1979; Tálos 1985:43; Pelinka 1981:106-114). This characterization is based not only on the cooperation – often deemed unique – between the large representative associations of interest, but also on their collaboration and agreement on contents and interlinkage with the authorities of public policy in the system of 'social partnership'. One of the system's roots is historic experience, with the fatal consequences of escalating conflicting interests in the period between the two World Wars. Above all, it stands for "balancing interests by consensus" (Gerlich 1985a:113) and giving priority to overall economic goals in the pursuit of particular interests. The voluntary and autonomous cooperation between the central associations of labor and industry is the core of social partnership. The strife for economic growth and social stability through the avoidance of conflicts constitutes the smallest common denominator of this expedient alliance. The expanded system of cooperation between these associations of interest, the representatives of public policy and the political parties constitutes the so-called non-autonomous part of social partnership. It is precisely the aggregate institutional, personnel, and functional interlock of actors and bodies in this quasi 'all-channel-network' that is considered an essential factor for the stability of economic and social partnership (see Tálos 1991:398-403).

This comprehensive system of cooperation is above all relevant for technology policy, due to its role as a political control pattern in balancing interests and its privileged position in the formulation and implementation of public – and, in particular, economic and social -policies. The implications of this concerted cooperation for public policy, and especially for technology policy, are briefly outlined in the four aspects described below:

First of all, the relationship between the 'cartel of associations' and the political system in Austria is characteristically interdependent, consequently limiting the scope of action on both sides (see Gerlich 1985a).[2] While this interdependence signifies a certain disentanglement of power and/or a balance of power between state and associations, it also implies a partial competition to the independent democratic formation of objectives at the parliamentary level. Secondly, the associations' voice in a multitude of institutions and an entire range of decisions increases the possibilities of political exchange, as well as the chances for compromises. Thirdly, a related implication concerns the focal aspect of political controllability. Oriented towards overall economic aims and towards avoiding social conflicts, the associations' cooperation definitely contributes to alleviating governmental policy from conflict management, thereby also bringing about a higher degree of overall steering capacity.[3] In this connection, the accent is rather on the positive effects achieved in the area of social policy, in the widest sense, and regarding social integration. Possible dysfunctional effects, visible in the form of trends towards immobility and an impairment of the innovative potential, are repeatedly mentioned as drawbacks (see Traxler 1986:345-347). Not least on account of the principle of consensus, the typical pattern of decision-making is characterized by incremental changes (as a consequence of the need to compromise) and concomitant high legitimacy (see Pelinka 1981:84-86).

A fourth aspect concerns the goals and implementation of technology policy in a more direct way. The basic consensus of social partnership is also operative in this area. The social partners' common basis of interest mainly consists of a growth and productivity coalition. The emanating influence can, however, shape the concepts and measures of public policy more in disfavor of incorporating competing demands (e.g. of social movements) than can the competitive democratic political process. Due to the differing logic of action in the area of public policy on the one hand, and of social partnership on the other – particularly in times characterized by an unfavorable economic development – the agreement mechanism might reach limits, especially regarding certain socially or ecologically delicate questions of technology policy. Social partnership definitely tends to become somewhat detached from and lose influence on government policy (see Gerlich 1985a:126), and some recent tensions in connection with important political endeavors are certainly tangible. These signs of crisis of social partnership have repeatedly evoked skepticism about this specifically Austrian type of a corporatist political style and have raised the question: "Can it (social partnership, the authors) also fulfill its task under more difficult conditions, or does it fail when faced with intensive challenges?" (Gerlich 1985b:356).

2.4 New Challenges

Various developments in society – such as pluralization of social structure, erosion of ideological affiliations, broadening topics through value changes and social movements, changed party competition (see Prisching 1991) – have all required tremendous adaptations in politics. New tasks notably also arise through the globalization of economy, the European integration process, and the abolition of the "iron curtain". Austria is among the group of small EFTA states that are substantially influenced by the development of the EU because of their strong economic interlinkage. There are fears of a "new discrimination" in case of non-participation once the EU has discarded its internal trade barriers. The application for membership in the EU, filed in 1989, and the expected adaptation pressure on the Austrian economy are hard tests, not only for Austria's competitiveness, but also for the established social partnership (see Martinsen 1991:513). At the same time, the "EU lever" is at times expected to facilitate the political enforcement of structural reforms (particularly in the protected sector). Moreover, considering its exposed geographical position and its historic relations, the opening of the East is of special importance for Austria. Eastern countries not only offer new chances as trading areas, but competing cheap labor also threatens entire branches, such as the cement and agricultural machine industries.

Like other small states, Austria is currently in a kind of 'sandwich position': on the one hand, it must try not to miss the connection to high-tech development (upward perspective); on the other hand (opposite, downward perspective), the opening of the East entails the danger that low-tech products can be produced more cheaply in Eastern countries. Besides, the intermediate position is getting more and more uncomfortable; this is reflected by the new industrialized countries' (NICs) growing diversification into medium technologies (see Hotz-Hart 1992:195). The global economic rivalry has become extremely dynamic and provoked the nations to attempt securing their competitiveness. Against this background, a (socially) integrated technology policy – making maximum use of the available economic potential by exploiting synergy effects and mobilizing human resources – is internationally propagated as the "mandate of the hour". Because of limited resources and existing technology deficiencies, this also – and in particular – applies to small states like Austria.

3 Development and Organization of Technology Policy

The development of an explicit technology policy is closely connected with Austria's overall economic development. After World War II, and supported by the so-called Marshall Plan, economic reconstruction started and yielded above-average economic growth rates and productivity in Austria until the 1950s and 1960s. The upward development was mainly based on the strategy of modernizing the domestic economy by purchasing and imitating foreign know-how in order to overcome the low technological level inherited from the period between the two World Wars. Relying on imitation, however, had negative consequences in the long run: qualified researchers went abroad, licences had to be purchased from competing nations, the country was dependent on the purchase of technology-intensive pre-products, etc. (see Goldmann 1985:194-195; Steindl 1977). The state only felt obliged to take an active role in R&D policy when the adopted growth path showed signs of exhaustion in the 1960s and when political attention was directed at the negligence of the previous years.

3.1 Initial Phase

In the mid-1960s, the OECD, the Austrian Chamber of Labor, and the Federal Economic Chamber for the first time analyzed the expenses for research and development in Austria. They identified a tremendous backlog; in Austria, the expenses for research and development only amounted to 0.3% of the gross domestic product, while they ranged between 1.5% and 2.5% in the rest of Western Europe (see Goldmann 1985:196). Austrian economists, as well as OECD experts, had repeatedly stressed the need of a future-oriented R&D policy, until a purposive R&D policy was finally initiated in the second half of the 1960s. An indicator of this development is the enactment of the Research Promotion Act in 1967. This law is the basis for the establishment of the Fund for the Promotion of Scientific Research and of the Commercial Research Promotion Fund. This first 'naive' phase of Austrian R&D policy lasted until the end of the 1970s and was characterized by the belief that R&D as such would contribute to higher economic growth and to a higher standard of living (see Zaruba 1985:499).

This phase is characterized by a general expansion of science and research and the establishment of a centralist research organization structure (see Melchior 1990). In the initial phase, political measures concentrated on science-oriented research. This includes the development of techno-economic study lines; opening science – especially at universities – to the demands and interests of industry[4]; incorporating criteria of industrial exploitability into research promotion, and information measures intended to facilitate "knowledge transfer" (symposia, scientific expositions, documentation on research institutions and research work, etc.). First results were achieved, as the next section will describe.

In the period 1966-1981, the actual expenditures for research and development tripled, amounting to as much as 12,300 million Austrian Shillings (AS) in 1981. Their share in the GDP rose from around 0.85% to 1.17%; this, however, is still lower than the OECD average (see Goldmann 1985:198; BMWF 1989b:9; Tichy 1992:30). In this period, around 50% of expenses were covered by the state, and around 50% by the industrial sector, with shares remaining constant over time (BMWF 1989b:9). During the 1970s, the number of persons employed in research grew from 7,665 to 18,599 employees in full-time equivalents (Goldmann 1985:198). Yet this strategy soon reached limits which required re-orientation:

- The intensified pressure of industrial exploitability endangered basic scientific research at the universities. It was feared that, in the long run, this would impair the university's efficiency (see OECD 1988:75-78). Resistance against too close a link between university research and the short-term interests of industry also arose from within the university (see Kneucker 1985:230).
- It became obvious that the problem of translating research results into new technologies had been neglected. Applied research and transfer to production require a special infrastructure (separate research institutions) and organization (cooperation and strategic planning), as well as substantial financial means (see Detter 1985).
- The ability of Austria's economy to absorb scientific research results was overrated. There is no automatic structure that would guarantee the transfer of new knowledge into technologies and incorporation into the production process. Moreover, direct public research promotion soon met (passive) resistance from

the economy, fearing "that a publicly-financed and -controlled technology policy might entail interventions in the domain of free entrepreneurial decisions" (Detter 1985a:434).

3.2 Integration Phase

The second phase in Austrian research and technology policy started at the beginning of the 1980s and has not yet been completed. The expectations regarding the direct economic effects of research policy have been adapted to the structural limitations, and new efforts have been made to establish an integrated technology policy. This is also the phase of intensified conceptual considerations and of the introduction of new institutions and measures within the framework of the proclaimed "technology offensive". Research Strategy 80 introduced the reorientation of Austrian R&D policy and reacted to the new challenges both by propelling selected focal programs in research promotion and by searching for new types of cooperation between state, economy, and science. Moreover, new orientations were followed, particularly regarding measures related with the economy: instruments of investment and tax policies were now increasingly applied to support and facilitate the industrial exploitation of R&D results. The Research Organization Act of 1981 introduced tax privileges for R&D expenses and investment promotion measures.[5]

The creation of the Innovation and Technology Fund (ITF) marked yet another step forward. The interests earned each year from the fund's invested capital of 8,000 million AS are used for research, development, and conversions in trade and industry.[6] More recent efforts to make the Innovation and Technology Fund a central instrument of coordination are connected with the elaboration of the 1989 Technology Policy Concept of the Federal Government, where a comprehensive technology policy strategy was formulated for the first time. How efficient it can be will, among other things, depend on the efficiency of established coordination and conflict management.

3.3 Distribution of Competences

In Austria, there is no central responsibility for technology policy. Research and development are considered as factors inherent in the respective administrative components. For this reason, each federal ministry is also responsible for matters of research and development connected with the respective ministry's assignments. As a consequence, competences and duties in technology policy tend to be fragmented. The complicated distribution of competences generates a promotional maze and competing claims by different ministries. The Federal Ministry of Science and Research, the Federal Ministry of Public Economy and Transport, and the Federal Ministry of Economic Affairs are the main actors in Austrian technology policy. Since no suitable authority was available, the federal government took the lead in formulating and initiating a purposeful, integrated research and technology policy.[7] The Research Strategies (1972, 1980), the Technology Policy Concept (1989), and the first technology programs (1985) approved by the federal government constitute the normative basis of public research policy.

At the federal level, Parliament is responsible for enacting and amending the Research Promotion Act (1967) and the Research Organization Act (1981). Though the latter constitutes the legal basis for establishing a research policy structure and for formulating principles and goals, it failed to comprehensively regulate competences. With the Research Organization Act, Parliament – to a great extent – detached itself from research and technology policy. The bodies it had created (in particular the independent funds) – originally meant to be instruments of coordination and cooperation, together with the Federal Ministry of Science and Research and the Federal Ministry of Public Economy and Transport – evolved as the actual vehicles and designers of Austrian research and technology policy.

3.4 Decision-making

Contrary to the division of responsibilities in matters of technology policy relating to industry, the competences and shaping instruments of scientific research are concentrated in one single ministry (Federal Ministry of Science and Research). For this reason, the Federal Ministry of Science and Research also plays a central role in technologically-oriented research policy. A special advisory structure was introduced by the Research Organization Act in 1981. Coordinate to the ministry, it consists of the Council for Science and Research and the Conference for Science and Research. While the former is considered a kind of "Council of Sages" (8-12 members), advising the federal government on general and specific matters of science and research and making suggestions, the latter (56 members) is hardly more than an acclaiming forum.

As the Austrian research strategies and the development of promotion programs show, it is mainly government representatives and the ministries who take the initiative in matters of research and development. The (internal) formulation of objectives is based on a large advisory system comprising more than 100 coordination, counselling, and project teams that have been established since 1972 by the Federal Ministry of Science and Research to develop research strategies and individual projects. There is a close institutional and personnel interlock between the two research promotion funds, the Research Promotion Council, the project teams, and the two advisory bodies of the Federal Ministry of Science and Research, i.e. the Council for Science and Research and the Conference for Science and Research. However, as the OECD country review points out (1988:73), there is no independent advisory body that could serve as a kind of national science council. Thus, all relevant functions (setting priorities, formulating policies, decision-making and implementation) rest with one ministry. Since the opinions and suggestions of the advisory bodies are not published, Parliament and the public are practically excluded from the formulation of objectives (OECD 1988:100).

The integration phase is characterized by two main thrusts: by the 'invention' of technology policy instruments[8] and by efforts to coordinate their application. Coordination and purposive application of the different instruments largely depend on the development of technology policy strategies in the different areas; these must be based on a broad consensus, which is a prerequisite for their successful implementation. The problem of having to select and possibly disturbing long-established client relationships is inherent in developing technology policy programs; however, in the case of strategy-finding, the problems become even more intricate when the technology policy behavior of powerful actors is to be influenced. The implementation of overriding technology policy goals is faced with the inactivity and the conservatism of established routines, institutional structures and distribution patterns.[9] In the past, the process of formulating objectives in technology policy typically displayed the following characteristics:

- The very articulation of the problem was considered an issue to be dealt with by research, and drafting policy concepts was regarded as a quasi-scientific endeavor. This particularly applies to the focal programs that are usually based on scientific studies about a possibly existing "demand", from which the respective programs should subsequently be deduced.[10]
- From a procedural point of view, decisions on the institutionalization of promotion programs were the result of an internal balance of interests of competing groups in science, economy, and central administration.
- Decisions in the area of technology policy were hardly public. In technology policy, the formulation of objectives usually took place without the participation of Parliament, the parties or the political public.

The 'consensual basis' of Austrian R&D policy embraced the representatives of employees and employers, be it in the afore-mentioned project teams dealing with strategic questions of science and research policy, or be it in departmental committees striving for common technology policy initiatives. The "social partners" (above all the Federal Economic Chamber and the Council of Austrian Chambers of Labor, and less frequently the Trade Union Federation and the Presidential Conference of the Chambers of Agriculture as well) are represented in most of the bodies relevant for technology policy, thus forming the link in the landscape of technology policy institutions characterized by heterogeneous interests.[11]

In 1988, the OECD, in its review of research and technology policy in Austria, advised the ministry to grant the representatives of employees and employers more influence and more possibilities to partake in decisions; and hence, the social partners continue to advance. The Innovation and Technology Fund is particularly important, as the social partners are prominently represented in its bodies and intend to make it the central instrument of technology policy (see ITF 1990:6, 9). Moreover, the social partners have observer status in the Departmental Committee on Technology Promotion and are represented in the Advisory Board on Technology Development established in 1989, working with the Minister of Science and Research. The social partners' close interlock with public bodies and institutions is typical for the way they exercise their influence in the area of technology policy, proving their 'state-sustaining' role in the sense of non-autonomous social partnership. Meanwhile, the interlock has developed to such an extent that it seems almost impossible (i.e. only by resorting to intrinsic analysis) to pinpoint the social partners' influence in individual cases. The social partners' firm embedding at the level of operational institutions is as much a prerequisite as it is a limit for conceptual considerations of a programmatic nature, such as the federal government's Technology Policy Concept of 1989.

4 Goals, Instruments and Resources of Technology Policy

4.1 Technology Policy Concept

The agreement of 1987, constituting the basis for the great coalition between the Socialist (now Social-Democratic) party and the Austrian People's party, envisaged the elaboration of a technology policy concept that would introduce an "efficient innovation and technology policy" as an "important prerequisite for the renewed modernization of Austrian economy" (BMWF 1989a:11). On April 11, 1989, the Technology Policy Concept of the federal government and a catalogue of operational technology policy measures were both adopted. For the first time, Austria thus had a comprehensive technology policy concept, defined as an "essential element of the overall economic, structural, environmental, and industrial policy" (BMWF 1989a:12). The concept comprises four strategic goals (BMWF 1989a:7-8):

- strengthening and modernizing applied research and development by adequate financing and the development of programs;
- strengthening the international competitiveness of Austrian economy by selective direct and indirect promotion;
- intensifying the cooperation between industry and science;
- improving the strategic position of public authorities in the process of planning and implementing large-scale technological projects with long-term consequences; reforming federally-owned institutions for research, development, and technical experiments; and plans for the medium- and long-term assessment of the consequences of technology.

The goals are elaborated in 16 items, supplemented by a catalogue containing technology policy measures and a catalogue containing operational measures. For the first time, the technology policy concept provides a yardstick by which technology policy can be legitimized and measured. Whether the lofty aims can be achieved also depends on an effective set of instruments.

4.2 Instruments

Even the measures listed in the Technology Policy Concept of the federal government cannot conceal the fact that a mature, clearly-defined set of instruments is not yet available for technology policy. The available instruments rather depend on the extent to which technology policy goals become orientation measures in all relevant policy areas. The available set of instruments hence changes with the significance which important political actors attach to technology policy. Primarily research-oriented, technology policy in Austria tends to be increasingly interpreted as an integral element of economic policy. Tracing this development, the following paragraphs contain an outline of the most important instruments of Austrian technology policy:

1. Technology Promotion: Direct promotion of technological development started with the foundation of the Commercial Research Promotion Fund in 1967. In the 1970s, the promotion of industrial research constituted the main instrument of public technology policy. In the 1980s, the expenditures of the Commercial Research Promotion Fund rose from 487 million AS (1981) to 1,037 million AS (1991), around 50% thereof being federal subsidies.[12] A second promotional instrument specially dedicated to the promotion of innovation and application, is the Innovation and Technology Promotion Fund, established in 1987.[13] The available funds go to space research (programs of the European Space Agency) and directly to domestic companies. While around 121 million AS were spent on space research in 1990, companies received around 390 million AS.[14] In 1989, the Federal Ministry of Public Economy and Transport introduced a five-year pilot program to promote new firms in the high-tech sector. This so-called seed-financing program promotes management planning, the establishment of firms and accompanying advice.[15] Despite all these measures, the state subsidies in 1991 only amounted to about 10% of all R&D expenses incurred by the firms (Tichy 1992:30). The state's contribution to industrial research and development expenses is hence rather low when compared with other OECD countries (except Japan and Switzerland) (see Passweg 1989:518, 527).

In order to make promotion more focused, the federal government has approved and implemented a number of different technology programs since 1985. The focus "microelectronics and information processing" was followed by the promotional focus "biotechnology and gene technology", and by corresponding programs in the areas "new materials", "environmental technology", "space technology", "laser technology", and – last but not least – "flexible computer-integrated production for small and medium-sized firms".[16]

It has already been mentioned that focusing involves difficult decisions. Instead of autonomously controlling technology, the selection of programs followed the international trend – which, in turn, rather led to an "imitation competition" with other countries (see van Tulder 1988:47). Though this strategy reduces the risk of becoming the sole failure, it also deprives Austria of the chance to benefit from competitive advantages (see Glatz 1992:61). Yet another problem is to actually commit the individual promotional agencies to the programs and/or to make them so concrete that the funds are actually concentrated. Unlike Holland, Finland or Sweden, Austria has no strong intermediary organizations that could guarantee the coordinated implementation of a coherent technology policy (see Glatz 1992:66). This only underlines the general view "that the way in which the aid is distributed is more important than its total amount" (Gaudin 1985:18).

2. Cooperation: Among the measures to promote the transfer of knowledge to the industrial sector are the establishment of technology and innovation centers, as well as the innovation agency initiated by the social partners. Established around the mid-1980s, the innovation agency is conceived as an interface between research and industrial commercial practice. Moreover, it functions as secretariat to the Federation of Austrian Technology Centers (established in 1989), and as executive body of the seed-financing program (BMWF 1989b:36-37; ITF 1990:27). At the beginning of the 1980s, the model projects 'Scientists for Industry' and 'Scientists Found Companies' were introduced in cooperation between the Federal Ministry of Science and Research and the Federal Economic Chamber, with the intention of transferring knowledge by transferring staff (see Bundeskammer 1981). In the 1980s, the number of experimental and research university stations that should increasingly engage in research cooperation and solicited research, grew to 35, i.e. 4% of all university institutes (see BMWF 1991b:21). The Christian Doppler Laboratories were founded in cooperation with the nationalized industry; they are financed by the *ÖIAG* (the central holding of the nationalized Austrian Industries) and should establish close cooperation with university research.[17] Yet another step towards a closer link between university research and industry was the expansion of the universities' partial legal capacity. Since 1988, the universities may use the funds they obtain from third parties for solicited research for their own purposes and for contracting employment under private law (see BMWF 1991b:19).

Intensifying international research and development cooperation within the framework of the EU should help Austria not to miss the technological boat. Austria's way into the European technology community started at the beginning of the 1970s, when it participated in 20 model projects of COST (Coopération Européenne dans le domaine de la recherche scientifique et technique) (see Pichler 1990:317-320). An intensification of technical research and development cooperation with the EU is to be noted only from 1985 onwards. Especially Austria's participation in EUREKA, a program designed for cross-national cooperation between firms and research institutions in the area of market-oriented research and technology development, has considerably expanded since 1985 (see BMWF 1991b:175-176). Austria's full participation in the so-called framework programs of the EU in the technology sector should ultimately be realized by the introduction of the European Economic Area. In order to be able to fully exploit the possibilities offered by this international cooperation, it will, however, in the future be necessary to match the research and development potentials of the nationalized industry and of small and medium-sized firms with the activities of the EU. So far, it is mainly university and research institutes – as well as subsidiaries of foreign multinational companies – that have profited from research cooperation with the EU (see Hutschenreiter/Leo 1992:457).

In a few isolated cases, international industrial cooperation was also supported by locating firms in Austria (e.g. Siemens-Villach; AMI-VOEST; see Detter 1985a:437, 444). Pre-competitive research cooperation is rare in Austria. An intensification of this type of industrial cooperation is, however, expected to occur with Austria's greater involvement in EU "framework programs".

3. Technology-Oriented Research Policy: In the 1980s, promotion policy underwent diversification. Above all in connection with the technology-oriented programs, this meant monetary and equipment privileges for the techno-scientific disciplines and for medical research, which is highly technologized in Austria. The establishment of new research institutes was coordinated with the new programs (see BMWF 1989:20). In research organization, the aim is to interlock science, state, and economy to an extent surpassing present forms and scopes (see Detter 1985a:439-455). In particular with regard to program formulation, firms are to be included at an early stage. The most recent promotion recommendations of the Austrian Research Promotion Council therefore aim at such a formulation of programs 'from below' (see Steinhöfler 1992:496).

4. Raising Venture Capital: In the 1980s, efforts to build up a risk-oriented funding system and a functioning capital market were stepped up; e.g. providing venture capital, creating new financial sources, and orientating traditional industrial promotion towards innovation criteria. In this connection, the resources of the *ERP-Fonds* were increasingly used to promote application-oriented research projects in the industrial sector (BMWF 1988:18). Through the 'TOP initiatives' started in 1981, the *Investitionskredit AG* granted long-term loans for investment projects of interest from a structural and innovative point of view.[18] The provision of venture capital was further facilitated by the tax reform in 1989 (see BMWF 1989b:8).

5. Government Procurement Policy: Conceptualizing a technology-oriented demand policy for large public and communal organizations (post, federal railways, electric industry, nationalized industry, hospitals) should help to overcome the structural limitations of Austrian economy by initiating new technological developments – with expected spin-off-effects – through orders placed with Austrian firms (see Goldmann 1985:208; Nowotny 1985:420-423). Apart from a few rather unsuccessful attempts and the creation of technology information centers for the postal service and the federal railways, selective government procurement in the area of technology policy is rather underdeveloped.[19] With the establishment of the European Economic Area and the approximation towards the EU, a positive discrimination of domestic firms when placing public orders will, however, become more and more difficult. In this respect, technology policy's scope of action is expected to become more limited (see Latzer 1991:621).

6. Legal Framework: The legal conditions to facilitate industrial restructurization were mainly improved by a tax policy directed towards growth and internal funding and by a liberalized foreign-trade policy "that, perhaps, was the most important impetus for the positive structural change in industry" (Goldmann 1992:465). When the East opened up, some Austrian branches of industry – producing on a high technological level due to more stringent environmental conditions – faced massive competition through cheap imports from neighboring Eastern countries. Pointing out the unequal competitive conditions, a "temporary protection of foreign trade" was hence propagated (Goldmann 1992:467). Yet Austria has, in principle, opted for liberalization, implying more intensive competition.[20] In line with the development in the OECD, industry will thus be the most important actor in technology policy. The state will rather be assigned the task of supporting the firms' adaptive endeavors, of removing any obstacles, and of improving the material and immaterial infrastructure (training, design, engineering, organization of production and labor relations, market research, software) (see Steinhöfler 1992:495).

Moreover, social and environmental perspectives in the development of technology can also be introduced through the legal framework. Making catalyzers obligatory, introducing an eco-point system for transit traffic, preparing a gene-technology bill and discussions about the so-called environmental impact assessment are first attempts of such a socially integrated technology policy.

7. Technology Assessment: In Austria, both industry and the social partners are often skeptical regarding the assessment of the consequences of technology, as this is usually considered an obstacle to innovation and a technology-impeding strategy (see Gottweis/Latzer 1991:606). Instruments of techno-economic technology evaluation (such as technology monitoring) have been employed to identify programs for technology promotion. The scientific assessment of the consequences of technology has not yet been institutionalized within the framework of the techno-political decision process. Accordingly, the respective instruments and institutions required to assess the consequences of technology on society as a whole are relatively young and just being developed.[21] New steps in technology assessment were taken in 1992 by establishing a Parliamentary commission for official inquiries on questions of gene technology that has elaborated pertinent recommendations. This is intended to revaluate the role of Parliament in matters of technology policy and to make the formulation of technology policy objec-

Table 1: Overall R&D expenditures in % of Gross Domestic Product (GDP), 1985-1991

Countries	1985	1987	1989	1991
USA	2.93	2.87	2.82	2.82
Japan	2.77	2.82	2.98	3.07*
Germany	2.72	2.88	2.88	2.81*
France	2.25	2.27	2.34	2.40*
Italy	1.13	1.19	1.24	1.35
Netherlands	2.09	2.33	2.16	–
Belgium	1.68	1.69	1.70	1.69*
Sweden	2.89	3.00	2.85	–
Switzerland	–	–	2.86	–
Denmark	1.25	1.42	1.54	–
Norway	1.62	1.82	1.85	1.87
Finland	1.57	1.73	1.80	1.88*
Austria	1.27	1.32	1.40	1.50

* Data from 1990

Source: OECD, STIID (STIU data base), April 1992

tives more democratic (see *Bericht der parlamentarischen Enquete-Kommission* 1992:128-131).

4.3 Expenditures

In 1991, the overall expenses for research and development were some 28,400 million AS, corresponding to 1.46% of Austria's gross domestic product (GDP) (see BMWF 1991b:202). In a comparison with OECD countries, Austria thus is at the lower end of the center range. In the EU, expenses for research and development amount to an average of 2%; in eight countries the expenses are even higher (see Table 1). Strikingly enough, Austria follows the OECD trend with a certain time-lag. Starting from a low level, the expenses for R&D showed a relatively strong increase in the 1970s. At the beginning of the 1980s, the Austrian growth rates lagged significantly behind international development. Above-average growth rates, indicating that the industrial sector is partly catching up, have been registered only since 1988 (see Aiginger et al. 1992:38). Industry's funding portion accordingly rose from 48% (1986) to around 53% (1991), while the state's share in the area of R&D dropped from 49% (1986) to 45% (1991). Nevertheless, the state's share in financing R&D is above the EU average (around 41%), but lower

Table 2: Overall R&D expenditures performed by sectors 1989 (share in %)

Countries	Industrial Sector	Public Sector	University Sector	Private Non-profit Sector	Total
USA	70.2	11.8	15.3	2.7	100
Japan	69.7	8.1	18.0	4.2	100
Germany	73.0	12.3	14.1	0.6	100
France*	59.5	24.9	14.8	0.9	100
Italy	57.1	23.7	19.2	–	100
Netherlands*	60.0	17.2	20.7	2.1	100
Belgium*	73.6	4.6	18.2	3.7	100
Sweden	66.2	4.1	29.6	–	100
Switzerland	74.8	4.3	19.9	0.9	100
Denmark	55.0	19.1	24.8	1.1	100
Norway**	62.0	15.8	21.2	1.0	100
Finland	60.8	18.4	20.2	0.5	100
Austria***	54.8	8.4	34.9	2.0	100

* Data from 1988
** Data from 1987
*** Data from 1985

Source: OECD, STI

than that of North America (around 48%) (see OECD 1992b:22).[22] The overwhelming part of expenditures for R&D is federal. All provinces together spent an estimated 1,500 million AS in 1990; this corresponds to a share of 6% of all R&D expenses and includes R&D expenses of the provincial hospitals.[23]

Problems regarding the economic efficiency of practiced research become evident by the fact that around 80% of public expenditures are spent on enlarging general scientific knowledge; 64% thereof go directly to university research (see Aiginger et al. 1992:45). Consequently, around 35% of all R&D expenses become effective in the university sector, an extremely high percentage when compared with other nations. However, a mere 55% become effective in the industrial sector, internationally the lowest value (see Table 2). This leads us to conclude that publicly-financed research and development in Austria is predominantly basis-oriented and less problem- or industry-oriented (see Aiginger et al. 1992:44). The figures show that Austria mainly has to fight two structural problems in the area of technology development: great deficiencies in the area of application and applied research (see Glatz 1992:60), and industrial research and development activities that are concentrated in a few large firms and receive below-average funds (see Passweg 1989:526-527).

5 State of Affairs and Problems Regarding the Implementation of an Integrated Technology Policy

Though adjusting the R&D share to the OECD level is still maintained as a goal, it is the redressing of the structural problems and the pronounced implementation weakness existing in the area of research and development that have become the priority of research policy. Concurrently, the innovative potential's dependence on the interlink with an active and comprehensive technology policy has become a political focus. Just under four years following the completion of the Technology Policy Concept of the Federal Government, the technology policy discussion in Austria hence concentrates on questions of a concrete strategic elaboration and implementation. Yet, the practically unanimous acceptance of the Policy Concept's basic orientation and of its value as a long overdue basis for a purposive technology policy, is occasionally pervaded by criticism for insufficiency regarding concreteness, binding character and monitoring of success, going hand in hand with skepticism regarding implementation (e.g. Goldmann 1990:54-55; Hutschenreiter/Leo 1992:459-460). To what extent the program of an offensive, yet socially-oriented technology policy has been realized, can partly be judged on the basis of systematic evaluation studies, and partly by having to resort to less comprehensive evaluations and indicators.

5.1 Policy Mix and Effects

Within the framework of the relatively modest Austrian technology policy so far, the main emphasis is on the offered promotional instruments – and in particular on the promotion of the application and diffusion of technology. In contrast, the demand from the public sector is hardly exploited as a technology and innovation-oriented instrument. Transfer of technology and knowledge through cooperation between science institutions and firms, but also international research cooperation, are being pushed forward. The assessment of the consequences of technology is just getting established on a firmer institutionalized basis, but so far has hardly had any relevance in decision-making processes. In the regulatory sector, measures relevant for technology policy were rather introduced in the areas of environmental and traffic policies, while legal regulations, e.g. regarding gene technology, are passed comparatively late from an international point of view.

Regarding technology promotion, Austria clearly differs from other small countries in its priorities by concentrating on the diffusion of process technologies, by an almost complete absence of application research on a broad basis, and by the dominance of basic research in the allocation of public research funds (see Glatz 1992:60). A goal since the introduction of an explicit technology policy, the reallocation of resources earmarked for traditional investment promotion to the promotion of research, technology, and environment is catching on rather slowly. The pertinent data required for any conclusive statements are just as unavailable as are the data on the development of the relation between all these forms of direct and the various indirect instruments of research and technology promotion (above all the different types of tax privileges). Not only the prevalent 'sprinkler principle', but also insufficiently binding guidelines, were the reason why, in the actual allocation of funds, promotion was not exhaustively utilized as a directive tool – an advantage in principle inherent to direct promotion. This fact, as well as the historically-developed split into various institutional representatives and types of promotion, continues to diminish the promotional system's efficiency with multi-tracked and uncoordinated operations (Hutschenreiter/Leo 1992:459; Tichy 1990:285). Efforts to rectify these weak points are currently under way as part of an overall reform of industrial promotion. Their goal is a centralized promotion system and the creation of a core 'center of call', perhaps at the Innovation and Technology Fund, without sacrificing a problem-adequate differentiation of instruments.

Though the introduction of promotion programs linked to certain focal points in new key technologies was primarily launched by international practice, it indicated a first technology policy convergence in the course of Austrian project promotion practice. The programs were primarily designed for small and medium-sized firms, focusing on microelectronics and information technologies. The rising resort to scientific success monitoring is expressed by the fact that the first three of these programs – i.e. microelectronics/information processing, CAD/CAM, and biotechnology/gene technology – were subjected to an exhaustive evaluation (see Hutschenreiter 1991). The incentive to establish cooperation between the applicant and pertinent research institutes was an innovation in the promotion concept; it was evaluated positively, though it was noted that this aspect had developed insufficiently in practice. Likewise, the achieved concentration on small and medium-sized firms, the considerable share of product innovations, and the favorable assessment of business effects were rated positively, while the following drawbacks were identified: an excessively broad coverage, and the fact that a more detailed specification of the promotion criteria had been dispensed with, which also contributed to relatively high pocketing. In addition, the advisory-informative component had been neglected, probably one of the reasons for the small overall reach. There was also a certain insecurity due to the stop-and-go policy exercised in several instances; and last but not least, there was insufficient coordination between the promotion agencies involved.

First improvements were made in more recent technology promotion programs like "flexible automation" and contribute to their apparently positive reception.

These programs meanwhile also offer a more extensive promotion of organizational and managerial functions in innovation projects, thus helping to overcome the overemphasis on the hardware component. In view of the organizational and implementational problems in the course of process innovations – an aspect frequently voiced by the firms – this constitutes an important improvement.

Progress is not only visible regarding the desired stronger focusing; other areas also display signs of some movement along these lines towards the aspired goals. This not only holds for the initiated reorganization of the promotion system, but also for the most recent developments in the areas of intensified transfer of knowledge and expansion of cooperation networks.[24] Likewise, the expansion of the information basis by evaluation studies, technology monitoring, and technology assessment has also progressed. Yet, all told, it seems right to conclude that the implementation of the Technology Policy Concept is slow in getting under way and faces quite a number of problems: these range from the limited interest in establishing programs, problems of project evaluation and lack of long-term planning (see Tichy 1990:289-290), to the still unsolved problems of dispersed powers of decision and administration of measures (Glatz 1992:62; Hutschenreiter/Leo 1992:459). Apart from industrial promotion, there is moreover "no cooperative interplay between industry and the state, in a conceptual and strategic sense that would also heed the national interest" (Goldmann 1992:465). The question is to what extent the Austrian technology policy, under these conditions, has approached the stipulated goal of achieving an innovation strategy oriented not only towards economy, but towards society as a whole.

5.2 Considering Social and Ecological Aspects

In the current discussion, technology's social embedding is repeatedly addressed from at least two different aspects: on the one hand, from the perspective of consequences; on the other hand, from the perspective of prerequisites. Meanwhile, it is generally acknowledged that social factors – and in particular the central 'human factor' – are a prerequisite for the productive exploitation of technological innovations. This is characteristic of a 'socio-economic' perception of innovation policy (see Badham, this volume). However, this model still contains a dualism, characterized by a disintegration of technology and innovation promotion on the one hand, and the assimilation or 'repair' of undesired social repercussions on the other. Anticipating and considering social repercussions by explicitly incorporating the social dimension into the goals within the framework of an integrative, overall social innovation policy, is a more extensive approach.

The program catalogue of the Austrian technology policy (BMWF 1989a) contains explicit elements of such an anticipatory integrative orientation. Economic and social progress are not only generally defined as goals of technology policy (BMWF 1989a:12); the program also states such aims as enacting the improvement of working conditions and the changing requirements of the protection of employees, but also preventing any negative effects by anticipatory measures. In view of the social repercussions expected as a result of the development and implementation of new technologies, it is stressed that technical innovations "require a design oriented towards society and those affected" (BMWF 1989a:13). At the same time, the social dimension has a central role, primarily when considered from the viewpoint that it might contribute towards safeguarding and raising "social productivity" and, from there, as feedback towards "technical productivity" (BMWF 1989a:13), and from the viewpoint of securing acceptance. Moreover, previous strategy documents on research policy, such as Research Strategy 80, have also attached great importance to the social shaping and repercussion potentials of technical innovations. Here, the sphere of work and the exploitation of information technologies are in the foreground, while transcending social repercussion potentials are also included, particularly by creating a special program on technology assessment within the framework of the first technology programs established in 1985.

When analyzing political practice beyond the programmatic level, the activities prior to the development of an explicit technology policy were more or less limited to initiatives such as the following: soliciting studies in the area of technology assessment, e.g. on the economic and social aspects of microelectronics in the form of alternative scenarios based on quantitative model calculations; the creation of a research focus "labor science" at the Fund for the Promotion of Scientific Research; and organization of inquiries and symposia, e.g. on the "humanization of the sphere of work". So far, none of the programs contains far-aimed shaping and innovative endeavors dedicated to the humanization of work in Austria, while a number of important other initiatives were taken by the trade unions and the Chamber of Labor. Yet even the representative bodies of workers and employees began considering technology policy in the narrow sense a special topic only as late as the mid-1980s (see Hochgerner 1988).

Against this background, the explicit incorporation of social aspects in technology promotion programs is an innovation. The program Microelectronics and Information Processing explicitly favors project applications containing socio-scientific accompanying measures and the assistance of the staff councils. The operationalization of this regulation was, however, characterized by a gradual loss of this emphasis on the social component (see Dimitz/Hartmann 1991:351-360). It should be added that the promotional regulations also referred to ecological aspects insofar as the required economic measures and any applicable proof for the "due removal of waste waters" had to be contained in the application (Dimitz/Hartmann 1991:355). Moreover, this aspect was taken into account only half-heartedly and practically without any effect.

Recently, a comprehensive quantitative evaluation scheme was elaborated for promotion applications; it also comprised differentiated social and ecological aspects.[25] The comprehensive catalogue of criteria was, however, already considerably reduced by a departmental committee. The creation of the focus "environmental technology" within the framework of technology promotion programs can be regarded as supporting the ecological orientation.[26]

Compared with these time-limited programs, a higher degree of institutionalization is certainly a stronger indicator for the attention paid to social and ecological interests connected with technological change. Consequently, an expansion of the role of technology assessment and more binding legal regulations are expected to be more effective. So far, only the rudiments are discernible: the already-mentioned tie into the Parliamentary Commission for Official Inquiries in connection with the preparation of a gene-technology law; the elaboration of a concept for the legal regulation of the environmental impact assessments; and, last but not least, the discussions about the problem of environmental protection taken up in core institutions of social partnership. So far, however, the maintenance of social compatibility has primarily rested on the social partners' basic consent. For this reason, the following statement applies to Austria just as it does to the majority of other countries: "The social reflection and co-determination in matters of technology development is in its initial phase everywhere. We currently still lack procedures to appropriately bring about this integration" (Glatz 1992:65).

6 Summary

The step from the programmatic incorporation of an integrated "welfare-oriented technology policy" (Nowotny 1985:427) to its practical implementation constitutes an enormous challenge for the cooperation between the state, economy, science and associations. In Austria, the established institutions and traditions of a cooperatively concerted policy offer a favorable premise. The social partners have always been linked with important research policy institutions – even though, for a long time, their presence did not lead to the formulation of a strategy but rather resulted in a marginal influence regarding the significance and orientation of research and technology policy, especially when compared with their influence in the traditional domain of regulating income policy. Nevertheless, one should not fail to see the impulses repeatedly given by the social partners, in particular through studies and concepts of their advisory body on economic and social affairs, both in the narrower sphere of technology policy (e.g. on industrial policy) and their ambient field (e.g. on structural, capital-market, educational, and environmental policies and on the internationalization of the economy). Concurrently with this "agenda-setting" function, the contents of public policy were also influenced to a certain degree (see Farnleitner 1991:100).

The social partners' active engagement in technology policy has become increasingly stronger since the mid-1980s, primarily taking place under the auspices of a structural policy aimed at higher international competitiveness. This endeavor also finds its expression in the recently-elaborated suggestions for a new orientation of industrial policy.[27] Cooperation between the social partners and the state has now changed. While regulation by demand was central in former times, a kind of 'supply-centered corporatism' is becoming more pronounced after the turn in economic policy (see Traxler/Unger 1990:216).

Regarding the consideration of ecological and social aspects in the modernization consensus of social partnership, a differentiation has to be made. Demands to consider criteria of environmental compatibility of economic and technological changes have so far been confronted with a limited assimilation capacity in the framework of decisions taken by the social partners. Though the obstruction of large projects in the energy-supply sector (Zwentendorf atomic power station, Hainburg hydraulic power station) by the pressure of social resistance already gave rise to the slogan of an expanded cooperation between the associations in the sense of an 'environmental partnership', this was rather connected with a defensive perspective. Though environmental problems have meanwhile become more and more topicalized, this process is still widely separated from the arena of technology policy. Impulses and activities to integrate social and ecological criteria into technology policy currently come rather from sections of science, from social movements, and recently from Parliament as well.

"Social absorption" remains the core element and is mostly understood as easing negative social repercussions following restructurization.[28] Far-reaching offensive calls for social compatibility, particularly in the sense of an anticipatory consideration of the social effects of technological change, are hardly to be found in practice so far. Even the social partners' efforts to achieve an integrated technology policy currently rather concentrate on redeeming coordinative deficiencies as well as on strengthening organizational and strategic capacities and not so much on the integration of non-economic factors. Nevertheless, the system of social partnership offers an important and, in principle, favorable supplement to the mechanisms of market and state policy in the striving towards a welfare-oriented innovation and technology policy. Its full realization, however, will greatly depend on the capacity of the overall political system in its current attempts to achieve a dual integration: both in terms of strategy coordination and in terms of a greater level of foresight with regard to social and ecological concerns.

Notes

[1] The decisive criterion for classifying a nation as a "small state" is the size of its population: according to Jacobs (1989:43), a population of less than 20 million constitutes the limit for this definition.

[2] This assessment is based on a differentiated empirical analysis of the type and intensity of relations, of the degree of corresponding interests, and of the available

instruments of power. It contradicts occasionally voiced theses of a one-sided dominance or instrumentalization ("collateral government").

[3] Also see Traxler (1986:34) and the effects mentioned there. These were empirically proved in international comparisons regarding individual indicators of success such as "controllability", "unemployment rate", and "economic development".

[4] Among others, mention should be made not only of the social partners' participation in drafting curricula, but also of the model projects Scientists for Industry and Scientists Found Companies; see Bundeskammer der gewerblichen Wirtschaft (1981); BMWF (1984).

[5] Promotion of investments improving the balance of goods and services and having a high R&D share; promotion of transfer to production. Institutions created exactly for this purpose, like the *Investitionskredit AG* and the *Finanzierungsgarantie-Gesellschaft*, offered favorable loans and guarantees in the framework of special initiatives (e.g. initiatives TOP) (Goldmann 1985:203).

[6] The capital of the ITF is due to sales returns of federal participation in special firms of the electric industry. Promotion applications are handled by the administrations of the *ERP-Fonds* and of the Commercial Research Promotion Fund (see ITF 1989:5-7).

[7] Moreover, the Federal Chancellor chairs the Board of the ITF, which also comprises the Federal Minister of Economic Affairs, the Federal Minister of Finance, the Federal Minister of Public Economy and Transport, the Federal Minister of Science and Research, one representative of the Council of Austrian Chambers of Labor, one representative of the Federal Economic Chamber, and two representatives of each of the two parties with the largest number of representatives in the *Nationalrat*.

[8] This is usually done by re-directing traditional economic measures and instruments towards the goals of innovation and technology policy.

[9] These problems are illustrated by two case studies on biotechnology policy (see Gottweis 1991) and on telecommunication policy (see Latzer 1991).

[10] A recent example is the study titled "Austrian Technology Monitoring System" (ATMOS), solicited in 1990, to supply "the innovation and technology fund (ITF) with a decision basis regarding the focus of its policy for the early 1990s". In line with the recommendations of the study, promotion focused on "environmental process technology", "laser technology", and "FlexCim". An evaluation of the study, however, raised doubts about the study's approach that was oriented towards direct policy formulation. It is suggested that the reduction to itemized studies and a few experts be substituted by a more process-oriented policy, which should contribute to broadening the formulation of objectives and lead to a "dialogue on innovation and technology policy" (see BMWF et al. 1991a:6-7; ITF 1990:9).

[11] The social partners are represented in the two major research promotion funds. They are represented in equal number in the *Finanzierungsgarantie-Gesellschaft*, established in 1977, and have shares in the Innovation Agency founded in 1984 (see Gottweis 1991:608). Moreover, the social partners have strengthened their engagement in the area of telecommunication policy (see Latzer 1991:620). In addition, the Advisory Board on Economic and Social Affairs founded the Working Committee Innovation in 1984. This committee drafted programs and elaborated research and technology policy suggestions that were incorporated into the federal government's Technology Policy Concept (see BMWF 1989a:12).

[12] Around 90% of the promotion funds are granted to firms, the remainder is given to study groups, joint research institutes and others.

[13] As the ITF is only a source of money, projects are administered via the Commercial Research Promotion Fund and the *ERP-Fonds* which have a 'management function'.

[14] On average, the promotional subsidies amounted to 23% of the project costs; this is around three times higher than the amount usually covered (see ITF 1990:16-17).

[15] In 1990, a total of around 43 million AS were pledged as promotional subsidies. In the same year, around 19 million AS were paid out, of which around 9 million AS were in the form of loans (ITF 1990:27).

[16] Each promotional program was projected for three years, but usually prolonged upon expiration. They are primarily intended for small and medium-sized Austrian firms, but also for pertinent research institutions (ITF 1990:9).

[17] Until 1991, 11 laboratories were set up at Austrian universities. Each laboratory employs 10-15 scientists, and five more are planned (see *Der Standard*, November 2/3, 1991:11; BMWF 1989b:20).

[18] From 1981-1984, the TOP initiatives financed a project volume of more than 19,000 million AS. Because they were rather successful, they were continued (Goldmann 1985:203).

[19] More steps in this direction have only been taken very recently, particularly regarding traffic concepts *('Neue Bahn')*, e.g. by an orientation towards stimulating interlacing across branches and lines in so-called umbrella projects.

[20] One example is the Factory Act, liberalized in Austria in anticipation of the Single European Market.

[21] Comprehensive and interdisciplinary assessment of the consequences of technology is merely carried out by two specialized institutes at the Austrian Academy of Sciences, even though a number of scientific institutes also deals with these questions in the broadest sense (see Braun et al. 1990; Bruckmann 1985).

[22] The high governmental share in North America is due to high expenditures on armaments in the USA.

[23] R&D expenditures of the federal provinces have, however, almost doubled in the 1980s (BMWF 1989b:9, 11).

[24] The number of innovation centers and technology transfer institutions has quickly risen to a total of 46 so far, with around 18 of them innovation centers in the narrow sense (Tödtling-Schönhofer/Laburda 1992:14-19).

[25] Among the criteria for social and structural adaptation are impacts on qualification and the sphere of work, effects in the context of the law on equal treatment, regional structural effects, etc. The ecological criteria comprise a longer list, ranging from the consumption of energy and amount of waste to the release of organisms alien to the ecosystem and the emission of various pollutants.

[26] Among others, it includes environmental processing technology, waste technology, water/waste water technology, and clean-air technology.

[27] One core element of this concept of an "industrial policy oriented towards the power of disposition" is "safeguarding and developing strong firms with a strategic center in Austria by protecting the strategic influence of 'strategic owners' located in Austria" (*Beirat* 1991:8).

[28] One example is the so-called structural billion which the Austrian Trade Union Federation got the federal government to accept for fighting unemployment.

References

Aiginger, K., N. Geldner, A. Guger, G. Hutschenreiter, G. Lehner, H. Leo, W. Puwein (1992) *Industriepolitik 2000. Studie des Österreichischen Instituts für Wirtschaftsforschung, vorläufige Fassung*, Vienna: Austrian Institute of Economic Research.

Beirat für Wirtschafts- und Sozialfragen (1991) *Industriepolitik III*, Vienna: Beirat für Wirtschafts- und Sozialfragen.

Bericht der parlamentarischen Enquete-Kommission (1992) Technikfolgenabschätzung am Beispiel der Gentechnologie, 740 der Beilagen zu den Stenographischen Protokollen des Nationalrats, XXVIII.GP, Vienna: Bibliothek des Parlaments.

BMWF (1984) *Hochschulbericht*, Vienna: Bundesministerium für Wissenschaft und Forschung.

BMWF (1988) *Mikroelektronik und Informationsverarbeitung. Forschungskonzept*, Vienna: Bundesministerium für Wissenschaft und Forschung.

BMWF (1989a) *Technologiepolitisches Konzept der Bundesregierung*, Vienna: Bundesministerium für Wissenschaft und Forschung.

BMWF (1989b) *Forschungsbericht*, Vienna: Bundesministerium für Wissenschaft und Forschung.

BMWF (1991a) *Technologiemonitoring – Technologiepolitik. Information über die Evaluierung der ATMOS-Studie und die geplante weitere Vorgangsweise*, Vienna: Bundesministerium für Wissenschaft und Forschung.

BMWF (1991b) *Forschungsbericht*, Vienna: Bundesministerium für Wissenschaft und Forschung.

Braun, E., M.Th. Nentwich, Ch. Rakos (1990) Technikbewertung in Österreich, research report, Vienna: Österreichische Akademie der Wissenschaften.

Bruckmann, G. (1985) Zur Institutionalisierung von 'Technology Assessment' in Österreich, in: H. Fischer (ed.) *Forschungspolitik für die 90er Jahre*, Vienna, New York: Springer, 105-115.

Bundeskammer der gewerblichen Wirtschaft (ed.) (1981) *Wissenschafter für die Wirtschaft*, Vienna: Österreichische Wirtschaftskammer.

Detter, H. (1985a) Strategische Planung von Forschungs- und Technologieschwerpunkten und ihre Umsetzung in die industrielle Praxis, in: H. Fischer (ed.) *Forschungspolitik für die 90er Jahre*, Vienna, New York: Springer, 434-455.

Detter, H. (1985b) Die österreichische Wirtschafts- und Forschungspolitik seit 1970 aus der Sicht des technischen Bereiches, in: BMWF (ed.) *15 Jahre Bundesministerium für Wissenschaft und Forschung*, Vienna: Bundesministerium für Wissenschaft und Forschung, 21-28.

Dimitz, E., B. Hartmann (1991) Soziale Auswirkungen/Qualifikation, in: G. Hutschenreiter (ed.) *Evaluierung der Technologieförderungsprogramme der Bundesregierung 1985/1987*, Vienna: Austrian Institute of Economic Research, 351-384.

Farnleitner, J. (1991) Die Zukunft der Sozialpartnerschaft, in: *Wirtschaftspolitische Blätter* 1: 96-105.

Fesselhofer, K. (1992) Wirtschaftsprognosen von WIFO und IHS 1992/93, in: *Report. Berichte und Analysen zur Wirtschaftslage* 3: 4-6.

Gaudin, Th. (1985) Definition of Innovation Policies, in: G. Sweeney (ed.) *Innovation Policy. An International Perspective*, New York: St. Martin's Press, 11-47.

Gerlich, P. (1985a) Sozialpartnerschaft und Regierungssystem, in: P. Gerlich, E. Grande, W.C. Müller (eds.) *Sozialpartnerschaft in der Krise. Leistungen und Grenzen des Neokorporatismus in Österreich*, Vienna, Cologne, Graz: Böhlau, 109-133.

Gerlich, P. (1985b) Sozialpartnerschaft in der Krise, in: P. Gerlich, E. Grande, W.C. Müller (eds.) *Sozialpartnerschaft in der Krise. Leistungen und Grenzen des Neokorporatismus in Österreich*, Vienna, Cologne, Graz: Böhlau, 353-366.

Glatz, H. (1991) Kleinstaaten im Wirtschaftlichen Strukturwandel. Industrie- und technologiepolitische Strategien ausgewählter Industrieländer. Studie im Auftrag des Bundesministeriums für öffentliche Wirtschaft und Verkehr, Vienna: Institut für Höhere Studien.

Glatz, H. (1992) Die Industrie- und Technologiepolitik kleiner europäischer Länder im Vergleich, in: *Wirtschaft und Gesellschaft* 1: 47-74.

Goldmann, W. (1985) Forschung, Innovation und Technologie in Österreich, in: H. Fischer (ed.) *Forschungspolitik für die 90er Jahre*, Vienna, New York: Springer, 187-208.

Goldmann, W. (1990) Industriepolitik in Österreich, in: *Wirtschaft und Gesellschaft* 1: 43-64.

Goldmann, W. (1992) Österreichische Industrie- und Technologiepolitik, in: *Wirtschaftspolitische Blätter* 4: 461-468.

Gottweis, H. (1991) Biotechnologiepolitik in Österreich, in: H. Dachs, P. Gerlich, H. Gottweis, F. Horner, H. Kramer, V. Lauber, W.C. Müller, E. Tálos (eds.) *Handbuch des politischen Systems Österreichs*, Vienna: Manz, 613-617.

Gottweis, H., M. Latzer (1991) Technologiepolitik, in: H. Dachs, P. Gerlich, H. Gottweis, F. Horner, H. Kramer, V. Lauber, W.C. Müller, E. Tálos (eds.) *Handbuch des politischen Systems Österreichs*, Vienna: Manz, 601-612.

Hochgerner, J. (1988) Zwischen Zielen und Zwängen. Überlegungen zur Positionsbestimmung der Gewerkschaften in der Technologiepolitik, in: *Österreichische Zeitschrift für Politikwissenschaft* 3: 263-274.

Hotz-Hart, B. (1992) Technik und Technologiepolitik in der Schweiz, in: *Wirtschaft und Gesellschaft* 2: 191-207.

Hutschenreiter, G. (ed.) (1991) *Evaluierung der Technologieförderungsprogramme der Bundesregierung 1985/1987*, Vienna: Austrian Institute of Economic Research.

Hutschenreiter, G., H. Leo (1992) Künftige Aufgaben österreichischer Technologiepolitik, in: *Wirtschaftspolitische Blätter* 4: 453-460.

ITF (Innovations- und Technologiefonds) (1989) *Jahresbericht*, Vienna: ITF.

ITF (Innovations- und Technologiefonds) (1990) *Jahresbericht*, Vienna: ITF.

Jacobs, D. (ed.) (1989) Small Countries' Opportunities for Participating in Science-based Development, in: R. van Tulder (ed.) Small Industrial Countries and the Economic and Technological Development, Working Document 9, The Hague: The Netherlands Organization for Technology Assessment, 43-51.

Katzenstein, P. (1985) *Small States in World Markets. Industrial Policy in Europe*, Ithaka, New York: Cornell University Press.

Kneucker, R. (1985) Förderungsentscheidungen in Österreich, in: H. Fischer (ed.) *Forschungspolitik für die 90er Jahre*, Vienna, New York: Springer, 211-237.

Kuntze, O.-E. (1990) *Vorbereitungen zehn westeuropäischer Industrieländer auf den EG-Binnenmarkt 1993*, Munich: IFO.

Latzer, M. (1991) Staatliche Technologiepolitik im österreichischen Telekommunikationssektor, in: H. Dachs, P. Gerlich, H. Gottweis, F. Horner, H. Kramer, V. Lauber,

W.C. Müller, E. Tálos (eds.) *Handbuch des politischen Systems Österreichs*, Vienna: Manz, 618-623.

Lehmbruch, G. (1979) Consociational Democracy, Class Conflict and the New Corporatism, in: Ph.C. Schmitter, G. Lehmbruch (eds.) *Trends Toward Corporatist Intermediation*, Beverly Hills: Sage: 53-61.

Martinsen, R. (1991) Technologiepolitische Strategien von Kleinstaaten im europäischen Kontext, in: *SWS-Rundschau* 4: 509-515.

Melchior, J. (1990) Zur österreichischen Forschungs- und Technologiepolitik: Entwicklungen und Probleme im Kontext internationaler Diskussionen, in: *Österreichische Zeitschrift für Politikwissenschaft* 3: 245-265.

Nick, R., A. Pelinka (1989) *Politische Landeskunde der Republik Österreich*, Berlin: Colloquium.

Nowotny, E. (1985) Technologiepolitik als wirtschaftspolitische Herausforderung, in: H. Fischer (ed.) *Forschungspolitik für die 90er Jahre*, Vienna, New York: Springer, 417-428.

OECD (1988) *Reviews of National Science and Technology Policy. Austria*, Paris: OECD.

OECD (1992a) *Economic Surveys. Austria*, Paris: OECD.

OECD (1992b) *Main Science and Technology Indicators*, Paris: OECD.

Passweg, M. (1989) Forschung und Entwicklung. Trends in der OECD und in Österreich, in: *Wirtschaft und Gesellschaft* 4: 511-536.

Pelinka, A. (1981) *Modellfall Österreich? Möglichkeiten und Grenzen der Sozialpartnerschaft*, Vienna: Braumüller.

Pichler, F. (1990) Österreichs Weg in die Europäische Technologiegemeinschaft, in: *Österreichische Zeitschrift für Politikwissenschaft* 3: 317-327.

Prisching, M. (1991) Bestandsaufnahme der Sozialpartnerschaft, in: *Wirtschaft und Gesellschaft* 1: 9-36.

Steindl, J. (1977) Import and Production of Know-how in a Small Country: The Case of Austria, in: *Industrial Policies and Technology Transfer between East and West* 3: 211-219.

Steinhöfler, K.H. (1992) Zur Diskussion Technologiepolitik und Wettbewerbsfähigkeit, in: *Wirtschaftspolitische Blätter* 4: 483-497.

Tálos, E. (1985) Sozialpartnerschaft: Zur Entwicklung und Entwicklungsdynamik kooperativ-konzertierter Politik in Österreich, in: P. Gerlich, E. Grande, W.C. Müller (eds.) *Sozialpartnerschaft in der Krise. Leistungen und Grenzen des Neokorporatismus in Österreich*, Vienna, Cologne, Graz: Böhlau, 42-83.

Tálos, E. (1991) Sozialpartnerschaft. Kooperation – Konzertierung – politische Regulierung, in: H. Dachs, P. Gerlich, H. Gottweis, F. Horner, H. Kramer, V. Lauber, W.C. Müller, E. Tálos (eds.) *Handbuch des politischen Systems Österreichs*, Vienna: Manz, 390-415.

Tichy, G. (1987) *Österreich und die Integration der europäischen Forschung*, Vienna: Signum.

Tichy, G. (1990) F&E Politik: Volkswirtschaftliche Bedeutung und Umsetzungsschwierigkeiten, in: *Österreichische Zeitschrift für Politikwissenschaft* 3: 281-291.

Tichy, G. (1992) Forschungs-, Technologie- und Industriepolitik, in: *Wirtschaft und Gesellschaft* 1: 29-45.

Tödtling-Schönhofer, H., A. Laburda (1992) *Innovationszentren und Technologietransfereinrichtungen in Österreich (Gutachten des Österreichischen Instituts für Raumplanung)*, Vienna: Österreichisches Institut für Raumplanung.

Traxler, F. (1986) *Interessenverbände der Unternehmer*, Frankfurt, New York: Campus.

Traxler, F., B. Unger (1990) Institutionelle Erfolgsbedingungen wirtschaftlichen Strukturwandels, in: *Wirtschaft und Gesellschaft* 2: 189-223.

Tulder, R. van (1988) Hochtechnologieförderung in Westeuropa als Hebel der Weltmarktkonkurrenz?, in: J. Becker, B. Wagner, K.-P. Weiner (eds.) *EUREKA. Technologiepolitik im Spannungsfeld wirtschaftlicher und sicherheitspolitischer Interessen*, Marburg: Interdisciplinary Working Group on Peace and Disarmament Research at the University of Marburg, 43-50.

Tulder, R. van (ed.) (1989) *Small Industrial Countries and the Economic and Technological Development*, The Hague: NOTA.

Zaruba, E. (1985) Überlegungen zur Forschungskoordination und Forschungskooperation, in: H. Fischer (ed.) *Forschungspolitik für die 90er Jahre*, Vienna, New York: Springer, 499-508.

Biographical Notes

Georg *Aichholzer* is working at the Institute of Technology Assessment (ITA) at the Austrian Academy of Sciences and is also Lecturer at the University of Economics in Vienna. He was formerly Assistant Professor in the Department of Sociology at the Institute for Advanced Studies, Vienna. His main subjects of research and publication are in the areas of Industrial Sociology, Labour Market Policy, Sociology of Technology and Technology Assessment.

Lahsen *Abdelmalki*, is Senior Lecturer at the University of Lyon 2 and Research Fellow at the CNRS-affiliated research centre *Economie des Changements Technologiques* (Economics of Technological Change). His research activities were focused on the building sector and housing in developing countries. He published a book entitled *Le logement au Maroc* in 1987 and several articles on this subject. He is now developing research in the field of environmental concerns.

Richard J. *Badham* is a Senior Lecturer in Science and Technology Studies at the University of Wollongong, Australia. He has written a book on *Theories of Industrial Society* and published numerous articles in international journals in the areas of industrial sociology, human factors, technology policy, and the sociology of technology. He has been employed as a technology policy consultant or adviser by over 15 different public and private associations in Australia and Europe, and is on the International Editorial Advisory Boards of the journals *AI and Society* and *Journal of Industrial Studies*.

Ernest *Braun*, Ph.D in physics, is now Visiting Professor at the Open University in London. He did research in industry for several years. He was at Bristol University, Professor in Physics in Aston University, Birmingham, and Head of Technology Policy Unit in Aston University, Head of the Department for Technology Assessment at the Austrian Academy of Sciences in Vienna. His main subjects are Technological Innovations and Technology Assessment. He has published several books and numerous articles on these subjects.

J.A.M. *van Boxsel* obtained his Ph.D. in biochemistry in 1976 and then worked for ten years in the staff of the Minister for Science Policy. In 1986 he joined NOTA, the newly formed Netherlands Organization for Technology Assessment. Since 1992 he has been affiliated with the unit for Medical Technology Assessment of TNO Health Research in Leyden. His main interests are medical TA, methodology of TA, ethical aspects of science and technology and technology dynamics. Recent publications include *Technology Assessment:*

To Adjust or to Channel? and *Ethische aspecten van wetenschap en technologie* (Ethical Aspects of Science and Technology).

Bent *Dalum* is Associate Professor at the Institute of Production, Aalborg University and member of the IKE research team. He teaches macroeconomics and international economics. His research fields are: technological development and international trade, industrial and technology policy and development of the electronics industry.

Gregory A. *Daneke* received his Ph.D from the University of California in 1976. He has taught at the University of Michigan, Stanford University, and the University of British Columbia and has been an advisor to the US Department of Energy, as well as to several major high-tech firms. He is the author of numerous works in technology and environmental policy and management. His current research focuses on corporate responses to "sustainable development" initiatives in the USA, Europe, and Canada.

Charles *Edquist* is a Professor at the Graduate School of Technology and Social Change at the University of Linköping (Sweden), where he coordinates a research programme entitled "The Anatomy of Development and Diffusion of New Technologies" (the ADD-TECH Research Programmme). During 1991-1992, he spent his sabbatical at the University of California at Berkeley. His fields of research are social and economic aspects of technological change, organizational innovations, technology and productivity, technology policy, technology and working life, and national systems of innovations.

Masahiro *Kawasaki* is currently Senior Vice-President of the Research Development Corporation, Japan (JRDC) since 1991. Before entering his present position, he has held several top executive positions at the Ministry of International Trade and Industry (MITI) and at the Science Technology Agency (STA), most recently as Director of the National Institute of Science & Technology Policy, STA.

Thierry *Kirat*, is Senior Lecturer of economics at the Institute of Political Studies (University of Lyon 2) and Research Fellow at the CNRS research centre *Economie des Changements Technologiques* (Economics of Technological Change). His research deals with theoretical and applied studies of technological change in the following fields: evolutionary approaches to technical change, technical standardization, and the technology and organization of work. He now develops research on the technological aspects of environmental protection.

Erich *Latniak* studied Political Sciences at the University of Heidelberg, where he became a member of the academic staff in 1987. In 1990 he joined the *Institut Arbeit und Technik* (Institute of Work and Technology) at the North Rhine-Westphalian Science Center at Gelsenkirchen. His main fields of research are technology policy/shaping of technology and studies in regional politics.

Torsti *Loikkanen*, M.Sc. (Econ.) from Helsinki University, is Senior Planning Officer at the Planning Unit of the Technical Research Centre of Finland (VTT). His main duties consist in developing the centre's strategic planning and R&D evaluation. Earlier, he worked in the Ministry of the Environment as researcher and planner responsible for environmental economics and promotion of environmental technology. He has participated in national and international working groups for environmental economics, as well as for technology assessment and future studies. His publications include articles on environmental economics, technology policy, technology assessment and R&D evaluation.

Renate *Martinsen* is Assistant Professor at the Department of Political Science, Institute for Advanced Studies, Vienna. She has published and researched on Political Regulation of Technology and Political Theory.

Josef *Melchior* is Research Fellow at the Department of Political Science, Institute for Advanced Studies, Vienna. Besides research and technology policy his other research interests include university and culture policy.

Emilio *Muñoz* is Research Professor in Biology and Biomedicine at the *Consejo Superior de Investigaciones Científicas* (CSIC), now in the Instituto de Estudios Sociales Avanzados (IESA), Research Unit on Science and Technology Policy (RUSTEP), Spain. From 1982 to 1990, he served different managerial positions in the Spanish R&D Administration. He has been member of the CREST European Committee and Chairman of the COST Programme. He has published more than 250 articles on biochemistry, molecular biology, science and technology policy, as well as on the public understanding of science. He is now involved in the analysis and evaluation of scientific and technological policies, with special emphasis on the problems of R&D Regional Policy and coordination among administrations as well as in the biotechnology field.

Frieder *Naschold* has been Director heading the Research Programme on Labour Politics at the Science Center Berlin since 1976. Before then, he was Professor of Political and Administrative Sciences at the University of Constance. His current activities within the Research Programme on Labour Politics concern projects dealing with questions of production policies in enterprises, as well as technology and social policies at the state level in an internationally comparative perspective. He has written numerous books and articles.

Luis *Sanz-Menéndez* is Research Fellow in Science and Technology Policy at the *Consejo Superior de Investigaciones Científicas* (CSIC), now in the *Instituto de Estudios Sociales Avanzados* (IESA), Research Unit on Science and Technology Policy (RUSTEP), Spain. He is also Associated Professor at the Sociology Department of the University of Madrid, Complutense. His background is in Political Science and Sociology and he has been a visiting scholar at the University of California, Berkeley. He is now involved in research on the impacts of political process and institutions in science and technology policy, and science, technology and regional development.

Gerd *Schienstock* is working at the Academy of Technology Assessment in Stuttgart and is also Lecturer at the University of Stuttgart, Department of Social Sciences. He was formerly Head of the Sociology Department at the Institute for Advanced Studies in Vienna. He studied Business Administration and Sociology in Berlin and Hamburg. His main fields of research are industrial relations, technical and social change, labour market, work organization and management. He has published numerous books and articles in these fields.

Esko-Olavi *Seppälä*, M.Sc., received his degrees from the University of Helsinki where he specialized in applied mathematics and theoretical physics. He is Chief Planning Officer of the Science and Technology Policy Council of Finland. Earlier, he worked at the Academy of Finland, first as Secretary for the Science Research Council and later as Head for Science Policy Planning. He is a member of the Finnish Society for Science Studies and on the Editorial Board of the Society's Scandinavian Journal *Science Studies*, member and Vice-Chairman of the Finnish Committee for IIASA (The International Institute for Applied Systems Analysis), and board member of TUTKAS (Finnish Association of scientists and MPs). He has written articles and reviews on S&T policy, evaluation and planning issues.

Georg *Simonis* earned his doctor's degree in Political Science at Berlin and is now professor at the University of Hagen. At the *Institut Arbeit und Technik* he is head of the Department of *Politische Steuerung* (Political Governance). His main fields of research are technology policy/shaping of technology and international politics.

Udo E. *Simonis*, Dr.sc.pol., is currently Full Professor of Environmental Policy at the Science Center Berlin, where he has been since 1988. He is a member of the Scientific Advisory Board of the federal government on Global Environmental Change and member of the United Nations' Committee for Development Planning. He is editor of *Jahrbuch Ökologie* and has published widely on environmental topics. Among several books, his most recent one is titled *Sustainability and Environmental Policy*.

Subject Index

Name Index

Bei Fragen zur Produktsicherheit wenden Sie sich bitte an:
If you have any questions regarding product safety,
please contact:

Walter de Gruyter GmbH
Genthiner Straße 13
10785 Berlin
productsafety@degruyterbrill.com